Few-Body Systems

Editor in Chief: W. Plessas, Graz

Supplement 11

N^* Physics and Nonperturbative Quantum Chromodynamics

Proceedings of the Joint ECT*/JLAB Workshop, Trento, Italy, May 18–29, 1998

Edited by
S. Simula, B. Saghai, N. C. Mukhopadhyay, V. D. Burkert

Springer-Verlag Wien GmbH

Dr. Silvano Simula
Istituto Nazionale di Fisica Nucleare
Rome, Italy

Dr. Bijan Saghai
Service de Physique Nucléaire, CEA/Saclay
Gif-sur-Yvette, France

Dr. Nimai C. Mukhopadhyay
Physics Department, Rensselaer Polytechnic Institute
Troy, NY, USA

Dr. Volker D. Burkert
Jefferson Laboratory, CEBAF
Newport News, VA, USA

© 1999 Springer-Verlag Wien
Originally published by Springer-Verlag Wien New York in 1999
Softcover reprint of the hardcover 1st edition 1999

Typesetting: Camera-ready by authors

Printed on acid-free and chlorine-free bleached paper
SPIN: 10695603

With 130 Figures

ISSN 0177-8811
ISBN 978-3-7091-7410-4   ISBN 978-3-7091-6800-4 (eBook)
DOI 10.1007/978-3-7091-6800-4

N^* physics and non-perturbative QCD

Trento, Italy

May 18 – 29, 1998

Organizing Committee

Volker D. BURKERT
Jefferson Laboratory, U.S.A.

Nimai C. MUKHOPADHYAY
Rensselaer Polytechnic Institute, U.S.A.

Bijan SAGHAI
Service de Physique Nucléaire, CEA/Saclay, France

Silvano SIMULA
Istituto Nazionale di Fisica Nucleare, Italy

Sponsored by

ECT*, Italy

Jefferson Laboratory, U.S.A.

CEA/Saclay, France

Site

European Centre for Theoretical Studies in Nuclear Physics and Related Areas,
Trento, Italy

Preface

The Workshop *N* Physics and non-perturbative QCD* was held at the European Center for Theoretical Studies and Related Areas (*ECT**) in Trento, Italy, during May 18-29, 1998. Previous workshops of the series on *N* Physics* took place at the Florida State University (1994), at *CEBAF* (1995), at the Institute for Nuclear Theory in Seattle (1996) and at the George Washington University (1997).

The Workshop was devoted to a summary of recent experimental and theoretical research on *N** phsyics and special emphasis was given to the information that photo- and electro-production of nucleon resonances can provide on the non-perturbative regime of Quantum Chromodynamics. The idea was to stimulate discussions among experimentalists and theoreticians in order to pursue the interpretation of the huge amount of forthcoming data from several laboratories in the world. It was therefore decided to have both experimental and theoretical lectures on the main topics, like ,among the others, single and double pion production, η- and K-meson production, the *GDH* sum rule, the spin of the proton, etc. Thanks to the unusual two-week extension of the Workshop, the allotted time for the lectures was extended up to one hour in order to allow the invited lecturers to give a detailed presentation of their topics. Finally, various short contributions were selected to sharpen the discussion about selected items.

This Volume contains both the invited lectures and the contributions to the Workshop and we feel very obliged to express our appreciation to all the speakers for providing us their manuscripts in due time. Thanks to the nice athmosphere at the *ECT**, it was possible to organize also panel discussions among the participants; the summary of one of them, prepared by N.C. Mukhopadhyay and R.M. Davidson, has been enclosed in this Volume.

A special acknowledgment is gratefully due to Richard Davidson, who has given us a decisive help for the preparation of this Volume.

The Workshop was attended by about 60 scientists from more than 40 universities and laboratories in 15 countries, with the largest delegations from USA (19), Germany (12) and Italy (12). We thank all of the participants for their valuable contribution to the high scientific quality achieved in the Workshop.

Many people and Institutions have contributed to the organization of the Workshop. We want to express our warmest thanks to the Director of the *ECT**, Prof. Ben Mottelson, to the Vice Director, Prof. David Brink, to the "unreplaceable" Scientific Secretary, Prof. Renzo Leonardi, and to all the members of the Scentific Board of the *ECT**, whose generous financial and logistic support made our two-week Workshop possible. The co-sponsorships from the Jefferson Laboratory in Newport News and the *CEA* in Saclay are also very much appreciated.

We are deeply indebted with the whole staff of the *ECT**, that before, during and after the Workshop has provided a decisive support to smooth out all the problems encountered in its organization; in particular, we want to warmly thank Mrs. Cristina Costa and Mrs. Ines Campo for their valuable administrative and pratical assistance to each of the participants as well as to the organizers.

The last, but not the least thank is for Nathan Isgur, who has encouraged and supported our idea to have a joint Workshop on N^* physics in Europe.

December, 1998

V.D. Burkert, N.C. Mukhopadhyay, B. Saghai, and S. Simula

An Introduction

I feel extremely fortunate to be able to pen these lines in the midst of my recovery from a potentially disastrous illness in India a few weeks ago, at the end of my physics sojourn at the Tata Institute, Bombay. Great luck (or heavily protection) prevented a more serious outcome at Calcutta. Moral from this episode!: please protect your health first and foremost, as you explore excited baryons and other good things in physics that we talked about in our last workshop in Trento, Italy; avoid too much stress!

Excited baryons (and mesons) are wonderful windows of Nature to explore the chromodynamics of their constituents, quarks and gluons. Now experimental devices, such as the continuous wave accelerator CEBAF at the J-Lab, improved accelerators such as those at Bates, Bonn and Mainz, light source at GRAAL (Grenoble, France) and LEGS (at Brookhaven, U.S.A.), along with excellent detectors (for example, the spectrometer at the Hall B at the J-Lab) and target-beam polarization techniques, provide us with powerful ways of looking through these windows. Naturally, we had at Trento a fair number of discussions on these technologies and the newest results coming from exploiting them, more on which you will have heard at the Bonn conference, which, unfortunately, I am not able to attend in my present state. On the technical side, we heard progress report on the lattice QCD and its ability to illuminate physics of quark model. On the later, we had a number of presentations, on the progress, for example, by the GRAZ approach, wherein mesons exchanged between the quarks play a vital role besides the gluon exchange, along with its traditional quark model and its strengths, weakness and extentions. On all of these, we had a lively panel discussions. Other approaches to the hadron structure were also reviewed.

I thank my fellow organizers, Volker Burkert, Bijan Saghai, Silvano Simula and all attendees for a wonderful time we had at Trento. We are grateful to Prof. Leonardi and the members of the supporting staff at the Institutions for their wonderful hospitality, special banquets and their nicities that made our workshop so memorable. We thank CEBAF and all other agencies for their generous supports. The great city of Trento had many attractions, such as wine testing, open to us.

We look forward to the next get-together like this. Hopefully, by then, we would be all wiser on the issues of hadron structure. Please stay healthy!

Nimai C. Mukhopadhyay,
Rensselaer Polytechnic Institute,
Troy, New York 12180-3590.

Contents

Few-Body Systems Suppl. 11, 1–9 (1999)

Few-
Body
Systems
© by Springer-Verlag 1999

Electromagnetic Excitation of Baryon Resonances and the CLAS N^* Program.

Volker D. Burkert*

Jefferson Laboratory, 12000 Jefferson Avenue, Newport News, Virginia 23606, USA

and the

CLAS collaboration

Abstract. The status of N* physics with electromagnetic probes is discussed. The experimental program with CLAS and first preliminary data are presented.

1 Introduction

The goal of the N^* program at CEBAF (N^* is here used for N and Δ states) is to probe the internal structure of light quark baryons. This will be accomplished in various way: (1) by studying the resonance transition form factors in a large Q^2 range and for a large number of states. This will allow detailed tests of baryon structure models and probe the contributions from hard and soft processes. (2) by searching for the so-called "missing resonances", states which are predicted in the symmetric quark model but have not been seen so far, and (3) by searching for gluonic excitations of the nucleon ("hybrid baryons").

In this lecture I will use the constituent quark model [1, 2] (CQM) as a guide. The model predicts a large number of resonant light quark (u,d,s) baryon states within the symmetry group $SU(3)_{flavor} \otimes SU(2)_{spin} \otimes O(3)_{space}$. The states fall into supermultiplets with fixed orbital angular momentum and energy excitation level. The mass degeneracy within one supermultiplet is broken by the color magnetic hyperfine coupling between the quark spins.

2 Total absorption cross section

The total photon absorption cross section exhibits 3 or 4 enhancements associated with the excitation of some of the prominent resonances - the $\Delta^+_{\frac{3}{2}}(1232)$, $N(1520)$, $N^+_{\frac{5}{2}}(1680)$, and $\Delta^+_{\frac{7}{2}}(1950)$. From inclusive measurements we obtain

---

*e-mail: burkert@jlab.org

only global information about the resonance form factors, and even this is only true in the region of the Δ resonance, a relatively isolated state. The higher mass states overlap, and since they have several hadronic decay channels open they are broad and do not appear as isolated states in the inclusive spectrum (Fig. 1). Three states are contributing to the enhancement around 1.5 GeV, at least seven to the bump at 1.7 GeV, and at least six are known in the mass region around 1.9 GeV. More are predicted with masses near 2 GeV.

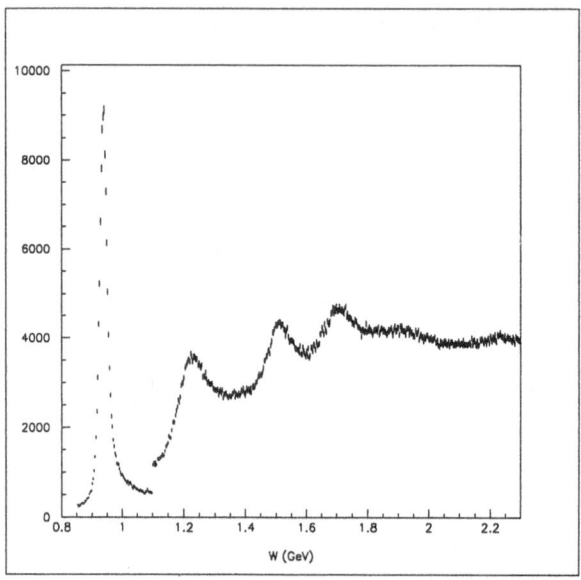

Figure 1. Inclusive electron scattering. The spike is due to elastic ep scattering, the enhancements indicate resonance production.

The various states can only be separated if the hadronic final states are measured, e.g. $N\pi$, $N\eta$, $N\pi\pi$. Spin, parity and isospin of the intermediate state can be identified, and, using information on the hadronic decay vertex, the transverse ($\lambda_{\gamma N} = \frac{1}{2}, \frac{3}{2}$) and scalar ($\lambda_{\gamma N} = \frac{1}{2}$) electromagnetic transition amplitudes $A_{\frac{1}{2}}$, $A_{\frac{3}{2}}$, and $S_{\frac{1}{2}}$ can be determined and compared to model predictions. So far only single pion and single eta photo- or electroproduction data have been used in analyses.

3 Results of previous experiments

3.1 The $\gamma_v p \Delta(1232)$ Transition

The lowest $I = \frac{3}{2}$ state is the $\Delta(1232)$, which may be excited by a magnetic dipole transition or through electric and Coulomb quadrupole transitions. In the symmetric quark model, a spin-flip in the ground state mediates the excitation. Dynamical model give small values for $R_{EM} = \frac{E_{1+}}{M_{1+}}$. However, chiral invariance requires $R_{EM} \to 1$ for $Q^2 >> M^2$.

Experimentally, such quadrupole contributions can be detected in distortions of the pion angular distribution. The electroproduction data (Fig. 2) have large systematic uncertainties. The $Q^2 = 0.5, 1.0, 3.2$ GeV^2 points [3] indicate that R_{EM} may rise slowly with Q^2. This is also seen in a recent dispersion relation analysis [4] of the same data, as well as of recent data at 2.8 and 4. GeV^2 from Jefferson Lab. However, the experimental analysis gives small negative values [6]. These discrepancies may be seen as a measure of model-dependencies in the analyses when only pπ^0 data are used. Despite of these shortcomings the results clearly show that hard processes are not important at $Q^2 < 4.0$ GeV^2 for this process.

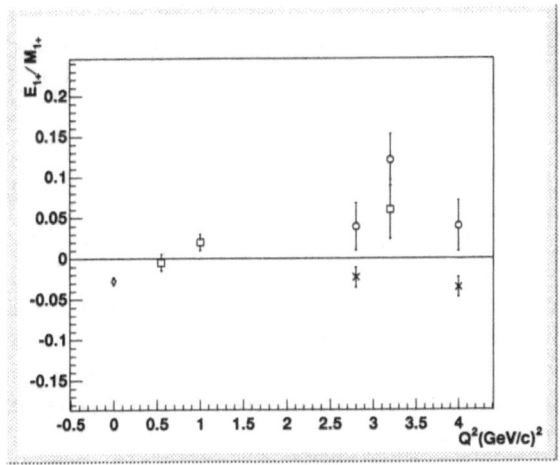

Figure 2. R_{EM} for the $\gamma N \Delta(1232)$. Symbols: squares - Ref. [3], open circles - Ref. [4], crosses - Ref. [6], photon point - Ref. [5]

3.2   Helicity Switch in $\gamma p \to N_{\frac{3}{2}}^-(1520)$, $N_{\frac{5}{2}}^-(1680)$

Radiative transitions to the higher mass states $N(1520)$, $N(1680)$ were found to be purely helicity $\frac{3}{2}$. In the quark model, this is explained by a cancellation between the spin-flip and the orbit-flip terms. However, in electroproduction such a cancellation is no longer possible, and the model predicts the $A_{\frac{1}{2}}$ amplitude dominates in both cases. In terms of the helicity asymmetry:

$$A_1 = \frac{A_{\frac{1}{2}}^2 - A_{\frac{3}{2}}^2}{A_{\frac{1}{2}}^2 + A_{\frac{3}{2}}^2} \tag{2}$$

a "switch" from $A_1 = -1$ at $Q^2 = 0$ to $A_1 \to +1$ at $Q^2 >> 1$ GeV^2 is predicted. This has been seen for the N(1520), however, the data for the N(1680) are too poor for a definite conclusion.

3.3   The Transition $\gamma_v p \to N_{\frac{1}{2}}^{-}(1535)$

Perhaps the most consistent data are available for the $N(1535)$ resonance, which is the $SU(6)$ partner of the $N(1520)$. This transition is most cleanly measured in the ηp channel which also isolates isospin $\frac{1}{2}$ states. Since no other near by state has a significant decay to ηp, and non-resonant amplitudes are small, this channel can be used to 'tag' the $N(1535)$ resonance. While photoproduction experiments indicate some D-wave contribution, electroproduction of $p\eta$ is consistent with s-wave only. This may be simply due to the fast drop of the $A_{\frac{3}{2}}$ amplitude for the $\gamma p N(1520)$ transition with Q^2.

3.4   Limits of the Constituent Quark Model

One way of testing the CQM is to study ratios of transition amplitudes of states belonging to the same supermultiplet. In the ratio

$$\frac{A_{\frac{1}{2}}(N(1520))}{A_{\frac{1}{2}}(N(1535))}$$

effects of the radial wave function drop out, and since both states have the same mass, kinematical effects should be small. Fig. 3 shows the data in comparison with model predictions. This ratio is rather insensitive to relativistic effects and other model ingredients. At high Q^2, the predictions are close to the $SU(6)$ limit for pure spin transitions, which are expected to dominate. While the data agree very well at low Q^2, the highest Q^2 point indicates a significant deviation. If this is a sign for a non-trivial breakdown of the CQM at small distances remains to be seen with improved data.

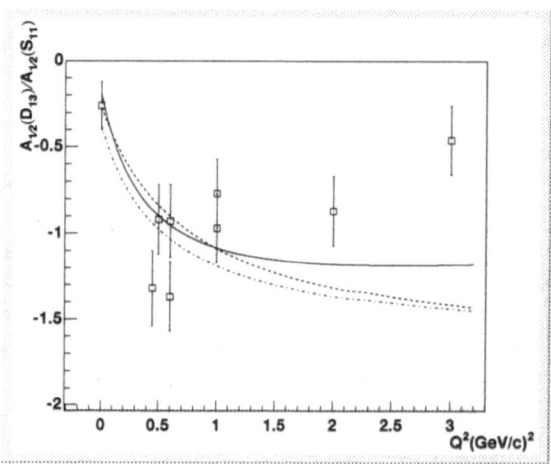

Figure 3. Test of the CQM symmetry structure; the curves are from Ref. [9].

Using experimental information on states belonging to the same supermultiplet, one can test symmetry properties of the electromagnetic transition between supermultiplets. In a single quark transition model, radiative transitions from the ground state to the $[70, 1^-]$ supermultiplet are described by a one-body current operator [8] :

$$J_+^{em} = A \cdot L_+ + B \cdot S_+ + C \cdot L_+ S_z \ , \tag{3}$$

corresponding to orbit, spin, and combined spin-orbit coupling, respectively. Deviations from the SQTM predictions may therefore yield information about multiple quark contributions. Using the data on the $N(1535)$ and $N(1520)$ we can determine A, B, C as a function of Q^2, and predict transition amplitudes to all other states belonging to $[70, 1^-]$. The predictions agree quite well with the data at the photon point, showing that SU(6) symmetry is approximately correct. However, this symmetry might break down at high Q^2. This can presently not be tested as electroproduction data on most of the higher mass states are lacking.

3.5 "Missing" Quark Model States

The symmetric quark model predicts many states with masses above 1.8 GeV which have not been observed in $\pi N \to \pi N$ reactions. Many of the "missing" states are predicted to decouple from the πN channel, however, they may couple to channels such as ρN, ωN, or $\pi \Delta$. Some may also couple to photons. Electromagnetic production of these channels may therefore be the only way to search for the "missing" states. Obviously, our picture of baryon structure could change dramatically if these states do not exist. Production of vector mesons or $\Delta^{++}\pi^-$ should provide a definite answer regarding the existence of at least some of these states [10]. For example, at forward angles the process $\gamma p \to p\omega$ is dominated by diffractive production, while resonance contributions would dominate at large angles. In electroproduction one expects the diffractive contribution to be reduced relative to the resonance contribution due to the increasing virtuality of the vector meson propagator involved.

3.6 Electro-Quenching of the Roper $N_{\frac{1}{2}}^+(1440)$

Existing data indicate a very fast drop of the transverse transition amplitude for the Roper resonance. There have been suggestions that this fast drop may be explained by pionic contributions, which die out fast with inceasing Q^2 and are not explicitly included in the quark calculations [9]. The non-relativistic quark model predicts for the neutron/proton ratio: $A_{\frac{1}{2}}^n / A_{\frac{1}{2}}^p = -0.67$. In chiral bag model calculations [11], contributions from the pion cloud of the proton bring this ratio closer to -1 at $Q^2 = 0$, while the averaged experimental value is -0.60 ± 0.15, making an interpretation in terms of pion contributions rather unlikely, at least within the CBM.

Progress in describing the transition amplitude of the Roper has been reported at this workshop using relativized models, e.g. the light cone model and by introducing quark form factors [17].

3.7 Gluonic Excitations of the Nucleon

In the past decade there have been speculations about the existence of gluonic baryon states $|q^3G>$ consisting of three constituent quarks and constituent glue [12, 13]. QCD lattice simulations indicate that such configurations must exist for mesons, whereas no such calculations have been performed for baryons. Estimates within the framework of QCD sum rules [15] yield masses for the lowest gluonic baryon state P_{11}^G around 1.5 GeV.

Unfortunately, gluonic baryons cannot be distinguished from ordinary $|q^3>$ states in hadronic production because they are, unlike some gluonic mesons, characterized by quantum numbers which are also possible for the normal $|q^3>$ baryon states. However, as their internal structure must be quite different from ordinary baryons, measurement of their transition form factors in electroproduction could be a powerful tool in these studies. A study of the Roper states $N_{\frac{1}{2}}(1440)$ and $\Delta_{\frac{3}{2}}(1600)$ as radial excitations of the 3-quark system or as gluonic excitation shows the transition form factors may be quite different for these alternative models. Moreover, because gluons have only transverse excitation modes, the longitudinal coupling is absent for gluonic excitations, and $S_{\frac{1}{2}}(Q^2) \equiv 0$.

4 The N^* program with CLAS

The study of electro-excitation of N^* resonances in the past has suffered from several shortcomings, such as: (1) lack of theoretical guidance, as most measurements were completed in the pre-QCD, (2) limitations of single pion production experiments with small solid angle spectrometers, and (3) the complete lack of polarization measurements.

The CLAS program [16] is aimed at improving this situation quantitatively as well as qualitatively. Several experiments focus on the precise measurement of the multipoles for the $\Delta(1232)$ in a large Q^2 range, by measuring the unpolarized differential cross section for charged and neutral pion production: $ep \rightarrow e'p\pi^0$, $ep \rightarrow e'n\pi^+$, and $en \rightarrow e'p\pi^-$, with high statistics, and by covering the full azimuthal and polar angular ranges. Also, detailed measurements of polarization observables are in preparation.

Another focus is the region of the second enhancement in the cross section. Of particular interest is the measurement of the transverse and longitudinal transition amplitudes of the Roper over a large Q^2 range. These measurements will help clarify the nature of this state.

The region of the third enhancement contains several states belonging to the $[70, 1^-]_1$ supermultiplet, whose photocouplings are virtually unknown for $Q^2 > 0$, such as $\Delta_{\frac{1}{2}}^-(1620)$, $N_{\frac{3}{2}}^-(1700)$, $\Delta_{\frac{3}{2}}^-(1700)$, and $N_{\frac{5}{2}}^-(1675)$. Also, there exist only few data on electroproduction of neutron resonances. Experiments to study channels such as $ep \rightarrow e'\Delta\pi$, $ep \rightarrow e'N\rho$ are expected to generate much improved data on several of these states, therefore allowing accurate tests of the SQTM assumption, as well as tests of the validity range of the constituent quark model. These channels, as well as $ep \rightarrow ep\omega$, and $\gamma p \rightarrow p\pi^+\pi^-$ are sensitive to many of the "missing" $|q^3>$ states, and will be extensively studied with CLAS.

The reactions $ep \rightarrow ep\eta$, $\gamma p \rightarrow p\eta$ will be used to study the $N_{\frac{1}{2}}(1535)$ and $N_{\frac{1}{2}}(1710)$ states, as well as to search for other isospin $\frac{1}{2}$ excitations. The $\gamma p \rightarrow p\eta'$ reaction might be sensitive to gluonic baryon excitations at higher masses.

Most of the experimental studies of N^* transitions have focussed on the vector structure of the hadronic current. Electroproduction of resonances in coincidence with a pion allow access to the axial structure of the hadronic current. For example, the process $ep \rightarrow e'\Delta^{++}\pi^-_{soft}$, where the π^- is produced near threshold, is related to the axial vector transition form factor G^A_A from the nucleon to the $\Delta(1232)$ via current algebra and PCAC theorem [19].

5   A first look at CLAS data

As of September '98 CLAS has accumulated about 5% of the expected N* data using electron beams at 1.6, 2.4, and 4 GeV, and about 10% of the expected photon data. Some of the electron scattering data have been analyzed using preliminary reconstruction software and calibration procedures. Although the current status of the analysis is still extremely preliminary, some features of the data will not change significantly in course of the analysis, and I will briefly discuss some of them.

An important features of CLAS is its capability to detect nearly the entire phase space of a reaction simultaneously. Identification of charged particles is provided by the combination of time-of-flight from the target to the scintillation counters with momentum measurements in the drift chamber/magnet system. Neutral particles can be identified by missing mass techniques or, in a limited phase space region, by detection in the electromagnetic calorimeters. Fig. 4 shows a missing mass distribution M_X for $ep \rightarrow epX$ and $ep \rightarrow e\pi^+ X$. The π^0, η and ρ,ω and neutron bands, respectively are clearly visible, indicating strong production of these channels. An important part is also due to two-pion production, e.g. $ep \rightarrow ep\pi^+\pi^-$, which requires detection of at least 2 hadrons in the final state. All these channels will be analyzed as part of the N^* program.

Fig. 5 shows the relative yields of various exlusive reactions as a function of the invariant hadronic mass W. Nearly all processes ($p\omega$ needs more statistics) show clear resonance contributions, and the isospin structure of the dominant resonances with masses < 1.75 GeV are revealed when comparing charged and neutral pion production - the $\Delta(1232)$ being isospin $\frac{3}{2}$ and dominantly decaying to $p\pi^0$, the $N(1520)$ and $N(1535)$ and $N(1680)$ with strong $n\pi^+$ as well as $p\pi^+\pi^-$ decays.

6   Conclusions and Outlook

The study of photo- and electroproduction of pseudoscalar and vector mesons from nucleons, provides rich information on the excitation of N^* and Δ reso-

8

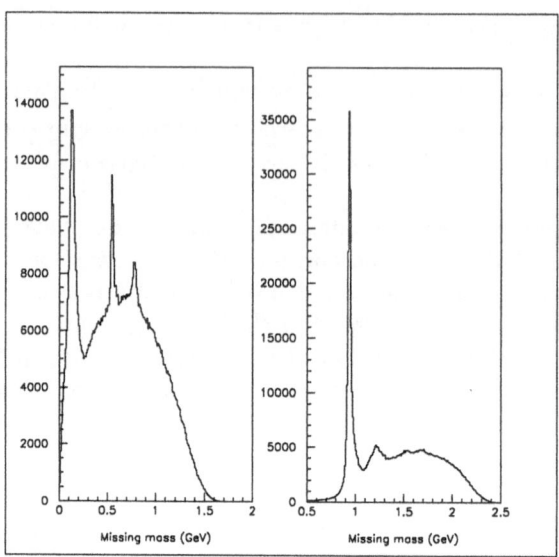

Figure 4. Missing mass for $ep \rightarrow epX$ (left), and $ep \rightarrow e\pi^+ X$ (right). The peaks correspond to the π°, η, ρ/ω, and neutron mass, respectively

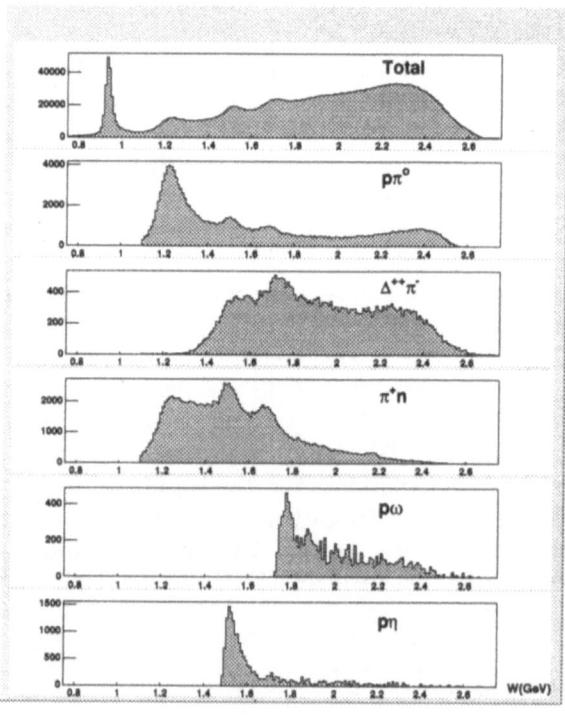

Figure 5. Some of the exclusive reactions recorded in CLAS at 4 GeV beam energy.

nances, their structure, and the strong interaction force in the non-perturbative regime. The utilization of state-of-the-art experimental equipment and data processing techniques has opened up the unique opportunity to collect the data needed for a comprehensive study of the structure of the nucleon in the confinement regime. CLAS has started to take data related to the N^* program in December 1997. The analysis is progressing rapidly and more data taking is scheduled for the year 1999.

References

1. N. Isgur, G. Karl: Phys. Lett. **72B**, 109 (1977); Phys. Rev. D23, 817 (1981)

2. S. Capstick, N.Isgur: Phys. Rev. **D36**, 2800 (1987)

3. V.D. Burkert and L. Elouadrhiri: Phys. Rev. Lett. **75**, 3614 (1995)

4. I. Aznauryan and S. Stepanyan: private communication

5. R. Beck et al.: Phys. Rev. Lett. **78**, 606 (1997)

6. P. Stoler: contribution presented at this workshop

7. F.E. Close, F.J. Gilman: Phys. Lett. **38B**, 541 (1972)

8. A.J.G. Hey, J. Weyers: Phys. Lett. **48B**, 69 (1974)

9. S. Capstick: Phys. Rev. **D46**, 2864(1992)

10. M. Ripani: lecture presented at this workshop

11. K. Bermuth et al.: Phys.Rev. **D37**, 89 (1988)

12. T. Barnes, F.E. Close: Phys. Lett. **123B**, 89 (1983)

13. E. Golowich, E. Haqq, G. Karl: Phys. Rev. **D28**, 160 (1983)

14. Z.P. Li, V. Burkert, Z. Li: Phys. Rev. **D46**, 70 (1992)

15. L. Kisslinger: talk presented at the CEBAF/INT Workshop on N^* *physics*, Seattle, Washington, September 9 - 13, 1996

16. See also: http://www.cebaf.gov/exp_prog/generated/meson.html

17. F. Cardarelli, E. Pace, G. Salme, S. Simula: Phys. Lett. **B371**, 7 (1996)

18. V. Burkert and B. Mecking: *Large Acceptance Detectors for Nuclear Physics*, in: Modern Topics in Electron Scattering, eds. B. Frois, I. Sick, World Scientific, Singapore, 1991

19. S. Adler and W. Weisberger: Phys. Rev. **169**, 1392 (1968)

Few-Body Systems Suppl. 11, 10–17 (1999)

Few-
Body
Systems
© by Springer-Verlag 1999

Perturbative QCD Applied to Baryons

Carl E. Carlson*

Nuclear and Particle Theory Group, Department of Physics,
College of William and Mary, Williamsburg, VA 23187-8795, USA

Abstract.

We review standard applications of perturbative QCD to baryon production, and argue by examining data that it is generally relevant at high but experimentally feasible momentum transfers. Then we consider some new initiatives, particularly meson photoproduction off baryons and the seeming quagmire of $\Delta(1232)$ electroproduction.

1 Introduction

This talk is a special one at a workshop dedicated to nonperturbative methods in baryon physics. It discusses the other side of things, namely perturbative QCD (pQCD) applied to baryons, with particular emphasis on applications to exclusive and semi-exclusive reactions.

We will start out in the next section discussing what I will call "standard old stuff," reviewing methods of calculation and scaling and normalization predictions that are well known to many, and seeing in what kinematic regime pQCD seems to work and how well it works there. I might say now that I am an optimist, thinking that pQCD results can be valid when momentum transfers are only a few GeV. The "standard old stuff" will come in three headings, namely the scaling behavior expected for amplitudes at high momentum transfer, with comparison to data, the polarization behavior expected for amplitudes at high momentum transfer, with comparison to data, and some review of results that have been gotten in the few cases where normalized calculations are possible.

To balance the old, section 3 will present a selection of new initiatives using pQCD, focusing on semi-exclusive reactions and connections between low and high momentum transfer behavior of $\Delta(1232)$ electroproduction.

---

*e-mail address: carlson@physics.wm.edu

2 Standard Old Stuff

2.1 Scaling—expectations and data

Perturbative QCD for exclusive reactions [1] begins by drawing all the relevant lowest order Feynman diagrams. There can be many for a given process and calculating all of them can be time consuming. However, the scaling behavior is generally the same for all the diagrams, and can be ferreted out relatively easily. The general categories of processes are form factors at high momentum transfer, or quasi-elastic reactions at high s at fixed large θ_{CM}. An example of the latter, specifically for $\gamma p \to \pi^+ n$, is given in the Figure below. The momentum transfer dependence comes from the internal propagators—a $1/Q^2$ for each gluon propagator (where Q is some momentum scale) and a $1/Q$ quark propagators—and a factor Q for each quark line [2, 3].

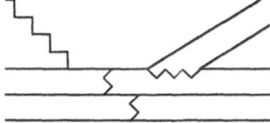

Figure 1. One lowest order diagram for $\gamma p \to \pi^+ n$.

The amplitude represented by this diagram has four quark lines and three each of internal quark and gluon propagators. Hence

$$\mathcal{M} \propto Q^4 Q^{-3}(Q^2)^{-3} = Q^{-5} \propto s^{-5/2}, \tag{1}$$

and the differential cross section is

$$\frac{d\sigma}{dt} = \frac{1}{16\pi s^2}|\mathcal{M}|^2 \propto s^7. \tag{2}$$

Does it work? Here is a plot of $s^7 d\sigma/dt$ vs. s for $\theta_{CM} = 90°$,

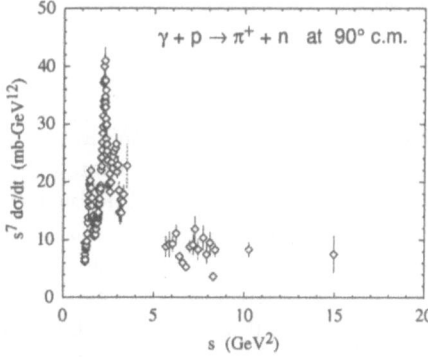

Figure 2. Scaled cross section for $\gamma p \to \pi^+ n$.

The bumps at low s are resonance excitations, and the pQCD expectation appears to succeed just above resonance region.

12

Form factors for electron elastic or quasi-elastic scattering from a hadron with N constituents generally go like,

$$F(Q^2) \propto 1/(Q^2)^{N-1}. \tag{3}$$

For baryon elastic or transition form factors this means $F \propto 1/Q^4$. (At least the leading form factor falls like this: there may be form factors that are zero to leading order, which then fall faster.)

Paul Stoler [4] has produced the following plots:

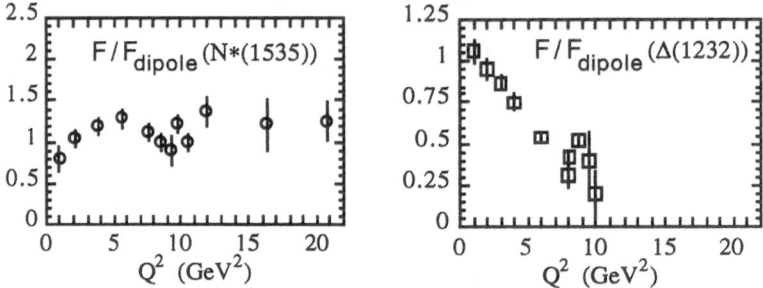

Figure 3. Form factors for two transition form factors, divided by F_{dipole}

For reasons of space, we have shown only the nucleon to $N(1535)$ and to $\Delta(1232)$ transition form factors. The dipole form is $(1 + Q^2/0.71\text{GeV}^2)^{-2}$, so a flat curve is what pQCD predicts. There are also plots for the elastic case and the $N(1688)$ region, which look rather like the $N(1535)$. Hence the pQCD results are successful, except for the $\Delta(1232)$.

The $\Delta(1232)$ falls faster than the others. There is a reason within the pQCD framework for this and a discussion will come in section 2.3. Also, there has been a suggestion that the $N(1535)$ is a ΛK bound state. This makes the minimum Fock component a 5 constituent state, with a faster form factor falloff according to Eqn (3). This is not supported by the data.

2.2 Polarization—expectations and data

The scaling rules tell us the leading scaling behavior, assuming nothing else suppresses the amplitude farther. In particular, there can be farther suppression if the helicity conservation rules are violated. The basic rule is that, neglecting quark mass and binding, the quark helicity is conserved in interactions with either gluons or photons. If all interactions are at close range, the orbital angular momentum of the quarks can be neglected, and then the helicity of the hadrons overall must be conserved. Each unit violation of the helicity conservation rule costs a factor of $O(m/Q)$ where m is some mass scale and Q is some momentum transfer scale [2, 3].

The nucleon electromagnetic form factors give a simple example. Thinking in the Breit frame, a transverse photon with helicity $+1$ hitting a nucleon with helicity $+1/2$ gives a final state nucleon also of helicity $+1/2$. Hadron helicity is conserved; The previous rules apply. The result in terms of G_M comes from

$$G_+ = \frac{1}{2m_N}\langle R, \lambda' = \tfrac{1}{2}|\epsilon_\mu^{(+)} \cdot j^\mu(0)|N, \lambda = \tfrac{1}{2}\rangle = \frac{Q}{m_N\sqrt{2}}G_M \propto \frac{1}{Q^3} \quad (4)$$

and so one gets $G_M \propto 1/Q^4$, which is well known to be true. However, bringing in a longitudinal photon leads to a final helicity of $-1/2$, and so the amplitude should be suppressed by a power of Q, and

$$G_0 = \frac{1}{2m_N}\langle R, \lambda' = \tfrac{1}{2}|\epsilon_\mu^{(0)} \cdot j^\mu(0)|N, \lambda = \tfrac{1}{2}\rangle = G_E \propto \frac{1}{Q^4}. \quad (5)$$

Thus for the Pauli form factor F_2 (using $\tau \equiv Q^2/4m_N^2$),

$$F_2 = \frac{G_M + G_E}{1 + \tau} \propto \frac{1}{Q^6}. \quad (6)$$

Comparing to F_1 in the figure ($F_1 = G_E + \tau G_M/(1+\tau) \propto 1/Q^4$), one sees that this prediction from hadron helicity conservation proves to be true in nature [5].

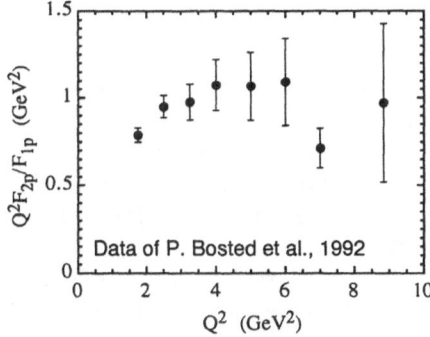

Figure 4. Checking the F_2 scaling behavior vs. data.

2.3 Normalized calculations

When normalized calculations can be done, they become the heart of the perturbative predictions for exclusive reactions. For example, for some typical form factor the whole high momentum transfer calculation is

$$F(Q^2) = \int [dx][dy]\phi(x, Q^2)T(x, y, Q^2)\phi(y, Q^2) \quad (7)$$

Here $\phi(x)$ is the distribution amplitude for the final baryon, simply related to its wave function, and describes finding three quarks with substantially parallel momenta, with a tolerance related to the scale Q, and with momentum fractions x_i; $\phi(y)$ is the same for the initial state. The distribution amplitudes are only weakly dependent on Q. The main, power law, Q dependence comes from the amplitude T, which describes one quark absorbing a large momentum transfer Q and sharing it with the other quarks so they are all parallel moving in the final state. It is calculated in perturbation theory.

The wave function or distribution amplitudes cannot be calculated in perturbation theory. One gets them using QCD sum rules to get moments of wave functions, which become constraints on model wave functions, and model wave functions have been offered by, for the nucleon, CZ and COZ (Chernyak, Oglublin, Zhitnitsky) and KS (King-Sachrajda) and GS (Gari-Stefanis).

These all lead to good results for proton G_M (of course),

$$Q^3 G_+(p \to p) \approx 0.75 \text{ GeV}^3, \tag{8}$$

with

$$Q^3 G_+(N \to \Delta) \approx 0.08 \text{ GeV}^3 \tag{9}$$

and

$$Q^3 G_+(p \to N^*(1535)) \approx 0.46 \text{ GeV}^3. \tag{10}$$

For definiteness, these use KS for the nucleon and CP (Carlson-Poor) for the Δ and S_{11} (with apologies to FOZZ (Farrar, Oglublin, Zhang, Zhitnitsky) and BP (Bonekamp-Pfeil)) [6].

The asymptotic Δ transition amplitude is small. Hence what we see in the data shown earlier is still the subleading part of the transition. A deep reason not known. Still, we can claim that the DDR (Disappearing Delta Resonance) is understood within pQCD.

A quick summary of this quick review is that pQCD has a decent record in explaining data at high but feasible momentum transfers, for single baryons

3 New Initiatives

3.1 Semi-exclusive reactions

A semi-exclusive reaction is one where one or a few, but not all, of the hadrons in a final are observed. We will focus on pion photoproduction [7, 8], $\gamma p \to \pi X$.

We will also suppose that the transverse momentum of the pion is high, and that the recoil mass m_X is high. These provisos ensure that perturbation theory can be used in the calculations.

We hope to learn or supplement what we know about:
- the polarized and unpolarized gluon distributions of the target,
- the quark distributions for high x, and
- the pion wave function at short range.

To proceed, let the transverse momenta be high enough (say $k_\perp > 2$ GeV) so that vector meson dominance is a small contributor. The pion in $\gamma p \to \pi X$ comes either from a parton emerging in some direction and fragmenting (so that the pion is part of a jet) or—at the very highest transverse momenta—directly as part of the short range process (whence the pion is kinematically isolated).

Where, fragmentation dominates, about 1/3 to 1/2 of rate comes from gluon targets in the proton. Note the importance of the high pion transverse momentum, and not just for allowing perturbative calculations. There has to be a recoiling particle, hence the process must be higher order. Then it is possible for the gluon target process to be of the same order of magnitude as a quark target process.

One quantity to consider is

$$E \equiv A_{LL} \equiv \frac{d\sigma_{R+} - d\sigma_{R-}}{d\sigma_{R+} + d\sigma_{R-}} \tag{11}$$

as a function of $k_\perp$. The R refers to the right handed polarization of the photon, and the "$\pm$" gives the helicity of the target proton. The corresponding quantity for the subprocess $\gamma g \to q\bar{q}$ is $(-)100\%$, so that there is a possibility of great sensitivity to the gluon polarization. This is borne out by actual calculations using a variety of proposed gluon in the proton distributions [8].

We will close this section with one more comment. As lower energies it is harder to find a fragmentation region between the direct pion production and VMD regions. Help may be available in fishing out gluon target events by looking two jets or two hadrons 180° apart in azimuth angle. Think of the two parton level diagrams,

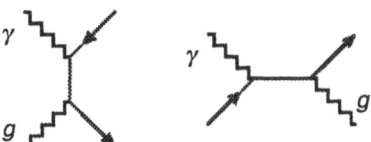

Figure 5. 'Gluon fusion' and 'quark Compton' subgraphs for pion photoproduction.

Fragmenting q's give faster hadrons than fragmenting glue. Perhaps observing two pions with some cut like each $k_\perp$ above 1.5 GeV suffices to ensure that gluon fusion dominates quark Compton [9] even at CEBAF with 12 GeV.

3.2 Approach to pQCD in $\Delta(1232)$ electroproduction

Electroproduction of the $\Delta(1232)$, $\gamma^* + N \to \Delta$, is a tough place to see pQCD at work for two reasons. One is that the low Q^2 starting point is so different from the asymptotic ending point. In terms of the multipole amplitudes, the quark model expectation, born out by data, is that the so-called electromagnetic ratio (EMR) or E_{1+}/M_{1+} is essentially zero at low Q^2, whereas the high Q^2 pQCD prediction is that same ratio is unity. The other is that the leading term asymptotically is unusually small, as we have already noted in section 2.3.

Since pQCD seems to work at a few GeV^2 in more normal cases, we [11] thought we should examine how the probably delayed approach to the pQCD

16

result might go as a function of Q^2. We did so by choosing simple forms that would give the correct results at low and high Q^2 and that obeyed a few principles. We worked using the language of helicity amplitudes, say the G_+ and G_- defined in section 2.2. The principles were basically three: the falloffs of G_+ and G_- should be $1/Q^3$ and $1/Q^5$ asymptotically; another is that there should be a kinematic zero in the amplitude at a (timelike) Q^2 where the Δ does not recoil when produced off a standing nucleon; and another is the high Q^2 normalization (with due regard for the uncertainties of the calculation) of G_+ that was quoted in section 2.3.

At the photon point, $Q^2 = 0$, the overall normalization of the two helicity amplitudes were fixed by comparing to existing data. The size of G_-, essentially given by the mass parameter governing its falloff in Q^2, was also determined from unseparated in helicity data on Δ electroproduction. Some tweaking of the G_+ mass parameter was also needed: there was some information about E_{1+}/M_{1+} at 3 GeV^2 even before the recent CEBAF data was released. Results of our fits are shown in the Figure below.

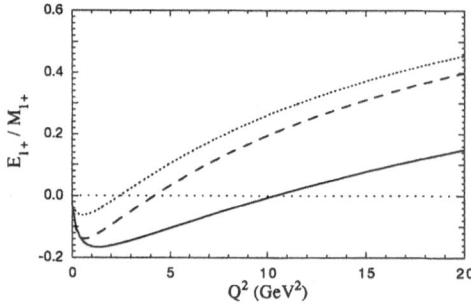

Figure 6. The electromagnetic ratio for Δ electroproduction

The solid curve is our preferred fit; the dashed curve is a naive fit that did not fit the unseparated data well, and the not so different dotted curve has a asymptotic G_+ that was in our opinion too large even given generous uncertainties in the calculated value. It appears that even in this tough situation there will be some push toward the pQCD result by 10 GeV^2 momentum transfer.

We have only mentioned two new initiatives because of space and time limitations. Others exist, notably [10] the idea of off-forward parton distributions and applications to deeply virtual Compton scattering and meson electroproduction and also including new work on inclusive/exclusive connections.

Acknowledgement.

I thank the organizers of this excellent workshop for their hard work, my collaborators on the projects described in the "new initiatives section," namely Andrei Afanasev, Nimai Mukhopadhyay, Chris Wahlquist, and the NSF for support under grant PHY-9600415.

References

1. G.P. Lepage and S.J. Brodsky: Phys. Rev. **D22**, 2157 (1980)

2. C.E. Carlson and F. Gross: Phys. Rev. Lett. **53**, 127 (1984)

3. T. Gousset and B. Pire: In Proceedings of the *"ELFE Summer School on Confinement Physics"*, ed. by S. Bass and P. Guichon, (Frontieres, Gif, 1996), hep-ph/9511274; C.E. Carlson, in *Correlations and Clustering Phenomena in Subatomic Physics*, Proceedings of a NATO Advanced Study Institute in Dronten, the Netherlands, August 1996, ed. by M. N. Harakeh, J. H. Koch, and O. Scholten, (Plenum, New York, 1997)

4. P. Stoler: Phys. Rev. Lett. **66**, 1003 (1991); Phys. Rev. **D44**, 73 (1991); Phys. Rep. **226**, 103 (1993); G. Sterman and P. Stoler, Ann. Rev. Nucl. and Part. Sci. **47**, 193 (1997)

5. P. Bosted et al.: Phys. Rev. Lett. **68**, 3841 (1992)

6. C.E. Carlson and J. L. Poor: Phys. Rev. **D38**, 2758 (1988); G. Farrar, A.A. Oglublin, H. Zhang, and I. R. Zhitnitsky: Nucl. Phys. **B311**, 585 (1989); J. Bonekamp: Thesis, Bonn Report No. BONN-IR-89-43, 1989

7. D. De Florian and W. Vogelsang: Phys. Rev. **D57**, 4376 (1998); B. A. Kniehl: Talk at Ringberg Workshop, hep-ph/9709261; M. Stratmann and W. Vogelsang: Talk at Ringberg Workshop, hep-ph/9708243.

8. A. Afanasev, C. Carlson and C. Wahlquist: Phys. Lett. **B398**, 393 (1997); Phys. Rev. **D58**, 054007 (1998)

9. A. Bravar, D. von Harrach, A. Kotzinian: Phys. Lett. **B421** 349 (1998); M. Amaryan: private communication

10. A. V. Radyushkin: Phys. Lett. **B380**, 417 (1996); X. Ji: Phys. Rev. Lett. **78**, 610 (1997); R. Vanderhaegen, P. Guichon, and M. Guidal: Phys. Rev. Lett. **80**, 5064 (1998)

11. Carl E. Carlson and Nimai C. Mukhopadhyay: Phys. Rev. Lett. **81**, 2646 (1998)

Few-Body Systems Suppl. 11, 18–24 (1999)

Few-
Body
Systems
© by Springer-Verlag 1999

Boson and Gluon Exchange and the Quark-Quark Interaction

D. O. Riska*

Department of Physics, 00014 University of Helsinki, Finland

Abstract. The roles of the Goldstone boson and gluon exchange interactions between quarks is discussed in the light of the empirical baryon spectrum. It is suggested that a qualitatively reasonable model for the hyperfine interaction between quarks is obtained by the combination of a weak gluon exchange interaction and the standard pseudoscalar meson exchange interaction with the irreducible pseudoscalar meson exchange + gluon exchange interaction.

1 Introduction

The spectra of heavy quarkonia - i.e. charmonium, bottomonium, and the recently discovered B_C^+ system may be qualitatively described by a combination of a linear confining interaction and an attractive single gluon exchange interaction, with a quark- gluon coupling with an inversely logarithmic dependence on momentum transfer. That model of the quark-interquark interaction is supported by construction of the interaction operator by lattice methods [1], although the full interaction operator is considerably more complicated than the simple combination of a linear confining interaction and the single gluon exchange interaction. It is noteworthy however, that such potential models – including those derived by lattice methods – fail to explain simultaneously the hyperfine splitting of the ground states (e.g. the η_C and the J/ψ) and the splitting between the S and P states in a quantitative way. This (small) imperfection may indicate a need to include higher order interaction mechanisms, an issue which will be returned to below.

The interaction between two-quarks is fundamentally different from the interaction between quarks and antiquarks. The former contains in addition the exchange interaction between quarks, which has a direct interpretation as $(q\bar{q})$ or meson exchanges. Among such meson exchange mechanisms the interaction mediated by the exchange of the light pseudoscalar meson octet plays a special

*E-mail address: riska@pcu.helsinki.fi

role, in that it has much longer range than other meson exchange interactions, but more importantly because of the role of the light pseudoscalar meson octet (π, K, η) as the Goldstone bosons of the spontaneously broken approximate chiral symmetry of QCD [2].

2 Symmetry Considerations

The gluon exchange interaction between between both quarks and antiquarks contains the $SU(3)$ color scalar operator $\lambda_C^1 \cdot \lambda_C^2$, whereas the pseudoscalar octet interaction between quarks contains the corresponding (somewhat broken) flavor $SU(3)$ scalar operator $\lambda_F^1 \cdot \lambda_F^2$. The most important components of these interactions for the lowest excitations in the baryon spectrum are respectively the color-magnetic hyperfine interaction, which has the color-spin structure $\lambda_C^1 \cdot \lambda_C^2 \sigma^1 \cdot \sigma^2$ and the spin-spin term of the meson exchange interaction, which has the flavor-spin structure $\lambda_F^1 \cdot \lambda_F^2 \sigma^1 \cdot \sigma^2$. This operator structure suffices for analyzing the effects of these interactions on the ordering of the lowest excitations of the baryon spectrum.

The ordering of the low lying positive and negative states in the baryon spectrum is anomalous. Any monotonically increasing confining interaction will organize the spectrum of the three-quark system in shells of alternating parity. Yet, the lowest excitations in all sectors of the baryon spectrum, without flavor singlet states, are positive parity states. So the lowest nucleon excitation is the $N(1440)$ $1/2^+$ state, the lowest Δ excitation is the $\Delta(1600)$ $3/2^+$ state, the lowest flavor nonsinglet Λ hyperon excitation is the $\Lambda(1600)$ $1/2^+$ state and that of the Σ hyperon the $\Sigma(1660)$ $1/2^+$ state.

The lowest excitations of the nucleon all have mixed color- spin symmetry $[21]_{CS}$, because that is the only possible symmetry of the combination of the antisymmetric color state ($[111]_C$) with the mixed symmetry spin state $[21]_S$ for total spin 1/2. As the color-spin operator $\lambda_C^1 \cdot \lambda_C^2 \sigma^1 \cdot \sigma^2$ is a Casimir operator of the decomposition of $SU(6)_{CS} \supset SU(3)_C \times SU(2)_S$ it cannot reverse the normal ordering implied by the confining interaction. In contrast the flavor spin-operator $\lambda_F^1 \cdot \lambda_F^2 \sigma^1 \cdot \sigma^2$, which is a Casimir operator in the decomposition $SU(6)_{FS} \supset SU(3)_F \times SU(2)_S$ may reverse the ordering, because, while the nucleon and the $N(1440)$ $1/2^+$ states both have the same completely symmetric flavor-spin symmetry $[3]_{FS}$, the $N(1535)$ $1/2^-$ state has to have mixed flavor-spin symmetry $[21]_{FS}$ in order to combine with the mixed symmetry orbital state to a completely symmetric state. The reversal of the ordering requires a negative sign for the flavor-spin hyperfine interaction. This argument carries over directly to the case of the ordering of the $\Lambda(1600)$ $1/2^+$ and the $\Lambda(1670)$ hyperon resonances.

This argument is even stronger in the case of the spectrum of the Δ, the lowest excitations of which are the $\Delta(1600)$ $3/2^+$ and $\Delta(1700)$ $3/2^-$ states. The color spin state of the Δ and the $\Delta(1600)$ is completely antisymmetric $[111]_{CS}$, whereas the $\Delta(1700)$ has mixed color spin symmetry $[21]_{CS}$. But with the negative sign of the color-magnetic hyperfine interaction, the latter is shifted downwards relative to the $3/2^+$ states, and hence worsens the spectral ordering problem. In this case the flavor-spin operator (with a negative sign) again, however, with a sufficiently strong orbital interaction matrix element, can achieve the required reversal of normal ordering. This is because while the Δ and the $\Delta(1600)$ have completely symmetric $[3]_{FS}$ flavor- spin states, the $\Delta(1700)$ $3/2^-$ has mixed flavor-spin symmetry $[21]_{FS}$.

3 Dynamical Considerations

There are several dynamical indications that the color-magnetic hyperfine interaction associated with one-gluon exchange cannot be the main source of the splitting of the ground state baryon states. Cooled lattice calculations suggest that the coupling of gluons to constituent quarks should be very weak [3]. The valence QCD approximation, which neglects $q\bar{q}$ terms that might be interpreted as meson exchange contributions, suggests the presence of a residual gluon exchange interaction, far too weak for explaining the splitting of the ground state baryons [4]. This suggests a decoupling of the gluons from or screening of the constituent quarks below the confinement scale Λ_{QCD} or the chiral restoration scale $\Lambda_\chi \sim 1$ GeV. That would be in line with phenomenological studies of the behavior of the running coupling constant of QCD in the infrared limit [5, 6].

The spin-spin component of the pseudoscalar Goldstone boson exchange interaction between constituent quarks has the form (neglecting flavor symmetry breaking):

$$V(\boldsymbol{r}) = -\frac{1}{12m^2}\frac{g_q^2}{4\pi}\{4\pi\delta(\boldsymbol{r}) - \mu^2\frac{e^{-\mu r}}{r}\}\boldsymbol{\lambda}^1 \cdot \boldsymbol{\lambda}^2 \boldsymbol{\sigma}^1 \cdot \boldsymbol{\sigma}^2. \tag{1}$$

Here μ is the pseudoscalar mass and m the quark constituent mass, and g_q is the quark-pseudoscalar meson coupling constant, which may be determined from the the πNN coupling constant $g_{\pi NN}$ –which is known–as

$$g_q = \frac{3}{5}\frac{m_q}{m_N}g_{\pi NN}. \tag{2}$$

The role of the δ function in (1) is to ensure that the volume integral of the interaction vanishes, which is a consequence of the chiral symmetry requirement that the coupling of Goldstone bosons to hadrons vanish with 4-momentum. An interaction, which vanishes with the exchanged momentum, has a vanishing volume integral.

In a numerical calculation the δ function should be smeared over the finite spatial distribution of the constituent quarks. This done, the interaction (1) leads to a remarkable description of the known part of the baryon spectrum up to the small spin-orbit splittings, when combined with a linear confining interaction and relativistic single quark kinetic energy operators [7]. The combination of long and short range terms of opposite sign in the interaction (1) moreover achieves the required splitting of the $N = 2$ $(SD-$shell) part of the spectrum.

4 Spin-Orbit and Tensor Interactions

The empirical spin-orbit splittings in the P shell of the baryon spectrum are small and consistent with zero, with the one exception of the flavor-singlet doublet $\Lambda(1405) - \Lambda(1520)$, which is split by 115 MeV. Because the $\Lambda(1405)$ is situated at threshold for $\bar{K}N$ decay, it is conventionally described as a $\bar{K}N$ molecular state [8] rather than a 3 quark state. This view is consistent with the Skyrme model description of the $\Lambda(1405) - \Lambda(1520)$ as bound states of a topological soliton and a $\bar{K}$, which predicts the size of the spin-orbit splitting correctly [9].

The tensor component of the Goldstone boson exchange interaction does not contribute to the splitting of this multiplet, so that even in an attempt to describe it as a three-quark system, its dynamical origin has to different. Apart from this multiplet the tensor component implies small, but mostly empirically contraindicated spin-orbit splittings among the other low lying spin-flavor multiplets with negative parity. Another dynamical mechanism is therefore required to cancel the effect of this tensor interaction. If vector mesons couple to constituent quarks the associated tensor interaction will tend to cancel that of pseudoscalar meson exchange while the associated spin-spin component adds to that of pseudoscalar meson exchange and therefore such a combination has the phenomenologically required features [2].

An alternative to balancing the pion exchange tensor interaction at short range by a corresponding vector meson exchange interaction of opposite sign is to consider a weak gluon exchange interaction along with the Goldstone boson exchange interaction, and in addition the irreducible π-gluon exchange interaction that is generated by the $\pi-$ gluon exchange loop diagrams as done in Ref. [10]. This is because the tensor component of the $\pi-$gluon exchange interaction is of the same order of magnitude as the pion exchange interaction, but has the opposite sign. As a result the net tensor interaction is very weak.

5 The Irreducible π-Gluon Exchange Interaction

The $\pi-$gluon exchange loop mechanisms allow combination of short range gluon exchange with long range pion exchange, and thus combines long and short range physics, even with the conventional assumption that the Goldstone bosons decouple from constituent quarks above the chiral restoration scale $\Lambda_\chi \simeq 4\pi f_\pi$.

The calculation of the $\pi-$gluon exchange interaction may be performed within the framework of the Blankenbecler-Sugar quasipotential framework, which allows a covariant extraction of the iterated single pion and gluon exchange interactions from the Bethe-Salpeter equation kernel [11]. The resulting interaction is then real and almost energy independent.

If the Goldstone boson exchange interaction between quarks is complemented with a fairly weak gluon exchange interaction along with the associated irreducible $\pi-$gluon exchange interaction the problem of the tensor component of the former is eliminated because the tensor component of the $\pi-$gluon exchange interaction is of the same order of magnitude as the pion exchange interaction, but with the opposite sign (Fig.1). The former dominates at short and the latter at large distances. As a result the net tensor interaction is very weak [10].

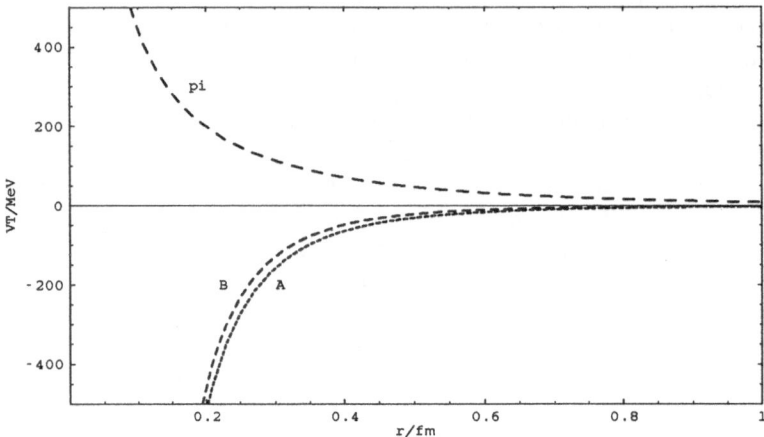

Figure 1. The tensor components of the pion ("pi") and the irreducible π−gluon exchange interactions with no (A) and with long range screening (B) of the gluon exchange interaction between the constituent quarks (From Ref.[10])

The π−gluon exchange interaction has an attractive spin-spin component, which adds to, but is somewhat weaker than that of single pion exchange at short range. Its detailed behavior at very short range is very sensitive to the high momentum behavior of the pion and gluon exchange interactions. Finally the spin-orbit and central components of the π−gluon exchange interaction turn out to be very weak. The π−gluon exchange interaction thus appears to provide part of the explanation for why the effective interaction between constituent quarks should have the form of an attractive flavor dependent spin−spin interaction and an at most very weak tensor interaction.

6 Discussion

This analysis suggests that the hyperfine interaction between light constituent quarks is made up of (at least) a (weak) single gluon exchange component, a pion exchange component with conventional strength and an irreducible π−gluon exchange interaction and possibly also vector meson exchange interactions. The single gluon exchange interaction is weak because the effective quark-gluon coupling is weak in the infrared limit [5, 6], and therefore the problem of a large gluonic spin-orbit interaction [12] is avoided. The pion exchange tensor interaction is in effect cancelled by the large tensor component of the π-gluon exchange interaction, and thus the incorrect (small) spin-orbit splitting of the low lying negative parity resonances is avoided. The π-gluon exchange and single pion exchange interactions combine to a strong attractive flavor dependent spin-spin interaction, which brings the low lying positive parity resonances below the lowest negative parity resonances in agreement with experiment.

But the results also implies that higher order mechanisms may in general be expected to contribute significantly to the effective interaction between constituent quarks. quarks. To these belong two-pion exchange, which mainly leads to a flavor independent attractive interaction, which adds to the strength of the effective confining interaction at short range. In the case of the quark-quark interaction the small values of the quark-meson coupling constants in comparison to the large meson-nucleon coupling constants suggests that the loop expansion may actually converge in the quark case.

The substantial size of the π-gluon tensor and spin-spin interaction components suggests that the interaction between constituent quarks may prove to be as complex as the nucleon-nucleon interaction proved to be, and that it – at least partly – has to be constructed phenomenologically as is the case with the latter [13]. Given this situation a purely phenomenological approach to the interaction may be well motivated, and then the main requirements are that the effective mass operator model satisfy the fundamental symmetries, as Poincaré invariance, in addition to the dynamical symmetries that are implied by the spectrum. An example of a simple mass operator model, which satisfies these requirements based on instant form kinematics, and which with a few adjustable parameters in the flavor spin dependent hyperfine interaction term is able to describe the presently known part of the baryon spectrum is given in Ref. [14].

References

1. G.S. Bali, K. Schilling and A. Wachter: Phys. Rev. **D56**, 2566 (1997)

2. L. Ya. Glozman and D. O. Riska: Phys. Rep. **268**, 263 (1996)

3. M.C. Chu et al.: Phys. Rev. **D49**, 6039 (1994)

4. K.-F. Liu et al.: hep-ph/9806491

5. A.C. Mattingly and P. M. Stevenson: Phys. Rev. **D49**, 437 (1994)

6. S.J. Brodsky et al.: Phys. Rev. **D57**, 245 (1997)

7. L.Ya. Glozman, W. Plesssas, K. Varga, R.F. Wagenbrunn: Phys. Rev. **D58**, 094030 (1998)

8. R.H. Dalitz and A. Deloff: J. Phys. **G17**, 289 (1991)

9. D. O. Riska and N. Scoccola: Phys. Lett. **B265**, 188 (1991)

10. C. Helminen and D. O. Riska: Phys. Rev. **C58**, 2928 (1998)

11. R. Blankenbecler and R. Sugar: Phys. Rev. **142**, 1051 (1966)

12. J. Carlson, J.B. Kogut and V.J. Pandharipande: Phys. Rev. **D28**, 2807 (1983)

13. M. Taketani, S. Nakamura and M. Sasaki: Prog. Theor. Phys. **6**, 581 (1951)

14. F. Coester, K. Dannbom and D. O. Riska: Nucl. Phys. **A634**, 335 (1998)

Few-Body Systems Suppl. 11, 25–28 (1999)

Few-
Body
Systems
© by Springer-Verlag 1999

Semirelativistic Constituent-Quark Model with Goldstone-Boson-Exchange Hyperfine Interactions

R.F. Wagenbrunn[1], L.Ya. Glozman[2], W. Plessas[1], K. Varga[3]

[1] Institute for Theoretical Physics, University of Graz, Universitätsplatz 5, A-8010 Graz, Austria
[2] Institute for Theoretical Physics, University of Tübingen, Auf der Morgenstelle 14, D-72076 Tübingen, Germany
[3] Theory Division, Argonne National Laboratory, Argonne, IL 60439, USA

Abstract. We address the semirelativistic constituent-quark model with linear confinement and a hyperfine interaction deduced from Goldstone-boson exchange. In the version where the latter is represented by pseudoscalar meson exchange (π, K, η, η') and restricted to the spin-spin component only a unified description of all light- and strange-baryon spectra is provided in close agreement with phenomenology. We discuss the extension of the model that includes also the tensor components of the pseudoscalar meson exchange and in addition vector-meson (ρ, ω, K^*, ϕ) and scalar-meson (σ) exchanges. A preliminary version thereof provides a description of the baryon spectra with similar good quality.

The spontaneous breaking of chiral symmetry in the u, d, s sector of QCD indicates that the relevant effective degrees of freedom for light and strange baryons at low energies are furnished by constituent quarks and Goldstone bosons, while the original QCD degrees of freedom, current quarks and gluons, are no longer active. The constituent quarks are to be viewed as quasiparticles with an acquired dynamical mass and a finite extension; the Goldstone bosons are manifested by the pseudoscalar mesons. Consequently, baryons should be considered as systems of three constituent quarks interacting by Goldstone-boson exchange and being subject to confinement [1].

In this spirit we developed a semirelativistic constituent-quark model [2, 3] based on the following Hamiltonian:

$$H = \sum_{i=1}^{3} \sqrt{{\boldsymbol{p}_i}^2 + m_i^2} + V_\chi + V_{\text{conf}}. \tag{1}$$

Here, $\boldsymbol{p}_i$ are the 3-momenta and m_i the masses of the constituent quarks. The relativistic form of the kinetic energy allows to avoid disturbing shortcomings of nonrelativistic quark models (see, e.g., the discussion in Ref. [2]). The confinement potential V_{conf} is taken in linear form. The chiral potential V_χ consists of the (most important) spin-spin interaction provided by the exchange of the pseudoscalar octet and singlet mesons. For further details and explanations see Ref. [3].

In Fig. 1 we show the spectra of the light and strange baryons produced by this version of the chiral constituent quark model. One observes a rather

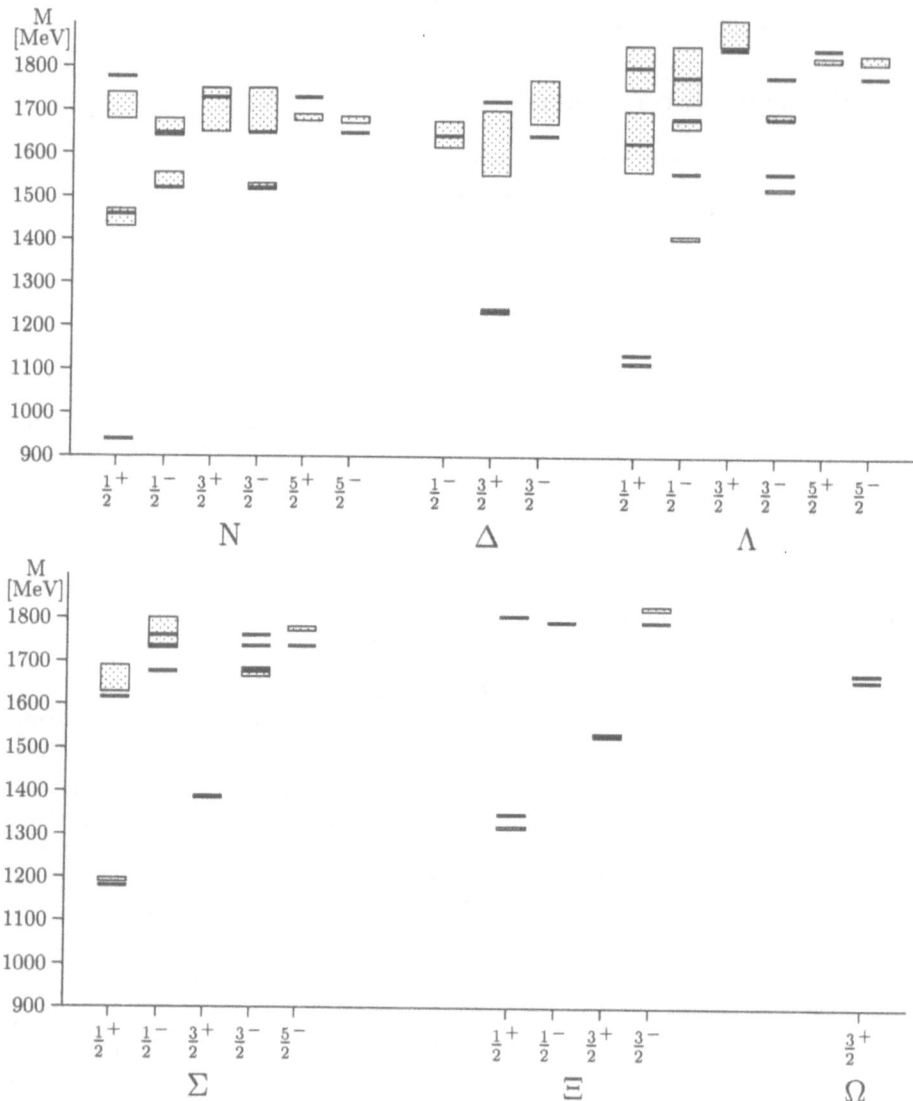

Figure 1. Energy levels of the lowest light- and strange-baryon states with total angular momentum and parity J^P for the model of Ref. [3].

satisfactory agreement with the phenomenological energy levels as given by the PDG [4]. In particular, the correct level orderings of positive- and negative-parity excitations are provided by the hyperfine interaction V_χ. Due to its specific flavor dependence this is achieved simultaneously in the light and strange sectors. Thereby a severe shortcoming of previous constituent quark models relying mostly on a one-gluon exchange hyperfine interaction is remedied.

In order to account for the further details in the hyperfine structure of

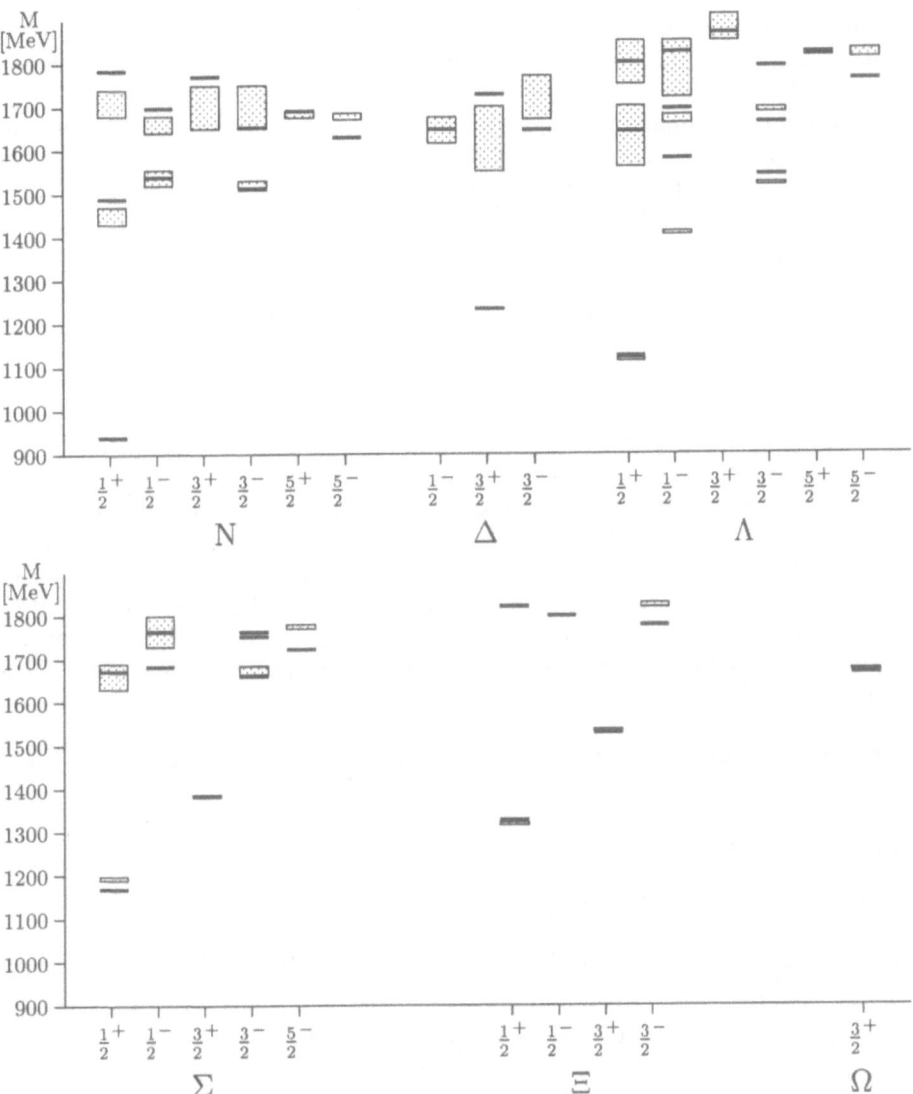

Figure 2. As in Fig. 1 but for a preliminary version of an extended model with the full interactions from pseudoscalar-, vector-, and scalar-meson exchanges included.

the baryon levels the chiral interaction must be extended beyond the spin-spin component, i.e., to include the full potential of the pseudoscalar meson exchange. At the same time it is advised to consider multiple Goldstone-boson exchanges. In this regard we introduce vector (ρ, ω, K^*, ϕ) and scalar (σ) meson exchanges. Thus, we end up with an extended chiral quark model whose chiral hyperfine interaction includes all possible central, spin-spin, tensor, and spin-orbit components of the pseudoscalar, vector, and scalar mesons. In a preliminary version thereof we have fixed all meson masses to their experimental values and kept all meson-quark coupling constants at the magnitudes as deduced from meson-nucleon phenomenology. In this way we have been able to avoid a proliferation of open parameters. In fact, only a few free parameters are needed to produce a spectrum as, for example, shown in Fig. 2.

The extended model meets further requirements of the baryon spectra. While the spin-spin components of the pseudoscalar- and vector-meson exchange interactions have the same sign for the short-distance interactions (providing the proper level orderings in the baryon spectra), the corresponding tensor components have opposite signs and largely cancel each other. The net tensor-force effects are thus small. This behavior is just required by the baryon spectra as the observed splittings within LS multiplets (like, e.g., $N(1535)$-$N(1520)$) are also small. Similar to the tensor-force contributions a partial cancellation of the spin-orbit effects from different meson exchanges occurs in the extended model.

At present we are still facing the problem of a proper description of the $\Lambda(1405)$ state. We expect this state to be affected to a large extent by higher Fock components other than just $\{QQQ\}$. A downshift of this particular state, which lies quite close to the $N\bar{K}$ threshold, by including mesonic components has already been observed, e.g., in Ref. [5]. It is one of our future aims to extend the chiral constituent-quark model to include $\{QQQ \text{ meson}\}$ states. This would constitute a natural and consistent generalization of the model and at the same time allow for a more realistic description of baryon resonances with finite widths.

Acknowledgement. R.F.W. wants to thank the ECT* and the Paul-Urban Foundation for financial support.

References

1. L.Ya. Glozman and D.O. Riska: Phys. Rep. **268**, 263 (1996)

2. L.Ya. Glozman, Z. Papp, W. Plessas, K. Varga, R. F. Wagenbrunn: Phys. Rev. **C57**, 3406 (1998)

3. L.Ya. Glozman, W. Plessas, K. Varga, R. F. Wagenbrunn: Preprint hep-ph/9706507, Phys. Rev. D (to appear)

4. Particle Data Group, C. Caso et al.: Eur. Phys. J. **C3**, 1 (1998)

5. M. Arima, S. Matsui, K. Shimizu: Phys. Rev. **C49**, 2831 (1994)

Few-Body Systems Suppl. 11, 29–32 (1999)

Few-
Body
Systems
© by Springer-Verlag 1999

Hadronic Decays of Baryon Resonances in the Goldstone-Boson-Exchange Constituent-Quark Model

W. Plessas[1], A. Krassnigg[1], L. Theußl[1,2], K. Varga[3], R.F. Wagenbrunn[1]

[1] Institute for Theoretical Physics, University of Graz, Universitätsplatz 5, A-8010 Graz, Austria
[2] Institut des Sciences Nucléaires, Université Joseph Fourier, Avenue des Martyrs 53, F-38026 Grenoble, France
[3] Theory Division, Argonne National Laboratory, Argonne, IL 60439, USA

Abstract. The recently proposed semirelativistic constituent-quark model whose hyperfine interaction relies on Goldstone-boson-exchange dynamics is applied to study hadronic decays of light and strange baryon resonances. Three-quark wave functions produced with the stochastic variational method are used to calculate partial decay widths of π and η decays within the elementary emission model.

The semirelativistic constituent-quark model (CQM) based on constituent-quark and Goldstone-boson degrees of freedom provides a unified description of all light and strange baryon spectra, much improved over previous CQMs relying on one-gluon-exchange hyperfine interactions only [1, 2, 3]. In particular, the Goldstone-boson-exchange (GBE) constituent-quark model with pseudoscalar coupling allows for a correct level ordering in the excitation spectra simultaneously for light (N and Δ) and strange (Λ, Σ, Ξ, Ω) baryons. The success essentially stems from the specific spin-flavor symmetry of the chiral potential V_χ coming with the new type of effective Q-Q interaction.

It is of immediate interest to test the GBE chiral quark model in problems beyond spectroscopy. As a first application of the three-Q wave functions produced by the pseudoscalar-exchange version of the model we have studied π and η decays of N and Δ resonances in Ref. [4]. These investigations have been carried out along the simplest possible approach, the elementary emission model (EEM) (see, e.g., Ref. [5]). No additional modifications of the decay vertex have been introduced. Rather, in this first step of our studies of hadronic resonance decays it has been our aim to have an unbiased test of the baryon wave functions generated by the stochastic variational method [6] in the solution for a semirelativistic three-Q problem. Here, we report further such results for the π and η decays of strange baryons.

The decay operator we employ derives from the interaction Lagrangian

$$\mathcal{L} \sim i\, g_{QM}\, \bar{\psi}\, \gamma_\mu \gamma_5\, \boldsymbol{\lambda}^F \cdot (\partial_\mu \boldsymbol{\Phi})\, \psi \,, \tag{1}$$

whith ψ and $\bar{\psi}$ the constituent-quark and $\boldsymbol{\Phi}$ the meson fields; g_{QM} is the quark-meson coupling constant, and $\boldsymbol{\lambda}^F$ represents the Gell-Mann flavor matrices. In the lowest-order nonrelativistic reduction the explicit form of the decay operator reads

$$\widehat{O}_\alpha = 3\, i\, \frac{g_{QM}}{2m_k}\, X^\alpha\, e^{-i\mathbf{q}\cdot\mathbf{r}_k} \left[\left(1 + \frac{\omega}{2m_k} \right) \boldsymbol{\sigma} \cdot \mathbf{q} - \frac{\omega}{m_k}\, \boldsymbol{\sigma} \cdot \mathbf{p}_k \right] \,, \quad (\alpha = 1, \dots, 8). \tag{2}$$

Here, ω and $\mathbf{q}$ are the meson energy and momentum, m_k, $\mathbf{p}_k$, and $\mathbf{r}_k$ are the mass, momentum, and position of quark k, $\boldsymbol{\sigma}$ are the Pauli spin matrices, and X^α is the flavor transition operator corresponding to a particular decay mode α. The operator in Eq. (2) describes the emission of a meson α (π: $\alpha = 1, 2, 3$; K: $\alpha = 4, 5, 6, 7$; η: $\alpha = 8$) from a single constituent quark k through a pointlike vertex. The first term in the square brackets is usually called "direct term", while the second one is referred to as "recoil term".

The calculation of the partial decay widths of resonance B going to the final-state baryon B' is performed in the rest frame of B. In order to get an idea of the structure dependence or independence of a certain decay one may compare the results with the direct term only and the full ones, i.e. with the recoil term included. While the first case just represents the overlap of the B and B' wave functions, the second one includes a derivative and thus introduces the spatial dependence of the wave functions more sensitively.

At the present stage, our studies of hadronic decays allow only for a gross insight into the wave function behaviour. Certainly the decay operator we employ is not realistic enough to explain the phenomenological decay widths. This is also why we have not attempted to introduce any phenomenological parametrization of the decay vertex, unlike, e.g., in Ref. [7]. The results we report are just straightforward predictions obtained with the wave functions yielded by the CQM as it stands. Also, we always employ theoretical resonance energies, as predicted by the CQM, rather than experimental ones.

Results of partial decay widths for the π- and η-decay modes of the light N^* and Δ^* resonances have already been given in Ref. [4]. Here, we add corresponding results for the π and η decays of the strange baryon resonances Λ^* and Σ^*. In Table 1 we show the partial decay widths for the transitions $\Lambda^* \to \Sigma\pi$, $\Sigma^* \to \Lambda\pi$, and $\Sigma^* \to \Sigma\pi$. The overall situation is similar as before [4]: in about half of the cases the decay widths come out right, at least by the order of magnitude, in the other cases one finds drastic differences to phenomenology. Still, it is useful to learn how much the direct and recoil terms contribute in a specific decay. Especially, whenever the latter plays a sensitive role, this hints to a pronounced wave function dependence.

For the η decays in Table 2 not much is known from experiment, as compiled by the Particle Data Group [8]. Some decay channels are not open in our model

Table 1. π partial decay widths for the GBE CQM calculated with the operator of Eq. 2. In all cases the theoretical energies of the CQM, as given in the second column, have been emlpoyed. The theoretical Σ and Λ ground-state masses entering the calculations are 1180 MeV and 1136 MeV, respectively. Experimental data are from the latest compilation of the Particle Data Group [8].

decaying resonance			Γ [MeV]		
$I(J^P_{LS})$	M_{th}	decay mode	direct	direct +recoil	Exp.
$0(\frac{1}{2}^-_{1\frac{1}{2}})$	1556	$\Lambda_{1405} \to \Sigma_{1189}$	3.1	289	50 ± 2
$0(\frac{3}{2}^-_{1\frac{1}{2}})$	1556	$\Lambda_{1520} \to \Sigma_{1189}$	3.1	4	6.5
$0(\frac{1}{2}^+_{0\frac{1}{2}})$	1626	$\Lambda_{1600} \to \Sigma_{1189}$	4.5	0.2	52 ± 37
$0(\frac{1}{2}^-_{1\frac{1}{2}})$	1683	$\Lambda_{1670} \to \Sigma_{1189}$	16	630	14 ± 7
$0(\frac{3}{2}^-_{1\frac{1}{2}})$	1683	$\Lambda_{1690} \to \Sigma_{1189}$	16	24	18 ± 6
$0(\frac{1}{2}^-_{1\frac{3}{2}})$	1779	$\Lambda_{1800} \to \Sigma_{1189}$	13	357	seen
$0(\frac{1}{2}^+_{0\frac{1}{2}})$	1800	$\Lambda_{1810} \to \Sigma_{1189}$	0.03	54	37 ± 22
$0(\frac{5}{2}^+_{2\frac{1}{2}})$	1842	$\Lambda_{1820} \to \Sigma_{1189}$	1.2	1.7	8 ± 3
$0(\frac{5}{2}^-_{1\frac{3}{2}})$	1779	$\Lambda_{1830} \to \Sigma_{1189}$	7.6	13	52 ± 19
$0(\frac{3}{2}^+_{2\frac{1}{2}})$	1842	$\Lambda_{1890} \to \Sigma_{1189}$	1.2	35	7 ± 3
$1(\frac{3}{2}^+_{0\frac{3}{2}})$	1389	$\Sigma_{1385} \to \Lambda_{1116}$	16	21	31 ± 1
$1(\frac{1}{2}^+_{0\frac{1}{2}})$	1616	$\Sigma_{1660} \to \Lambda_{1116}$	1.7	13	seen
$1(\frac{3}{2}^-_{1\frac{1}{2}})$	1678	$\Sigma_{1670} \to \Lambda_{1116}$	3.3	5.3	6 ± 3
$1(\frac{1}{2}^-_{1\frac{1}{2}})$	1678	$\Sigma_{1750} \to \Lambda_{1116}$	3.3	40	seen
$1(\frac{5}{2}^-_{1\frac{3}{2}})$	1736	$\Sigma_{1775} \to \Lambda_{1116}$	5.9	9.7	20 ± 4
$1(\frac{3}{2}^+_{0\frac{3}{2}})$	1389	$\Sigma_{1385} \to \Sigma_{1189}$	3.9	4.9	4 ± 1
$1(\frac{1}{2}^+_{0\frac{1}{2}})$	1616	$\Sigma_{1660} \to \Sigma_{1189}$	0.3	10	seen
$1(\frac{3}{2}^-_{1\frac{1}{2}})$	1678	$\Sigma_{1670} \to \Sigma_{1189}$	13	20	27 ± 9
$1(\frac{1}{2}^-_{1\frac{1}{2}})$	1678	$\Sigma_{1750} \to \Sigma_{1189}$	13	468	< 7
$1(\frac{5}{2}^-_{1\frac{3}{2}})$	1736	$\Sigma_{1775} \to \Sigma_{1189}$	1.5	2.4	4 ± 2

because the corresponding thresholds lie too high (we always use theoretical model energies). Except for the $\Lambda^* \to \Lambda\eta$ decay all partial widths are rather small. It will be interesting to see if this trend is confirmed by experiment.

As already said before [4], a realistic description of hadronic decays needs a number of decisive improvements both in the CQM description of baryons and in the decay model. Most important will be the introduction of explicit mesonic

Table 2. Same as Table 1 for the η decays.

decaying resonance			Γ [MeV]		
$I(J^P_{LS})$	M_{th}	decay mode	direct	direct +recoil	Exp.
$0(\frac{1}{2}^-_{1\frac{1}{2}})$	1683	$\Lambda_{1670}\to\Lambda_{1116}$	below	threshold	9 ± 4
$0(\frac{3}{2}^-_{1\frac{1}{2}})$	1683	$\Lambda_{1690}\to\Lambda_{1116}$	below	threshold	
$0(\frac{1}{2}^-_{1\frac{3}{2}})$	1779	$\Lambda_{1800}\to\Lambda_{1116}$	0.2	129	
$0(\frac{1}{2}^+_{0\frac{1}{2}})$	1800	$\Lambda_{1810}\to\Lambda_{1116}$	0.04	2.2	
$0(\frac{5}{2}^+_{2\frac{1}{2}})$	1842	$\Lambda_{1820}\to\Lambda_{1116}$	0.02	0.04	
$0(\frac{5}{2}^-_{1\frac{3}{2}})$	1779	$\Lambda_{1830}\to\Lambda_{1116}$	0.1	0.3	
$0(\frac{3}{2}^+_{2\frac{1}{2}})$	1842	$\Lambda_{1890}\to\Lambda_{1116}$	0.02	4.8	
$1(\frac{1}{2}^-_{1\frac{1}{2}})$	1678	$\Sigma_{1750}\to\Sigma_{1189}$	below	threshold	32 ± 18
$1(\frac{5}{2}^-_{1\frac{3}{2}})$	1736	$\Sigma_{1775}\to\Sigma_{1189}$	0	0	

components into the CQM wave functions; this is naturally possible in the GBE CQM. Furthermore, in the decay model an appropriate description of the decay mechanism taking into account the final extension of the quark-meson vertices is called for.

References

1. L.Ya. Glozman, Z. Papp, W. Plessas, K. Varga, and R.F. Wagenbrunn: Phys. Rev. **C57**, 3406 (1998)

2. L.Ya. Glozman, W. Plessas, K. Varga, R.F. Wagenbrunn: Preprint hep-ph/9706507, *to appear in* Phys. Rev. D

3. R.F. Wagenbrunn, L.Ya. Glozman, W. Plessas, K. Varga: These Proceedings

4. L.Ya. Glozman, W. Plessas, L. Theußl, K. Varga, R.F. Wagenbrunn: Proceedings of the 4th N^* Workshop, Washington, DC, 1997. πN Newsletter **14**, 99 (1998)

5. A. Le Yaouanc, L. Oliver, O. Péne, J.-C. Raynal: *Hadron Transitions in the Quark Model*; Gordon and Breach, New York, 1988

6. K. Varga, Y. Suzuki: Phys. Rev. **C52**, 2885 (1995); K. Varga, Y. Ohbayasi, Y. Suzuki: Phys. Lett. **B396**, 1 (1997)

7. R. Koniuk, N. Isgur: Phys. Rev. **D21**, 1868 (1980)

8. Particle Data Group: Eur. Phys. J. **C3**, 1 (1998)

Few-Body Systems Suppl. 11, 33–36 (1999)

Few-
Body
Systems
© by Springer-Verlag 1999

Multiquark States in a Goldstone Boson Exchange Model

Fl. Stancu*

Institute of Physics, B.5, University of Liege, Sart Tilman, B-4000 Liege 1, Belgium

Abstract. We discuss the stability of multiquark systems containing heavy flavours. We show that the Goldstone boson exchange model gives results at variance with one-gluon-exchange models.

1 Introduction

The study of exotic hadrons formed of more than three quarks and/or antiquarks ($q^m \bar{q}^n$ with $m + n > 3$) is a natural development of QCD inspired models. Both theoretical and experimental interest has been raised so far by particles described by the colour state $[222]_C$. These are the tetraquarks $q^2 \bar{q}^2$ [1], the pentaquarks $q^4 \bar{q}$ [2, 3] and the hexaquarks q^6 [1]. From theoretical general arguments [4, 5], one expects an increase in stability of multiquark systems if they contain heavy flavours $Q = c$ or b.

In the heavy sector, experiments are being planned at CERN and Fermilab to search for new heavy hadrons and in particular for doubly charmed tetraquarks [6]. Recently , the first search for pentaquarks with the flavour content $uuds\bar{c}$ and $udds\bar{c}$ has just been reported [7]. Within the confidence level of the analyzed experiments, no convincing evidence for the production of the above strange pentaquarks has been observed so far.

The theoretical predictions are model dependent. Here we are concerned with constituent quark models which simulate the low-energy limit of QCD. We compare results from models where the spin-dependent term of the quark-quark interaction is described by the chromomagnetic part of the one gluon exchange (OGE) interaction [8] with results we obtained from the Goldstone boson exchange (GBE) model [9, 10, 11, 12]. In this model the hyperfine splitting in hadrons is due to the short-range part of the Goldstone boson exchange interaction between quarks, instead of the OGE interaction of conventional

* *E-mail address:* fstancu@ulg.ac.be

models. The GBE interaction is flavour-dependent and its main merit is that it reproduces the correct ordering of positive and negative parity states in all parts of the considered spectrum. Moreover, the GBE interaction induces a strong short-range repulsion in the Λ-Λ system, which suggests that a deeply bound H-baryon should not exist [13]. This is in agreement with the high-sensitivity experiments at Brookhaven [14] where no evidence for H production has been found.

In the stability problem we are interested in the quantity

$$\Delta E = E(q^m \overline{q}^n) - E_T \qquad (1)$$

where $E(q^m \overline{q}^n)$ represents the multiquark energy and E_T is the lowest threshold energy for dissociation into two hadrons: two mesons for tetraquarks, a baryon + a meson for pentaquarks and two baryons for hexaquarks. A negative ΔE suggests the possibility of a stable compact mutiquark system.

According to Ref. [9] there is no meson exchange interaction between quarks and antiquarks. It is assumed that the $q\overline{q}$ pseudoscalar pairs are automatically included in the GBE interaction. Therefore the light quark and the heavy antiquark interact via the confinement potential only and the model Hamiltonian contains GBE interactions only between light quarks.

2   The Hamiltonian

The GBE Hamiltonian considered below has the form [10] :

$$H = \sum_i m_i + \sum_i \frac{p_i^2}{2m_i} - \frac{(\sum_i p_i)^2}{2\sum_i m_i} + \sum_{i<j} V_{\text{conf}}(r_{ij}) + \sum_{i<j} V_\chi(r_{ij}), \qquad (2)$$

with the linear confining interaction :

$$V_{\text{conf}}(r_{ij}) = -\frac{3}{8}\lambda_i^c \cdot \lambda_j^c \, C \, r_{ij}, \qquad (3)$$

and the spin–spin component of the GBE interaction in its $SU_F(3)$ form :

$$\begin{aligned} V_\chi(r_{ij}) &= \left\{ \sum_{F=1}^3 V_\pi(r_{ij})\lambda_i^F\lambda_j^F \right. \\ &\quad + \left. \sum_{F=4}^7 V_K(r_{ij})\lambda_i^F\lambda_j^F + V_\eta(r_{ij})\lambda_i^8\lambda_j^8 + V_{\eta'}(r_{ij})\lambda_i^0\lambda_j^0 \right\} \sigma_i \cdot \sigma_j, \quad (4) \end{aligned}$$

with $\lambda^0 = \sqrt{2/3}\,\mathbf{1}$, where $\mathbf{1}$ is the 3×3 unit matrix. The interaction (2) contains $\gamma = \pi, K, \eta$ and η' meson-exchange terms and the form of $V_\gamma(r_{ij})$ is given as the sum of two distinct contributions : a Yukawa-type potential containing the mass of the exchanged meson and a short-range contribution of opposite sign, the role of which is crucial in baryon spectroscopy. For a given meson γ, the

Table 3.1. Results for ΔE , Eq.(1), for charmed exotic hadrons

System	OGE	GBE
$u\,u\,\bar{c}\,\bar{c}$	19 MeV [15]	-185 MeV [16]
$u\,u\,d\,d\,\bar{c}$ $(P=+1)$	unbound	-76 MeV [17]
$u\,u\,d\,s\,\bar{c}$ $(P=-1)$	-51 MeV [18]	488 Mev [19]
$u\,u\,d\,d\,s\,c$	-7.7 MeV [20]	625 MeV [21]

exchange potential is

$$V_\gamma(r) = \frac{g_\gamma^2}{4\pi}\frac{1}{12 m_i m_j}\{\theta(r-r_0)\mu_\gamma^2 \frac{e^{-\mu_\gamma r}}{r} - \frac{4}{\sqrt{\pi}}\alpha^3 \exp(-\alpha^2(r-r_0)^2)\} \qquad (5)$$

For the Hamiltonian (2)-(5), we use the parameters of Refs.[10, 13]. These are
:

$$\frac{g_{\pi q}^2}{4\pi} = \frac{g_{\eta q}^2}{4\pi} = \frac{g_{Kq}^2}{4\pi} = 0.67, \quad \frac{g_{\eta' q}^2}{4\pi} = 1.206,$$

$$r_0 = 0.43\,fm, \quad \alpha = 2.91\,fm^{-1}, \quad C = 0.474\,fm^{-2},$$

$$m_{u,d} = 340\,MeV, \quad m_s = 440\,MeV, \qquad (6)$$

$$\mu_\pi = 139\,MeV, \quad \mu_\eta = 547\,MeV, \quad \mu_{\eta'} = 958\,MeV, \quad \mu_K = 495\,MeV.$$

3   Results

Values of ΔE , Eq. (1), for charmed systems are presented in the table both for OGE and GBE models. Details of our calculations based on the GBE model can be found in refs. [16, 17, 19, 21] together with results for $Q = b$.

One can see that the OGE and the GBE interactions predict contradictory results for the charmed exotic systems presented here: while the GBE interaction stabilizes a given system, the OGE interaction destabilizes it and vice versa. The following remarks are in order:

- As the $u\,u\,d\,d\,\bar{c}$ $(P = -1)$ pentaquarks are predicted to be unbound by a chromomagnetic interaction [18], the same system but with positive parity is expected to be even more unstable due to the increase in the kinetic energy produced by the excitation of a quark to the p-shell. While the OGE interaction favours negative parity pentaquarks with strangeness, the best candidates predicted by the GBE interaction have positive parity and are nonstrange [17].

- The GBE interaction destabilizes the hexaquarks in the presence of one or even two heavy quarks [21].

Acknowledgement. I am grateful to my collaborators J.-M. Richard, L. Glozman, S. Pepin and M. Genovese who were involved with me in parts of this work.

References

1. R. L. Jaffe: Phys. Rev. **D15**, 267 (1977)

2. C. Gignoux, B. Silvestre-Brac, J.-M. Richard: Phys. Lett. **B193**, 323 (1987)

3. H. J. Lipkin: Phys. Lett. **B195**, 484 (1987), Nucl. Phys. **A625**, 207 (1997)

4. A. V. Manohar, M. B. Wise: Nucl. Phys. **B399**, 17 (1993)

5. J.-M. Richard: Phys. Rev. **A49**, 3573 (1994)

6. M. A. Moinester: Z. Phys. **A355**, 349 (1996); see also G. Baum et al., COMPASS Collaboration, CERN-SPSLC-96-14, March 1996

7. E. M. Aitala et al.: Phys. Rev. Lett. **81**, 44 (1998)

8. A. de Rujula, H. Georgi, S. L. Glashow: Phys. Rev. **D12**, 147 (1975)

9. L. Ya. Glozman, D. O. Riska: Phys. Rep. **268**, 263 (1996)

10. L. Ya. Glozman, Z. Papp, W. Plessas: Phys. Lett. **B381**, 311 (1996)

11. L. Ya. Glozman, Z. Papp, W. Plessas, K. Varga, R. Wagenbrunn: Nucl. Phys. **A623**, 90c (1997)

12. R. Wagenbrunn, these proceeding

13. Fl. Stancu, S. Pepin, L. Ya. Glozman: Phys. Rev. **D57**, 4393 (1998)

14. R. W. Stotzer et al.: Phys. Rev. Lett. **78**, 3646 (1997)

15. B. Silvestre-Brac, C. Semay: Z. Phys. **C57**, 273 (1993), *ibid* **C59**, 457 (1993)

16. S. Pepin, Fl. Stancu, M. Genovese, J.-M. Richard: Phys. Lett. **B393**, 119 (1997)

17. Fl. Stancu: Phys. Rev. **D58**, 111501 (1998)

18. J. Leandri, B. Silvestre-Brac: Phys. Rev. **D40**, 2340 (1989)

19. M. Genovese, J.-M. Richard, Fl. Stancu, S. Pepin: Phys. Lett. **B425**, 171 (1998)

20. J. Leandri, B. Silvestre-Brac: Phys. Rev. **D47**, 5083 (1993)

21. S. Pepin, Fl. Stancu: Phys. Rev. **D57**, 4475 (1998)

Few-Body Systems Suppl. 11, 37–42 (1999)

Few-
Body
Systems
© by Springer-Verlag 1999

Hypercentral Constituent Quark Model *

M.M. Giannini[†] and E. Santopinto

Dipartimento di Fisica dell'Università di Genova, via Dodecaneso 33, 16164 Genova, Italy

Abstract. We have formulated a constituent quark model based on a hypercentral approach, taking into account three-body force effects and standard two-body potential contributions. We report our results on the baryon spectrum, the helicity amplitudes and the transition form factors, comparing them with experimental data. The effect of introducing relativistic corrections at a kinematical level is also briefly discussed.

Constituent Quark Models have been recently widely used for the description of the internal structure of baryons, leading to a satisfactory account of the spectrum [1, 2, 3, 4]. The study of hadron spectroscopy is very useful, however, it is not sufficient to distinguish among the various forms of quark dynamics. To this end one has to study in a consistent way all the physical observables of interest, in particular, besides the spectrum, the photocouplings, the electromagnetic form factors and the strong decay amplitudes. Such a systematic study of baryon properties is better performed within a general framework, and in this respect a hypercentral approach to quark dynamics can be used [5]. This method is sufficiently general to investigate new dynamical features, such as three-body mechanisms, and also to reformulate and/or include the currently used two-body potential models.

Before illustrating the results, we briefly remind the theoretical framework proposed in [5]. The internal quark motion is described by the Jacobi coordinates ρ and λ:

$$\rho = \frac{1}{\sqrt{2}}(r_1 - r_2), \quad \lambda = \frac{1}{\sqrt{6}}(r_1 + r_2 - 2r_3),\tag{1}$$

or equivalently, ρ, Ω_ρ, λ, Ω_λ. In order to describe three-quark dynamics it is convenient to introduce the hyperspherical coordinates, which are obtained

---

*Partially supported by EC-contract number ERB FMRX-CT96-0008
[†]e-mail: giannini@genova.infn.it

substituting the absolute values ρ and λ by

$$x = \sqrt{\rho^2 + \lambda^2} \ , \qquad \xi = arctg(\frac{\rho}{\lambda}), \tag{2}$$

where x is the hyperradius and ξ the hyperangle. In this way one can use the hyperspherical harmonic formalism [6].

The quark potential, V, is assumed to depend on the hyperradius x only, that is to be hypercentral. Therefore, $V = V(x)$ is in general a three-body potential, since the hyperradius x depends on the coordinates of all the three quarks. In this case, in the three-quark wave function one can factorize the angular and hyperangular parts, which are given by the known hyperspherical harmonics [6]. The Schrödinger equation is then reduced to a single hyperradial equation:

$$[\frac{d^2}{dx^2} + \frac{5}{x} \frac{d}{dx} - \frac{\gamma(\gamma + 4)}{x^2}] \ \psi_{[\gamma]}(x) = -2m \ [E - V(x)] \ \psi_{[\gamma]}(x) \ , \tag{3}$$

where $\psi_{[\gamma]}(x)$ is the hyperangular wave function and the grand angular quantum number γ is given by $\gamma = 2v + l_\rho + l_\lambda$; l_ρ and l_λ are the angular momenta associated with the ρ and λ variables and $v = 0, 1, \dots$.

There are at least two hypercentral potentials which can be solved analytically. First, the h.o. potential, which has a two-body character, turns out to be exactly hypercentral, since

$$\sum_{i<j} \frac{1}{2} k \ (r_i - r_j)^2 = \frac{3}{2} k \ x^2 = V_{h.o}(x) \tag{4}$$

The second one is the 'hypercoulomb' potential [7, 8, 9]

$$V_{hyc}(x) = -\frac{\tau}{x}. \tag{5}$$

This potential is not confining, however it has interesting properties. In fact it leads to a power-law behaviour of the proton form factor [8] and of all the transition form factors [10] and has a perfect degeneracy between the first 0^+ excited state and the first 1^- states [8, 9], which can be respectively identified with the Roper resonance and the negative parity resonances. This degeneracy seems to be in agreement with phenomenology and is typical of an underlying $O(7)$ symmetry [9]. This feature cannot be reproduced in models with only two-body forces and/or harmonic oscillator bases since the excited $L = 0$ state, having one more node, lies above the $L = 1$ state.

The dynamic symmetry $O(7)$ of the hyperCoulomb problem can be used to obtain the eigenvalues using purely algebraic methods, similarly to what is done in the hydrogen atom case with the $O(4)$ symmetry. In fact, the hyperCoulomb Hamiltonian can be rewritten as [9]

$$H = -\frac{\tau^2 m}{2 \ [C_2(O(7)) + \frac{25}{4}]} \ , \tag{6}$$

where $C_2(O(7))$ is the quadratic Casimir invariant of $O(7)$. The eigenvalues of $C_2(O(7))$ in the totally symmetric representations of $O(7)$ are $\omega(\omega + 5)$, with $\omega = 0, 1, ...\infty$, and introducing the principal quantum number $n = \omega + \frac{5}{2}$, the energy eigenvalues can be written as

$$E = -\frac{\tau^2 m}{2n^2}. \tag{7}$$

We shall present results obtained with two types of interactions: a) The hyperCoulomb potential plus a "small" confinement and a hyperfine interaction with spin-spin and tensor terms [10]; the calculation of all quantities of interest are performed analytically. b) The hypercentral potential of the form [5]

$$V(x) = -\frac{\tau}{x} + \alpha x, \tag{8}$$

to which a hyperfine interaction of the standard form is added. The hypercentral equation is solved numerically, using a computer code, which has been tested by comparison with the known analytical solutions. Model b) gives a reasonable description of the spectrum with $\tau = 4.59$ and $\alpha = 1.61\ fm^{-2}$. With these parameters (and a standard strength of the hyperfine interaction) the model has been used to calculate the helicity amplitudes for the photoexcitation of the nucleon resonances [11] and the transition form factors to the negative parity resonances [12]. The helicity amplitudes are in qualitative agreement with data and are comparable with the results of various models proposed in the literature, including those which take into account some relativistic effects [13, 14, 11]. This may due to the fact that the states in the different models have the same $SU(6)$-spin-isospin structure.

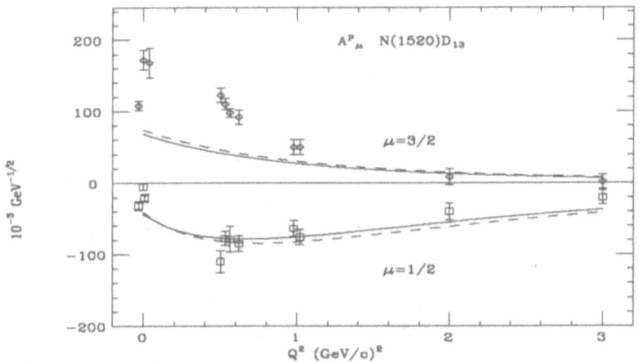

Figure 1. Comparison between the experimental data for the helicity amplitudes $A^p_{3/2}, A^p_{1/2}$ for the $D_{13}(1520)$ resonance and the calculations with the potentials a) (dashed curve) and b) (full curve). The data are from the compilation of ref. [16].

As for the proton transition form factors, in Figs. 1-2 we give the results for the $D_{13}(1520)$ and $S_{11}(1535)$ resonances [12]. The medium Q^2 behaviour is very well reproduced, while serious problems are present at small Q^2, specially in the $A^p_{3/2}$ amplitude of the transition to the $D_{13}(1520)$ state. It should be noted that in the calculations the hyperfine mixing has been taken into account for all states; this is a crucial aspect for the $S_{11}(1650)$-amplitude, which would be identically zero in pure $SU(6)$- symmetry.

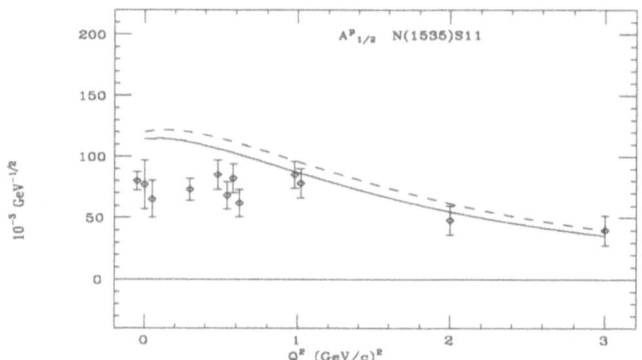

Figure 2: The same as in Fig. 1, for the helicity amplitude $A^p_{1/2}$ of the $S_{11}(1535)$ resonance.

The impossibility of reproducing the low-Q^2 part of the helicity amplitudes is a common feature of all constituent quark models. This discrepancy could be ascribed to the non-relativistic character of the model. To this end, we have included relativistic corrections [15], performing the calculations in the equal velocity frame and applying Lorentz boosts to the nucleon and resonance states. Expanding the transition matrix elements in quark momenta, one gets for any helicity amplitude:

$$A_{rel} = F \, A_{non-rel}(q_{eff}), \qquad (9)$$

where F is a known kinematical factor and $q_{eff} = q \, \frac{M_N}{E_N}$ is an effective momentum transfer. In Figs. 3 and 4 we report the calculations of the helicity amplitudes including the relativistic corrections. The modifications are not important in the whole range, thereby not destroying the agreement at medium-high Q^2 but giving no improvement at small Q^2.

The hypercentral three-body approach supplies a general framework, allowing a consistent and systematic analysis of various dynamical mechanisms. The various electromagnetic properties can be obtained by means of a parameter-free evaluation. The inclusion of relativistic corrections seems beneficial, but not sufficient to explain the discrepancy with the experimental data. This means that an improvement is needed, in particular one should include explicitly other mechanisms, such as the $q\bar{q}$-pair creation, which increases in importance in the outer region and is expected to give contributions to the excitation strength in the small Q^2-region.

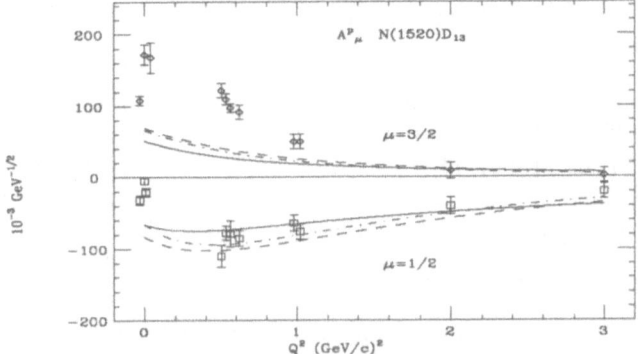

Figure 3. Comparison between the experimental data for the helicity amplitudes $A^p_{3/2}, A^p_{1/2}$ for the $D_{13}(1520)$ resonance and the calculations with the potential b). The full curve is the calculation including the relativistic corrections described in the text. The dashed and dot-dashed curves are the non-relativistic results in the Equal Velocity and in the Breit frames, respectively. Data are from the compilation of ref. [16].

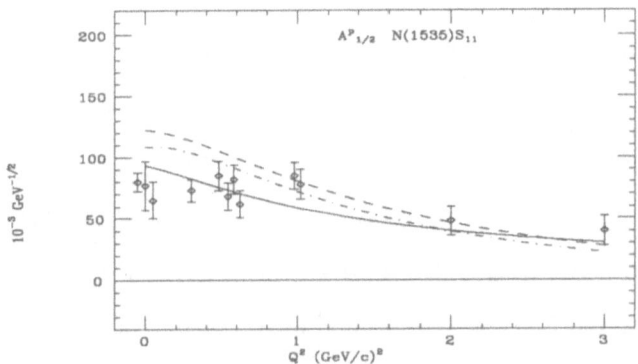

Figure 4. The same as in Fig. 3, for the helicity amplitude $A^p_{1/2}$ of the $S_{11}(1535)$ resonance.

References

1. N. Isgur and G. Karl: Phys. Rev. **D18**, 4187 (1978); **D19**, 2653 (1979); **D20**, 1191 (1979); S. Godfrey and N. Isgur: Phys. Rev. **D32**, 189 (1985); S. Capstick and N. Isgur: Phys. Rev. **D 34**, 2809 (1986)

2. M.M. Giannini: Rep. Prog. Phys. **54**, 453 (1991)

3. D.O. Riska: : In these Proceedings

4. F. Iachello: lecture presented at this Workshop

5. M. Ferraris, M.M. Giannini, M. Pizzo, E. Santopinto and L. Tiator: Phys. Lett. **B364**, 231 (1995)

6. G. Morpurgo: Nuovo Cimento 9, 461 (1952); Yu. A. Simonov: Sov. J. Nucl. Phys. 3, 461 (1966); J. Ballot and M. Fabre de la Ripelle: Ann. of Phys. (N.Y.) **127**, 62 (1980); M. Fabre de la Ripelle: in *"Models and Methods in Few-Body Physics"* (L.S. Ferreira, A. C. Fonseca and L. Streit eds.) *Lecture Notes in Physics 273*, Springer (Berlin), 1987, p. 283

7. M. Fabre de la Ripelle and J. Navarro: Ann. Phys. (N.Y.) **123**, 185 (1979)

8. H.J. Lipkin: Rivista Nuovo Cimento **I** , 134 (1969); J. Leal Ferreira and P. Leal Ferreira, Lett. Nuovo Cimento **III**, 43 (1970); M.M. Giannini: Nuovo Cimento **A76**, 455 (1983); D. Drechsel, M.M. Giannini and L. Tiator: in *"The Three-Body Force in the Three-Nucleon System"*, eds. B.L. Berman and B.F. Gibson, **Lecture Notes in Physics 260**, 509 (1986); Few-Body Syst. Suppl. **2**, J.-L. Ballot and M. Fabre de la Ripelle eds. , 448 (1987)

9. E. Santopinto, M.M. Giannini and F. Iachello: in *"Symmetries in Science VII*, ed. B. Gruber, Plenum Press, New York, 445 (1995); F. Iachello: *ibid* 213 (1995)

10. E. Santopinto, F. Iachello and M.M. Giannini: Nucl. Phys. **A623**, 100c (1997); Eur. Phys. J. **A1**, 307 (1998)

11. M. Aiello, M. Ferraris, M.M. Giannini, M. Pizzo and E. Santopinto: Phys. Lett. **B387**, 215 (1996)

12. M. Aiello, M. M. Giannini, E. Santopinto: J. Phys. **G24**, 753 (1998)

13. L. A. Copley, G. Karl and E. Obryk: Phys. Lett. **29**, 117 (1969)

14. R. Koniuk and N. Isgur: Phys. Rev. **D21**, 1868 (1980)

15. M. De Sanctis, E. Santopinto, M.M. Giannini: Eur. Phys. J. **A1**, 187 (1998); Eur. Phys. J. **A2**, 403 (1998)

16. V. Burkert: private communication.

Few-Body Systems Suppl. 11, 43–46 (1999)

Few-
Body
Systems
© by Springer-Verlag 1999

The Nucleon Wave Function in Light-Front Dynamics

V.A. Karmanov*

Lebedev Physical Institute, Leninsky Prospekt 53, 117924 Moscow, Russia

Abstract.
   The general spin structure of the relativistic nucleon wave function in the $3q$-model is found. It contains 16 spin components, in contrast to 8 ones known previously. The explicitly covariant form of the wave function automatically takes into account the relativistic spin rotations, without introducing any Melosh rotation matrices. It also reduces the calculations to the standard routine of the Dirac matrices and of the trace techniques. The importance of investigating the influence of the extra components on the observables is emphasized.

1   Introduction

Many calculations of the nucleon properties (magnetic moments, form factors, etc.) are carried out in the framework of the light-front dynamics (LFD) with the nucleon wave function in the $3q$ model containing one or a few spin components. Inspite of the fact that the average quark momenta in the nucleon exceed their masses, the nonrelativistic wave function is often used. It is implied that in general the full basis contains eight components found in the paper [1].

The aim of the present paper, based on [2], is two-fold. First of all, we show that eight components do not exhaust the relativistic nucleon wave function. We will find another eight components, so the total number of them is sixteen. In general, this is related to the fact known long ago [3] that in a many-body system the parity conservation does not reduce the number of the spin components. This is so for a relativistic three-body system and for any n-body system for $n \geq 4$ (both relativistic and nonrelativistic one). Hence, for the relativistic nucleon we get

$$N = (2S_1 + 1)(2S_2 + 1)(2S_3 + 1)(2S_N + 1) = 2 \times 2 \times 2 \times 2 = 16.$$

These 16 components are forming the full basis for the nucleon wave function. Their relative magnitude is determined by dynamics. In nonrelativistic limit their number is reduced to 8.

*e-$mail$: karmanov@sci.lebedev.ru

Secondly, we represent the nucleon wave function in the $3q$ model in the explicitly covariant form. This will allow one to use in calculations the standard Dirac-matrices algebra and the trace techniques. In particular, we will see that there is no any need to introduce explicitly any Melosh rotation matrices: the covariant approach incorporates automatically the spin rotation effects. In examples of the proton magnetic moment and of the axial nucleon form factor, we reproduce by this way the results which in the standard approach are obtained due to averaging the Melosh matrices.

The above problems is solved in the explicitly covariant version of LFD developed in a series of papers (see for review [4]). In the standard LFD the wave functions are defined on the light-front plane $t + z = 0$. In the covariant approach the wave functions are defined on the light-front plane given by the invariant equation $\omega \cdot x = 0$, where ω is the four-vector with $\omega^2 = 0$. This provides all the advantages of the explicit covariance, similar to advantages of the Feynman graph techniques. The standard approach is obtained as a particular case at $\omega = (1, 0, 0, -1)$. However, we would like to emphasize that this particular choice of ω does not reduce the number of the components of the nucleon wave function, it is 16 in any version of LFD.

2   Constructing the basis states

Eight basis states were constructed by Dziembowsky [1]:

$$
\begin{array}{ll}
(1) & [\bar{u}(k_1)U_c\bar{u}(k_2)][\bar{u}(k_3)\gamma_5 u(p)] \\
(2) & [\bar{u}(k_1)\gamma_5 U_c\bar{u}(k_2)][\bar{u}(k_3)u(p)] \\
(3) & [\bar{u}(k_1)\hat{p}\gamma_5 U_c\bar{u}(k_2)][\bar{u}(k_3)u(p)] \\
(4) & [\bar{u}(k_1)\hat{p}U_c\bar{u}(k_2)][\bar{u}(k_3)\gamma_5 u(p)] \\
(5) & [\bar{u}(k_1)\gamma^\mu U_c\bar{u}(k_2)][\bar{u}(k_3)\gamma_\mu\gamma_5 u(p)] \\
(6) & [\bar{u}(k_1)\gamma^\mu\gamma_5 U_c\bar{u}(k_2)][\bar{u}(k_3)\gamma_\mu u(p)] \\
(7) & [\bar{u}(k_1)\sigma^{\mu\nu}p_\nu U_c\bar{u}(k_2)][\bar{u}(k_3)\gamma_\mu\gamma_5 u(p)] \\
(8) & [\bar{u}(k_1)\sigma^{\mu\nu}U_c\bar{u}(k_2)][\bar{u}(k_3)\sigma_{\mu\nu}\gamma_5 u(p)]
\end{array}
\tag{1}
$$

However, it turns out that this list is not full. One can construct the pseudoscalar:

$$
C_{ps} = \epsilon^{\mu\nu\rho\gamma}k_{1\mu}k_{2\nu}k_{3\rho}p_\gamma
\tag{2}
$$

It is not zero, since the bound quarks are off-energy-shell: $k_1+k_2+k_3 = p+\omega\tau \neq p$. In ordinary approach this corresponds to the well known conservation law:

$$
\boldsymbol{k}_{1\perp} + \boldsymbol{k}_{2\perp} + \boldsymbol{k}_{3\perp} = \boldsymbol{p}_\perp, \quad k_{1+} + k_{2+} + k_{3+} = p_+,
$$

but $k_{1-} + k_{2-} + k_{3-} \neq p_-$. Using this pseuduscalar, we construct another eight basis states:

$$
(9) \qquad C_{ps}[\bar{u}(k_1)U_c\bar{u}(k_2)][\bar{u}(k_3)u(p)]
$$

$$(10) \quad C_{ps}[\bar{u}(k_1)\gamma_5 U_c \bar{u}(k_2)][\bar{u}(k_3)\gamma_5 u(p)]$$

$$(11) \quad C_{ps}[\bar{u}(k_1)\hat{p} U_c \bar{u}(k_2)][\bar{u}(k_3) u(p)]$$

$$(12) \quad C_{ps}[\bar{u}(k_1)\hat{p}\gamma_5 U_c \bar{u}(k_2)][\bar{u}(k_3)\gamma_5 u(p)]$$

$$(13) \quad C_{ps}[\bar{u}(k_1)\gamma^\mu U_c \bar{u}(k_2)][\bar{u}(k_3)\gamma_\mu u(p)]$$

$$(14) \quad C_{ps}[\bar{u}(k_1)\gamma^\mu \gamma_5 U_c \bar{u}(k_2)][\bar{u}(k_3)\gamma_\mu \gamma_5 u(p)]$$

$$(15) \quad C_{ps}[\bar{u}(k_1)\sigma^{\mu\nu} p_\nu U_c \bar{u}(k_2)][\bar{u}(k_3)\gamma_\mu u(p)]$$

$$(16) \quad C_{ps}[\bar{u}(k_1)\sigma^{\mu\nu} U_c \bar{u}(k_2)][\bar{u}(k_3)\sigma_{\mu\nu} u(p)] \tag{3}$$

The structures (9-16) are obtained from (1-8) by deleting or adding γ_5, that changes their parity, and multiplying then by C_{ps}, that restores their parity. So, the parity of (9-16) and (1-8) is the same. The structures (9-16) disappear on the energy shell $p = k_1 + k_2 + k_3$, since in this case $C_{ps} = 0$.

The wave function is represented in the form of the sum of the structures (1) and (3) with the scalar coefficients, properly symmetrized [2]. These coefficients depend, in general, on the scalar products of the quark momenta. Due to that they take into account the quark angular momenta.

3 Covariant wave function

Consider, for example, the nucleon wave function in c.m.-system with fully symmetrical S-wave spin structure:

$$\Psi_S = \frac{\psi_S}{\sqrt{72}}[3 + (\boldsymbol{\sigma}_{12} \cdot \boldsymbol{\sigma}_{3N})(\boldsymbol{\tau}_{12} \cdot \boldsymbol{\tau}_{3N})] \tag{4}$$

where $\boldsymbol{\sigma}_{12} = (w_1^\dagger \boldsymbol{\sigma} \sigma_y w_2)$, $\boldsymbol{\sigma}_{3N} = (w_3^\dagger \boldsymbol{\sigma} \sigma_y w_N)$ and similarly for the isospin matrices $\boldsymbol{\tau}_{12}, \boldsymbol{\tau}_{3N}$. In arbitrary system it is multiplied by the Melosh rotation matrices.

Our aim is to represent it in the explicitly covariant, four-dimensional form, in terms of the usual Dirac spinors, avoiding any Melosh matrices. For this aim we introduce the projection operators:

$$\Pi_+ = \frac{\mathcal{M} + \hat{\mathcal{P}}}{2\mathcal{M}}, \quad \Pi_- = \frac{\mathcal{M} - \hat{\mathcal{P}}}{2\mathcal{M}} \tag{5}$$

where $\mathcal{P} = k_1 + k_2 + k_3 = p + \omega\tau$, $\mathcal{M}^2 = \mathcal{P}^2$. We emphasize that $\mathcal{M}$ here is the effective mass of the free quarks. Then the wave function (4) is covariantly represented as:

$$\Psi_S = \frac{\psi_S}{\sqrt{72}} c_1 c_2 c_3 c_N \{3[\bar{u}(k_1)\Pi_+\gamma_5 U_c \bar{u}(k_2)][\bar{u}(k_3)\Pi_+ u(p)]$$
$$- [\bar{u}(k_1)\Pi_+\gamma^\mu \Pi_- U_c \bar{u}(k_2)][\bar{u}(k_3)\Pi_+\gamma_\mu \gamma_5 \Pi_+ u(p)](\boldsymbol{\tau}_{12} \cdot \boldsymbol{\tau}_{3N})\}, \tag{6}$$

where $c_{1,2,3} = 1/\sqrt{\epsilon_{1,2,3} + m}$, $c_N = 1/\sqrt{\epsilon_N + M}$ and, e.g., $\epsilon_1 = \sqrt{k_1^2 + m^2}$ is the energy of the quark 1. In the system where $k_1 + k_2 + k_3 = 0$ this wave function *exactly* coincides with (4). There is no nonrelativistic approximation. The wave function (6) can be decomposed in terms of the structures (1), (3). Other states are represented similarly.

4 Conclusion

We have established the general spin structure of the nucleon wave function in 3q-model. This structure is more complicated than the one used in the calculations of the nucleon properties. It contains sixteen spin components forming the full basis for decomposition of the wave function. The wave function is represented in the explicitly covariant form. The calculations of the form factors are then straightforward, through the standard routine of the Dirac matrices and of the trace techniques, like in the case of the Feynman approach. This fact is the strong advantage of the explicitly covariant version of LFD.

Among four spin structures independent of momenta there is one symmetric structure, eq. (4). Multiplied by the Melosh matrices, with the function ψ_S approximated by the S-wave, it is often used for calculating the nucleon properties (form factors, magnetic moments, etc.). Being represented in the covariant form by eq.(6) it does not require any Melosh matrices. Our calculations [2] by this way of the proton anomalous magnetic moment and of the nucleon axial form factor reproduce the results found previously by the standard methods. However, in view of the existence of other 15 componets, this wave function seems oversimplified.

We believe that in order to make comparison with experiment one should investigate the magnitude of other components and their influence on the observables. Some of these components may be important. Even in the deuteron wave function we found the component (so called f_5) which dominates in relativistic region over all other components, including usual S- and D-waves [4]. The nucleon is much more relativistic system than the deuteron. Therefore the same can take place in the wave function of the nucleon and of the nucleon resonances.

References

1. Z. Dziembowski: Phys. Rev. **D37**, 768 (1988)

2. V.A. Karmanov: Nucl. Phys. **A644**, 165 (1988)

3. V.M. Kolybasov: Nucl. Phys. **68**, 8 (1965)

4. J. Carbonell, B. Desplanques, V.A. Karmanov and J.-F. Mathiot: Phys. Rep., **300**, 215 (1998)

Few-Body Systems Suppl. 11, 47–52 (1999)

Few-
Body
Systems
© by Springer-Verlag 1999

Lorentz Covariant Spin–Grouping of Baryon Resonances

M. Kirchbach *

Institute for Nuclear Physics, University Mainz, D-55099 Mainz, Germany

Abstract. A well pronounced spin–grouping of baryon resonances to O(4) partial waves is found in baryon spectra and shown to be well interpreted in terms of Lorentz group representations of the type $\{\frac{1}{2} + l', \frac{1}{2} + l'\} \otimes [\{\frac{1}{2}, 0\} \oplus \{0, \frac{1}{2}\}]$ with l' integer. In this way the relativistic description of finite dimensional resonance towers containing higher-spin states becomes possible. It is further argued that the nucleon excitations into the $l' = 1$ and $l' = 2$ multiplets are chiral phase transitions.

The relativistic description of resonances with higher-spins is one of the oldest unsolved problems of baryon spectroscopy. Attempts to handle it in the mid to late 60's failed through the improper choice for the irreducible representations of the Lorentz group considered to embed the baryon excitations known at that time. For example, Kleinert considered in [1] unitary representations of the Lorentz group describing a spin-tower of states without upper end starting for the nucleon $P_{11}(939)$ with the $D_{13}(1520)$, $F_{15}(1680)$, $G_{17}(2190)$ etc. resonances. Alternatively, Regge phenomenology required in the t-channel near origin (i.e. $t_\mu \approx 0$) a grouping of approximately mass degenerate baryon Regge poles into the finite dimensional non-unitary representations of the Lorentz group of the type $\{l_1, l_2 \pm \frac{1}{2}\}$ with integer l_i's (so called Toller poles [2]). Unfortunately, both the schemes mentioned above are in disagreement with data, in particular because the parity duplication of the resonances required by them is absent from the spectra. On the other side, Weinberg described baryon resonances as isolated $\{J, 0\} \oplus \{0, J\}$ higher–spin states. In order to obtain the corresponding relativistic equations of motion Weinberg put forward the idea in [3] to embed the $\{J, 0\} \oplus \{0, J\}$ state, considered as pointlike, within the Rarita–Schwinger multiplet $\{\frac{1}{2} k, \frac{1}{2} k\} \otimes [\{\frac{1}{2}, 0\} \oplus \{0, \frac{1}{2}\}]$ (with k integer and related to J via $k = J - 1/2$) and to eliminate all the lower–spin components,

---

*E-mail address: mariana@kph.uni-mainz.de

considered as *redundant unphysical states* by means of a set of suitably choosen auxiliary conditions. Within Weinberg's scheme a spin–J resonance of mass M and momentum p_μ is described by a totally symmetric traceless rank–k Lorentz tensor with Dirac spinor components $\Psi_{\mu\mu_1\ldots\mu_{k-1}}$ satisfying the conditions

$$(\gamma \cdot p + M)\,\Psi_{\mu\mu_1\ldots\mu_{k-1}} \;=\; 0\,, \tag{1}$$

$$p^\mu\Psi_{\mu\mu_1\ldots\mu_{k-1}} = 0\,, \qquad \gamma^\mu\Psi_{\mu\mu_1\ldots\mu_{k-1}} = 0\,. \tag{2}$$

Here, Proca's equation in Eq. (2) eliminates the spin–0 and preserves the spin–1 components from the $\{1/2, 1/2\}$ representations as associated with the Lorentz indices. The second set of auxiliary conditions in (2) eliminates then the longitudinal components of the spin–1 field. It ensures that the state surviving the elimination procedure carries the maximal allowed helicities and describes, therefore, the highest–spin state of the Rarita–Schwinger multiplet. A serious disadvantage of the Rarita–Schwinger scheme for treating higher–spin states is that for off–shell particles the auxiliary conditions are not any longer valid and the appearance of arbitrary unphysical parameters in the Lagrangian describing the couplings of the states under consideration to external fields is inevitable. Because of all these difficulties, the requirement for relativistic and thereby for field theoretical description of higher–spin resonances was gradually given up by hadron spectroscopy thus opening space for pragmatic evidently non-relativistic schemes such as the $O(3)_L \otimes SU(6)_{SF}$ hadron book–keeping symmetry. Through this group the trivial spin-flavor (SF) correlation between three quarks in the 1s-shell was naively extended to arbitrary orbital angular momenta. The problems raised by symmetry groups of strong interaction based on tight correlations between the spin- and flavor degrees of freedom of quarks not only have conceptual disadvantages, but also their predictions are not satisfactory. For example, the $O(3)_L \otimes SU(6)_{SF}$ classification scheme predicts a substantial excess of baryon excitations known as 'missing resonances'. Finally, the basic idea of the multiplets as well separated families of particles of different internal but same space–time properties, seems quite inappropriate here, where the spacing between the $O(3)_L \otimes SU(6)_{SF}$ multiplets is much smaller as compared to the mass splitting within the multiplets.

On the other side, speed plot analysis of the pole positions on the complex energy plane of various baryon resonances $(L_{2I,2J})$ with masses below ~ 2500 MeV performed by Höhler [4] revealed a well-pronounced spin–clustering in baryon spectra. This is quite a surprising result as it was not anticipated by any model or theory. Independently, the symmetry of all reported N, Δ and Λ baryon excitations with masses below 2500 MeV was re-analyzed in our previous work [5] and shown to be governed by $O(1,3)_{LS} \otimes SU(2)_I$ rather than by $O(3)_L \otimes SU(6)_{SF}$ as predominantly used since the invention of the naive three flavor quark model. The $O(1,3)_{LS} \otimes SU(2)_I$ symmetry indicates that the spin-orbital correlation between quarks is much stronger than the spin-flavor one. Indeed, it was demonstrated in [5] that Höhler's poles correspond to $O(4)$ partial waves as they are identical to Lorentz multiplets of the type $\{\frac{1}{2} + l', \frac{1}{2} + l'\} \otimes [\{1/2, 0\} \oplus \{0, 1/2\}]$ with l' integer. All orbital angular momenta (denoted here by l) contained within a $\{\frac{1}{2}+l', \frac{1}{2}+l'\}$ multiplet have either natural or unnatural parity [1]. Coupling a Dirac particle to l is then standard and leads to covariantly transforming spin–groups of states with $J = l \otimes \frac{1}{2}$ and $l = 0, ..., 2l'+1$ all having either natural or unnatural parities according to $J = \frac{1}{2}^+, \frac{1}{2}^-, \frac{3}{2}^-, ..., (k+\frac{1}{2})^\pi$ with $k = 2l'+1$, $\pi = (-1)^l$ for natural, and $\pi = (-1)^{l+1}$ for unnatural parities. Note that the $\{\frac{1}{2}+l', \frac{1}{2}+l'\}$ multiplets are well known from the Coulomb problem, where they correspond to even principal quantum numbers $n = k+1$. These finite dimensional unitary representations of $O(4)$, the maximal compact group of the Lorentz group, contain only once each $O(3)$ representation corresponding to an intrinsic orbital angular momentum $l = 0, ..., k$. The Coulomb multiplets yield in the $O(1,3)$ group finite dimensional *non–unitary* representations. Because of the possibility of restricting $O(1,3)$ to $SU(2) \otimes SU(2)$ for the $\{\frac{1}{2}+l', \frac{1}{2}+l'\}$ representations, however, the latter may be used for the construction of relativistic equations of motion. The baryon resonances are now collected into the following representations [5]:

$$l' = 0: \quad \Psi_\mu \quad : \quad P_{2I,1}; S_{2I,1}, D_{2I,3}, \quad \text{for} \quad I = 0, \frac{1}{2}, \frac{3}{2}, \quad \text{and}$$

$$l' = 1: \quad \Psi_{\mu\mu_1\mu_2} \quad : \quad S_{2I,1}; P_{2I,1}P_{2I,3}; D_{2I,3}, D_{2I,5}; F_{2I,5}, F_{2I,7},$$

$$l' = 2: \quad \Psi_{\mu\mu_1...\mu_4} \quad : \quad S_{2I,1}; P_{2I,1}P_{2I,3}; D_{2I,3}, D_{2I,5}; F_{2I,5}, F_{2I,7};$$

$$G_{2I,7}, G_{2I,9}; H_{2I,9}, H_{2I,11}, \quad \text{for} \quad I = \frac{1}{2}, \frac{3}{2}. \quad (3)$$

Note that the states with the highest spins $J^\pi = (k+\frac{1}{2})^\pi$ from the Lorentz multiplets under consideration and the corresponding isolated $\{J^\pi, 0\} \oplus \{0, J^\pi\}$ states may belong to different Fock spaces due to different origins for their parities. Consider, as an illustration, a spin–3/2$^-$ resonance to emerge as the highest spin state of the Lorentz multiplet $\{1/2, 1/2\} \otimes [\{1/2, 0\} \oplus \{0, 1/2\}]$. The negative parity of this resonance signals an internal P wave state. Such a natural parity state resides within a Fock space built upon a scalar vacuum. On the contrary, the negative parity of an isolated $\{3/2^-, 0\} \oplus \{0, 3/2^-\}$ state arises necessarily from the coupling of the fermion degree of freedom, when considered as fundamental, to an internal 0^- state. Such an unnatural parity state necessarily resides within a Fock space built upon a pseudoscalar vacuum. This aspect of mapping isolated higher–spin states onto multi–spinor representations of the Lorentz group is of crucial importance for selecting baryons as genuine parity partners and, therefore, for establishing the scale of chiral symmetry restoration (see below).

Considering resonances as approximate $O(4)$ partial waves brings numerous advantages relative treating them as $O(3)$ partial waves. First of all, the number of the 'missing resonances' is strongly reduced. Indeed, within that scheme

---

[1] The parity $(-1)^{L+1}$ of a single πN resonance $L_{2I,2J}$ in standard notation [8], where L stands for the relative angular momentum in the decay channel, is determined in the present classification scheme by either $(-1)^l$ or $(-1)^{l+1}$, depending on whether the parity of the intrinsic orbital angular momentum is natural or unnatural. In the present notation, L takes the values of either $L = |l-1|, (l+1)$ for natural, or $L = l$ for unnatural parities.

the Δ spectrum below 2000 MeV appears complete as all observed states fit into the $l' = 0$, and $l' = 1$ Lorentz multiplets in Eq. (2). As the F_{37} state from the $l' = 1$ Lorentz multiplet has to be paralleled in the nucleon sector by a (not reported) F_{17} resonance with a mass around 1700 MeV, only that latter state has to be viewed as a 'missing resonance' among the non-strange baryon excitations with masses below 2000 MeV. In continuing by comparing the states from the third nucleon and Δ clusters with $l'=2$, four more 'missing resonances' are predicted. These are the $H_{1,11}$, P_{31}, P_{33}, and D_{33} states with masses between 2200 and 2400 MeV. In summary, five new, still unobserved non-strange resonances have been predicted in [5]. Remarkably, the spacing of about 200 MeV among the relativistic multiplets with masses below 2000 MeV appears now much larger as compared to the maximal mass splitting of 50-70 MeV between the corresponding multiplet members (see Fig. 1). The baryon classification scheme in Eq. (3) shows that the parities of the resonances from the first spin–clusters with $l' = 0$ are always natural. This means that they can be considered to emerge within a Fock space built upon a vacuum of positive parity as spontaneously selected in the course of the realization of chiral symmetry in the hidden Nambu–Goldstone mode in the low energy regime. On the contrary, the resonances of the second and third spin–clusters

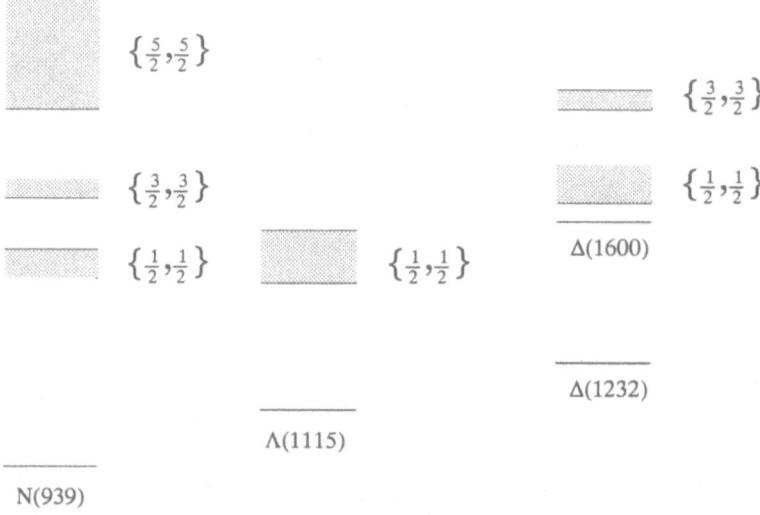

Fig. 1 Baryon spectra in terms of the $\{j,j\} \otimes [\{\frac{1}{2},0\} \oplus \{0,\frac{1}{2}\}]$ representations. Only the $\{j,j\}$ assignments are given. Each dashed area contains $4j+1$ states.

with $l' = 1$, and $l' = 2$, respectively, have always unnatural parities and can be considered to reside within a Fock space built upon a vacuum of negative parity. Both the $\psi_N \rightarrow \Psi_{\mu\mu_1}$, and $\psi_N \rightarrow \Psi_{\mu\mu_1\mu_2}$ transitions are, therefore, in contrast to the $\psi_N \rightarrow \Psi_\mu$ excitation, chiral phase transitions. A further advantage of the $O(1,3)_{LS} \otimes SU(2)_I$ classification scheme for baryons is that it opens a new possibility for the relativistic description of the higher-spin resonances as members of covariant representations. Indeed, the relativistic propagators of the spin–clusters can be directly read off from the representation theory of the Lorentz group as

$$S_{\mu\mu_1\ldots\mu_{k-1};\nu\nu_1\ldots\nu_{k-1}} = \frac{(\gamma\cdot p + M)\bigotimes_{n=1}^{n=k}(g_{\mu_n\nu_n} - \frac{1}{M^2}p_{\mu_n}p_{\nu_n})}{2M(p^2 - M^2)}, \qquad (4)$$

with M standing for the mass of the degenerate resonances. For example, the $(S_{2I,1}, D_{2I,3})$ cluster is described in terms of the Lorentz vector with Dirac spinor components Ψ_μ from Eq. (3) and its propagator is given by [5]

$$S_{\mu\nu} = \frac{(\gamma\cdot p + M)(g_{\mu\nu} - \frac{1}{M^2}p_\mu p_\nu)}{2M(p^2 - M^2)}. \qquad (5)$$

In noting that, say, the first S_{11} and D_{13} states appear separated by only 15 MeV, one sees that the relativistic contribution of these states to the amplitude of processes like meson photoproduction at threshold, can easily be calculated. Along the line of the representation theory of the Lorentz group, the construction of higher-cluster propagators and interactions with external fields is straightforward. For example, for the case of a $\mathcal{B} \to N + V$ process, where $\mathcal{B}$ stands for a Lorentz covariant spin-cluster, while V is a vector meson, a possible Lagrangian can be written as

$$\mathcal{L}_{\mathcal{B}VN} = \bar{\Psi}^{\mu\mu_1\ldots\mu_{k-1}}\Big(\frac{f_k}{m_\pi^{k-1}}\partial_{\mu_1}\ldots\partial_{\mu_{k-1}}A_\mu \;+\; \frac{f'_k}{m_\pi^k}\partial_\mu\ldots\partial_{\mu_{k-1}}\slashed{A}\Big)\psi_N, \quad (6)$$

where $k = 2l' + 1$, A_μ denotes the vector meson field, while the coupling strengths f_k and f'_k can be extracted from data.

The Lorentz covariant spin-clusters of baryon resonances introduced above are in fact nothing else but the special case of the Rarita–Schwinger representations with odd k's, the essential difference being that all the lower-spin states are no longer redundant components that need be eliminated, but *physically observable resonances* related to the composite character of the baryons. The Lorentz–Dirac index notation for the spin–clusters introduced in Eq. (3) indicates that the first resonance-tower will predominantly couple to the pion–nucleon (or η–nucleon) system carrying each one Dirac and one Lorentz index. The second and third spin–clusters will prefer couplings to multipion–nucleon final states (one Dirac- and several Lorentz indices) in agreement with the empirical observations. According to that, the reason for the suppression of the $S_{11}(1650)\to N+\eta$ decay channel as compared to the $S_{11}(1535)\to N+\eta$ one, can be a simple re–distribution of decay strength in favor of the new opened $S_{11}(1650)\to N+\pi+\pi$ channel in accordance with data. Now, the existence of the Lorentz covariant spin-grouping of baryon resonances can be interpreted to emerge via the coupling of a quark to approximately mass–degenerated Di-quarks of spins $l = 0, \ldots, 2l'+1$ having all either natural or unnatural parities. Indeed, the covariant quark-Di-quark model based on solving the Bethe-Salpeter equation reveals an internal O(4) symmetry [7] as visible from the rapid convergence of its solutions in the basis of the orthogonal (Gegenbauer) polynomials of the group O(4) corresponding to four dimensional partial waves.

52

Thus the major advantage of our new relativistic spectrum generating algebra for baryons is that it reconciles such seemingly contrary ideas of the baryon dynamics like the constituent quark model on the one side, and the multi–spinor representations of the Lorentz group corresponding to structureless (pointlike) particles, on the other side. Now the apparent analogy between the spectrum of the hydrogen atom and the baryon spectra leads to the question whether the positions of the Lorentz covariant spin–clusters is determined by the inverse squared $1/n^2$ of the principle quantum number and follow a sort of modified Balmer-series pattern. The answer to that question is positive [5]. Below quite a simple empirical recursive relation is suggested that describes with quite an amazing accuracy the reported mass averages of the resonances from the Lorentz multiplets only in terms of the cluster quantum numbers and the masses of the ground state baryons:

$$M_n - M_{n-2} = \left(\frac{1}{(n-1)^2} - \frac{1}{n^2} \right) J_{n-2}^{max} \left(J_{n-2}^{max} + 1 \right) M_{n-2},$$

$$n = 2l' + 2, \qquad l' > 0, \qquad J_{n-2}^{max} = n - \frac{1}{2}. \tag{7}$$

Here, M_n denotes the mass of the respective Lorentz multiplet. The position of the first excited nucleon spin–cluster with $l' = 0$ is given instead by (7) rather by $M_{n=2} - M_{n=1} = (\frac{1}{1^2} - \frac{1}{4})(1^2 - \frac{1}{4})M_{n=1}$ with $M_{n=1} = M_N$. For the Δ and Λ baryons, the mass scale $M_{n=1}$, has to be replaced by $(M_N + M_\Delta)/2$, and $(M_N + M_\Lambda)/2$, respectively (compare Table 2 of Ref. [6] for details). To conclude, baryon resonances group to O(4) rather than to O(3) partial waves, and the baryon spectra are evidently generated by the relativistic o(1,3)$_{LS}$⊗su(2)$_I$ group algebra rather than by the algebra of the non–relativistic O(3)$_L$⊗SU(6)$_{SF}$ group

Acknowledgement. Work supported by the Deutsche Forschungsgemeinschaft (SFB 201).

References

1. H. M. Kleinert: Fortschritte der Physik, **16**, 1 (1968).

2. M. Toller: Nuovo Cim. **54A**, 295 (1968)

3. S. Weinberg: Phys. Rev. **133**, B1318 (1964)

4. G. Höhler: In: *Physics with GeV- Particle Beams* eds. H. Machner and K. Sistemich (World Scientific, Singapore) 1995, p. 198

5. M. Kirchbach: Mod. Phys. Lett. **A12**, 2373 (1997); ibid. 3177

6. M. Kirchbach: Mod. Phys. Lett. **A13**, 823 (1998)

7. K. Kusaka *et al.:* Phys. Rev. **D55**, 5299 (1997)

8. Particle Data Group: Phys. Rev. **D54**, 1 (1996)

Few-Body Systems Suppl. 11, 53–56 (1999)

Few-
Body
Systems
© by Springer-Verlag 1999

Baryon Properties in Soliton Models *

Norberto N. Scoccola [†] [**]

Physics Dept., CNEA, Av. Libertador 8250, (1429) Buenos Aires, Argentina.

Abstract. We report on the results of some calculations of the radiative decays of decuplet baryons in the context of topological chiral soliton models. Such results are compared with those of alternative baryon models.

1 Introduction

At present, only few data are available on the electromagnetic decays of the $\frac{3}{2}^+$ baryons. Although recently the reaction $\Delta \to N\gamma$ has carefully been analyzed at MAMI, the decay parameters are still unknown for those $J = \frac{3}{2}$ to $J = \frac{1}{2}$ transitions which involve strange baryons. Upcoming experiments at JLAB [1] are expected to provide some data on these radiative decays soon and thus give more insight in the pattern of flavor symmetry breaking. Theoretical studies of these decays have been performed in a number of quark-based models like i.g. the non–relativistic quark model [2, 3], a quenched lattice calculation [3], etc. Here, we complete those studies by presenting the predictions of the Skyrme-type soliton models. In these models, baryons emerge as solitons configurations of the pseudoscalar mesons. The corresponding non-linear chiral action describes the main features of QCD at low energies. Since we are interested in hyperon properties, and in order to account for the non-negligible strange quark mass, convenient chiral symmetry breaking terms have to be included. One way to treat the kaon fields in the presence of these symmetry breaking terms is to assume that they correspond to small amplitude fluctuations off the soliton. In this picture, called "bound state approach" (BSA), hyperons appear as kaon-soliton bound states[4]. Alternatively, the "rigid rotator approach" (RRA) starts from a flavor symmetric formulation wherein non–vanishing kaon fields arise from a rigid rotation of the classical pion field. The associated collective coordinates, which parametrize these large amplitude fluctuations off the soliton, are canonically quantized to generate states which

---

*Work supported in part by a grant of Fundación Antorchas.
[†]Fellow of the CONICET, Argentina.
[**] *E-mail address:* scoccola@tandar.cnea.edu.ar

possess the quantum numbers of physical hyperons. It turns out that the resulting collective Hamiltonian can be diagonalized exactly even in the presence of flavor symmetry breaking [5]. In this contribution we will basically concentrate on this second method although results for the BSA will be also presented.

2 The model

Our starting point is a gauged effective chiral action with appropriate symmetry breaking terms written in terms of pseudoscalar octet ϕ and the photon field A_μ. The former is non–linearly represented by the chiral field $U = \exp(i\phi)$. Such action reads

$$\Gamma = \int d^4x \left\{ \frac{f_\pi^2}{4} \text{Tr} \left[D_\mu U (D^\mu U)^\dagger \right] + \frac{1}{32\epsilon^2} \text{Tr} \left[[U^\dagger D_\mu U, U^\dagger D_\nu U]^2 \right] \right\} + \Gamma_{an} + \Gamma_{sb}. \quad (1)$$

Here, $f_\pi = 93\text{MeV}$ is the pion decay constant and ϵ is the Skyrme parameter. The covariant derivative $D_\mu U = \partial_\mu U + ie\, A_\mu\, [Q, U]$ is defined via the usual electric charge matrix Q. Moreover, Γ_{an} is the Wess-Zumino action gauged to contain the photon field and Γ_{sb} a suitable symmetry breaking term that accounts for the finite pion and kaon masses as well as for the difference between the kaon and pion decay constants. The explicit expressions of Γ_{an} and Γ_{sb} in terms of U and the electromagnetic field can be found in i.g. Ref.[6].

To generate baryon states of good spin and flavor quantum numbers we consider the $SU(3)-$rotating hedgehog configuration,

$$U(\boldsymbol{r}, t) = A(t) \exp\left[i\boldsymbol{\tau} \cdot \hat{\boldsymbol{r}}\, F(r) \right] A(t)^\dagger. \quad (2)$$

Inserting this ansatz and following the standard canonical quantization one obtains the collective Hamiltonian

$$H = M_{sol} + \left(\frac{1}{2\Theta_\pi^2} - \frac{1}{2\Theta_K^2} \right) \boldsymbol{J}^2 + \frac{1}{2\Theta_K^2} C_2\left[SU(3) \right] - \frac{3}{8\Theta_K^2} + \Phi\, (1 - D_{88}). \quad (3)$$

In Eq. (3), Θ_π and Θ_K are the moments of inertia in the pionic and kaonic directions, respectively and Φ the symmetry breaking strength (see Ref.[7] for details). $\boldsymbol{J}$ denotes the spin operator while $C_2\left[SU(3) \right]$ refers to the quadratic Casimir operator of $SU(3)$. This collective Hamiltonian can be diagonalized exactly. The eigenfunctions $\Psi_B(A) = \langle A|B\rangle$ of the collective Hamiltonian (3) are identified as the wave–functions corresponding to baryon B. These are distorted $SU(3)$ D–functions since the resulting baryon eigenstates contain sizable admixtures of baryon states with appropriate spin and flavor quantum numbers in higher dimensional representations of $SU(3)$.

For our numerical calculations we take the meson masses and decay constants at their physical values and adjust the Skyrme parameter to $\epsilon = 3.9$. The observed mass differences for the $\frac{1}{2}^+$ and $\frac{3}{2}^+$ baryons are quite accurately reproduced for this set of values[7]. On the other hand, the absolute values of the baryons masses turn out to be all too large. Recently, it has been shown that even within the SU(3) Skyrme model this problem can be cured when quantum corrections of $\mathcal{O}(N_c^0)$ are included [8]. As we will see below these quantum corrections might also play an important role in the hyperon decay amplitudes.

3 Electromagnetic Decay Widths

For the radiative decays of the decuplet hyperons we are interested in both $M1$ and $E2$ allowed transitions. The corresponding partial widths can be expressed in terms of the baryon e.m. current J_μ^{em} as

$$\Gamma_{E2} = \frac{675}{8}e^2q\left|\langle\Psi_{J=\frac{1}{2}}|\int d^3r j_2(qr)\left(\frac{z^2}{r^2}-\frac{1}{3}\right)J_0^{em}|\Psi_{J=\frac{3}{2}}\rangle\right|^2, \quad (4)$$

$$\Gamma_{M1} = 18e^2q\left|\langle\Psi_{J=\frac{1}{2}}|\frac{1}{2}\int d^3r\, j_1(qr)\,\epsilon_{3ij}\,\hat{r}_i\,J_j^{em}|\Psi_{J=\frac{3}{2}}\rangle\right|^2. \quad (5)$$

where q is the photon momentum and $j_l(qr)$ are the spherical Bessel functions.

In our model, the general form of J_μ^{em} can be directly obtained from the gauged effective action (Eq. (1)). Using the ansatz (Eq. (2)), one can then get the explicit expression of the integrands in Eqs. (4,5). These expressions can be found in Ref.[6]. It should be noticed that in deriving Eq.(4) some approximations according to the Siegert's theorem have been made. The impact of such approximations on the $E2-\Delta N$ transition has been recently studied in the two flavor reduction of the model[9]. It turns out that the kinematical corrections coming from the low-momentum expansion of the Bessel functions are of the order of 5 %. In addition to this, pion fluctuations off the rotating soliton have to be included to consistently satisfy the continuity equation $\partial_\mu J^\mu = 0$ at subleading order in $1/N_C$. These induced fields account for shortcomings in the collective quantization and give corrections of the order of ($\sim 25\%$) for the ΔN case. Unfortunately such an inclusion of induced fields seems to be unfeasible in the three flavor model with symmetry breaking included. In any case, since the decuplet decays are almost completely M1–dominated these uncertainties do not affect the total decay widths in any significant way.

Results for the total decay widths are given in Table 1. There we list the corresponding widths normalized to the ΔN value. As in the case of magnetic moments, the calculated ΔN transition amplitude turns to be roughly 30% smaller than the empirical value. It has been recently shown, however, that the inclusion of next-to-leading order quantum corrections not only solves nucleon mass problem as already mentioned, but also that of the magnetic moments[10]. Preliminary results indicate that satisfactory ΔN decay amplitudes can be obtained in that case. From Table 1 we observe that the predictions of the present model (RRA) are very similar to those of the alternative approach to SU(3) Skyrme model (BSA) and also to those of quark-based models. In particular, in all models the $U-$spin selection rules that predict small $\Sigma^{*-} \rightarrow \gamma\Sigma^-$ and $\Xi^{*-} \rightarrow \gamma\Xi^-$ are rather well satisfied.

Table 1. Total decay widths normalized to that of the $\Delta \to \gamma N$ transition in various models. The results for RRA correspond to the model describe in the text while the BSA results have been taken from Ref.[11]. The data for the quark model (QM) and lattice calculation (Lat) are taken from Refs.[2, 3].

Transition	RRA	BSA	QM	Lat.
$\Sigma^{*0} \to \gamma\Lambda$	0.653	0.765	–	0.703
$\Sigma^{*-} \to \gamma\Sigma^-$	0.007	0.010	0.007	0.006
$\Sigma^{*0} \to \gamma\Sigma^0$	0.035	0.037	0.040	0.055
$\Sigma^{*+} \to \gamma\Sigma^+$	0.210	0.233	0.233	0.303
$\Xi^{*-} \to \gamma\Xi^-$	0.011	0.039	0.009	0.012
$\Xi^{*0} \to \gamma\Xi^0$	0.313	0.412	0.300	0.415

4 Conclusions

In this contribution we have discussed the hyperon radiative decay in the context of the $SU(3)$ soliton models. We have found that, when normalized to that of the ΔN transition, the corresponding predictions are qualitatively similar to those of alternative quark based models.

References

1. R.A. Schumacher: Contribution to this workshop.

2. J.W. Darewych, M. Horbatsch and R. Koniuk: Phys. Rev. **D28**, 1125 (1983)

3. D.B. Leinweber, T. Draper and R.M. Woloshyn: Phys. Rev. **D48**, 2230 (1993)

4. C.G. Callan and I. Klebanov: Nucl. Phys. **B262**, 365 (1985)

5. H. Yabu and K. Ando: Nucl. Phys. **B301**, 601 (1988)

6. T. Haberichter, H. Reinhardt, N.N. Scoccola and W. Weigel: Nucl. Phys. **A615**, 291 (1997)

7. H. Weigel: Int. J. Mod. Phys **A11**, 2419 (1996)

8. N.N. Scoccola and H. Walliser: Phys. Rev. **D58**, 094037 (1998)

9. H. Walliser and G. Holzwarth: Z. Phys. **A357**, 317 (1997)

10. F. Meier and H. Walliser: Phys. Rep. **289**, 383 (1997)

11. C.L. Schat, C. Gobbi and N.N. Scoccola: Phys. Lett. **B356**, 1 (1995)

Few-Body Systems Suppl. 11, 57–65 (1999)

Few-
Body
Systems
© by Springer-Verlag 1999

QCD Sum Rules and Soft-Hard Interplay for Hadronic Form Factors

A. V. Radyushkin*

Physics Department, Old Dominion University, Norfolk, VA 23529, USA and Jefferson Lab, Newport News,VA 23606, USA

Abstract. We discuss two types of contributions to hadronic form factors in QCD: hard gluon exchange and soft wave function overlap. Within the QCD sum rule approach, the hard contribution has strong numeric suppression by factor $(\alpha_s/\pi) \sim 0.1$ for each exchange. For this reason, the soft contribution dominates at accessible momentum transfers. The "humpy" distribution amplitudes used to enhance hard terms cannot be derived from QCD sum rules in a self-consistent way. The estimates of soft terms obtained within the local quark-hadronic duality approach in all cases are close to existing data, providing an experimental evidence that hard terms are small.

1 Soft *vs.* Hard

It is still a matter of controversy whether hard scattering [1] or the soft wave function overlap mechanism [2]) is responsible for the experimentally observed power-law behaviour of elastic hadronic form factors. At sufficiently large momentum transfer, the soft mechanism is dominated by configurations in which one of the quarks carries almost all the momentum of the hadron. On the other hand, the hard scattering term is generated by the valence configurations with small transverse sizes and finite light-cone fractions of the total hadron momentum carried by each valence quark. For large Q^2 in QCD, this difference results in an extra $1/Q^2$-suppression of the soft term compared to the hard scattering one.

The hard term can be written in a factorized form [3],[4],[5] as a product of a perturbatively calculable hard scattering amplitude and two distribution amplitudes (DAs) describing how the large longitudinal momentum of the initial and final hadrons is shared by their constituents. This mechanism involves exchange of virtual gluons, each exchange bringing in a noticeable suppression

*Also at: Laboratory of Theoretical Physics, JINR, Dubna, Russia

factor $(\alpha_s/\pi) \sim 0.1$. As a result, to describe existing data by the hard contribution alone, one should increase somehow the magnitude of the hard scattering term.

This is usually achieved by using the DAs with a "humped" shape [6]. However, the passive quarks in this situation carry a rather small fraction of the hadron momentum and, as pointed out in ref.[7], the "hard" scattering subprocess, even at rather large momentum transfers $Q^2 \sim 10\,GeV^2$, is dominated by rather small gluon virtualities. This means that the hard scattering scenario heavily relies on the assumption that the asymptotic pQCD approximations (e.g., the $1/k^2$-behaviour of the gluon propagator $D^c(k)$) are accurate even for momenta k smaller than $300\,MeV$, i.e., in the region strongly affected by finite-size effects, nonperturbative QCD vacuum fluctuations, etc. Including these effects decreases the magnitude of the gluon propagator $D^c(k)$ at small spacelike k converting $D^c(k)$ into something like $1/(k^2 - \Lambda^2)$ and shifts the hard contributions significantly below the data level even if one uses the humpy DAs and other modifications increasing the hard term (see, e.g., [8]).

An instructive illustration of possible modifications due to finite size or transverse momentum effects is given by the light-cone calculation of the $\gamma^*\gamma\pi^0$ amplitude [9, 10] in which hard propagator of a *massless* quark is convoluted with the two-body wave function $\Psi(x, k_\perp)$. Assuming the Gaussian dependence $\Psi(x, k_\perp) \sim \exp[-k_\perp^2/2x\bar{x}\sigma]$ on transverse momentum, one can easily calculate the $k_\perp$ integral to see that the pQCD propagator factor $1/xQ^2$ is substituted by the combination $(1 - \exp[-xQ^2/2\bar{x}\sigma])/xQ^2$ which monotonically tends to a finite limit $1/2\sigma$ as $x \to 0$. Hence, the effective virtuality is always larger than 2σ. The suppression of low virtualities has a simple explanation: propagation of quarks and gluons in the transverse direction is restricted by the finite size of the hadron. Numerically, $2\sigma \approx 1.35\,GeV^2$ in that case. However, even a milder modification of the "hard" propagators by effective quark and gluon masses $1/k^2 \to 1/(k^2 - M^2)$ with $M^2 \sim 0.1\,GeV^2$ or model inclusion of transverse momentum effects strongly reduces the magnitude of hard contributions [8], especially when the humpy DAs are used. For these reasons, a scenario with humpy DAs and bare $\sim 1/x_i y_j t$ propagators (which amounts to ignoring finite-size effects) considerably overestimates the size of hard contributions.

2 Lessons from Pion Studies

The relative smallness of hard contributions can be easily understood within the QCD sum rule context. The soft contribution is dual to the lowest-order diagram while the gluon exchange terms appear in diagrams having a higher order in α_s which results in the usual $\alpha_s/\pi \sim 1/10$ suppression factor per each extra loop. In particular, the α_s/π suppression factor is clearly visible in the expression for the hard contribution to the pion form factor [11, 12, 4, 13]

$$F_\pi^{\text{hard}}(Q^2)|_{\varphi_\pi = \varphi_\pi^{as}} = \frac{8\pi\alpha_s f_\pi^2}{Q^2} = 2\left(\frac{\alpha_s}{\pi}\right)\frac{s_0}{Q^2}. \tag{1}$$

Here, the combination $s_0 = 4\pi^2 f_\pi^2 \approx 0.67\,\text{GeV}^2 \sim m_\rho^2$ is what is usually called the "typical hadronic scale" in the case of the pion. At asymptotically high Q^2, the $O(\alpha_s/\pi)$ suppression of the hard terms is more than compensated by their slower decrease with Q^2. However, such a compensation does not occur in the subasymptotic region where the soft contributions, as we have seen, may have the same effective power behavior as that predicted by the asymptotic quark counting rules for the hard contributions. In ref. [14], both the soft contribution and the $O(\alpha_s)$ corrections for the pion form factor were calculated together within a QCD sum rule inspired approach. The ratio of the $O(\alpha_s)$ terms to the soft contribution was shown to be in full agreement with the expectation based on the α_s/π per loop suppression.

The use of the humpy DAs is usually motivated by the QCD sum rule analysis for the $\langle x^N \rangle$ moments of DAs [6]. However, applications of the QCD sum rules to DAs $\varphi(x)$ and form factors $F(Q^2)$ require a more detailed information about the nonperturbative QCD vacuum than those for the simpler classic cases [15] of hadronic masses and decay widths. The main problem is that the coefficients of the operator product expansion (OPE) for the relevant correlators now depends on an extra parameter, $e.g.$, on the order of the moment N for $\langle x^N \rangle$ or momentum transfer Q^2 for form factors. In particular, the higher condensates $\langle \bar{q}(D^2)^n q \rangle$ are accompanied in the $\langle x^N \rangle$ sum rule by large N^n-factors. Since for any reasonable shape of $\varphi(x)$ the moments $\langle x^N \rangle$ should decrease with growing N, the appearance of N^n-dependence is an artifact of the expansion procedure. Calculationally, the N^n-factors appear from the Taylor expansion of the nonlocal condensate $\langle \bar{q}(0)q(z) \rangle$. In this situation, one is forced to make, explicitly or implicitly an assumption about the structure of the OPE in higher terms. In the approach of ref.[6] only the lowest condensates were taken into account. A simple alternative is to model $\langle \bar{q}(0)q(z) \rangle$ by a smooth function with the width suggested by existing estimates $\langle \bar{q}D^2 q \rangle / \langle \bar{q}q \rangle \approx 0.4\,GeV^2$. This model gives a QCD sum rule in which all terms decrease for large N. It produces the pion DA close to a smooth "asymptotic" form (see [16]).

It was also observed that the sum rules with nonlocal condensates have the property that the humps in the relevant correlator (corresponding to a sum over all possible states) get more pronounced when the relative pion contribution $decreases$ (see ref.[17]). This means that the humps of the correlator are generated by oscillations in the DAs of the higher states rather than by the humps in the pion DA. The oscillatory behaviour of DAs of the radial excitations found in ref.[17] (see also [18]) is supported by the studies in two-dimensional QCD [19, 20].

An independent evidence in favour of the narrow form of the pion distribution amplitude $\varphi_\pi(x)$ is provided by the result of ref.[21], where it was found that $\varphi_\pi(1/2) \approx 1.2 f_\pi$, to be compared with $\varphi_\pi^{as}(1/2) = 1.5 f_\pi$ for the asymptotic distribution amplitude [4],[5] and $\varphi_\pi^{CZ}(1/2) = 0$ for the CZ form [3]. Furthermore, the lattice calculation of ref.[22] gives a rather small value $\langle \xi^2 \rangle \approx 0.11$ for the second moment of the pion DA incompatible with the humpy form (compare with $\langle \xi^2 \rangle^{CZ} = 0.43$ and $\langle \xi^2 \rangle^{as} = 0.2$). The statement that the pion DA is close to its asymptotic form even at a low normalization point is also supported by calculation of the pion DA in the chiral soliton model [23] and by a direct QCD sum rule calculation of the large-Q^2 behavior of the $\gamma^*\gamma\pi^0$ form factor [24]. Within the light-cone QCD sum rule approach one can relate the pion DA to the pion parton densities [25] known experimentally. According to the analysis performed in [26], existing data favor the asymptotic shape. Finally, the humpy pion DA advocated in [27, 3] is now ruled out by recent experimental data [28] on the $\gamma^*\gamma\pi^0$ form factor. The data are fully consistent with the next-to-leading pQCD prediction calculated using the asymptotic DA [29, 10, 30].

If the pion DA is narrow, the hard contribution to the pion form factor is small. On the other hand, in many models, the soft term calculated as an overlap of model wave functions $\Psi(x, k_\perp)$ is comparable in size with the data [7, 31, 32, 33]. It should be noted that the relevant distribution amplitudes (obtained from $\Psi(x, k_\perp)$ by integration over $k_\perp$), are narrow and the hard term is small. Moreover, if one intends to increase the hard term by using wave functions enhanced in the end-point regions, one also increases the soft term (see, $e.g.$, [7, 33]), since the latter is dominated by the regions where one of the quarks has small momentum.

The pion form factor was also studied within the QCD sum rule approach, which is applicable in that case both in the region of moderately large [34],[35] and small momentum transfers [36]. In the whole region $0 \leq Q^2 \lesssim 3\,GeV^2$, the QCD sum rule result for the contribution due to the Feynman mechanism is sufficiently large to explain the magnitude of existing data. In the region $Q^2 \gtrsim 4\,GeV^2$, the OPE is ruined by $O(Q^2/M^2)^n$ enhancement of condensate contributions. This phenomenon has exactly the same nature as the $O(N)$ enhancement of the lowest condensate contributions in the CZ sum rule for $\langle x^N \rangle$ [37]. As a result, if we would assume that the higher condensate corrections can be neglected, we would get a very large soft contribution, marginally exceeding the data. Alternatively, using the nonlocal condensates [16] we would get the soft term comparable in size with the data [38]. As mentioned above, the same model produces a rather narrow pion DA [16] generating a small hard contribution which is subdominant up to $Q^2 \sim 10\,GeV^2$. Since the QCD sum rules for $\langle x^N \rangle$ and $F^{soft}(Q^2)$ have similar structure, if one uses the same model for the condensates in both sum rules, the results for the hard term $F^{hard}(Q^2)$ (whose magnitude is determined by the shape of $\varphi(x)$) and soft term $F^{soft}(Q^2)$ are strongly correlated. Just like in quark model calculations, it is impossible to get a large hard term without getting a huge soft term. The existence of such a correlation is also supported by the light-cone QCD sum rules [39]. Just like in quark model calculations, it is impossible to get a large hard term without getting a huge soft term.

3 Nucleon Case

Since the structure of OPE in the pion and nucleon cases is very similar, there is no reason to expect a significant deviation of the nucleon DA from its asymptotic form. In particular, an evidence against humpy nucleon DAs is provided by a lattice calculation [40] which does not indicate any significant asymmetry. One may argue that the proton DA must be asymmetric to reflect the fact that the u-quarks carry on average a larger fraction of the proton momentum than the d-quarks. As shown in ref. [41], to accomodate this observation one needs only a moderate shift of the DA maximum from the center point $x_1 = x_2 = x_3 = 1/3$. Such a shift does not produce a drastic enhancement of the hard contribution provided by the humpy DAs. However, with the asymptotic DA, the leading twist hard contribution completely fails to describe the data: it gives zero for the proton magnetic form factor and a wrong-sign (positive) contribution for the neutron magnetic form factor, with the absolute magnitude of the latter being two orders of magnitude below the data [42].

In the case of the baryon form factors, the standard SVZ-Borel version of the QCD sum rule approach works only in the region of small momentum transfers $Q^2 \lesssim 1\, GeV^2$ [43]. Beyond this region, the OPE explodes because of $O(Q^2/M^2)$-enhancements in condensate contributions, and a regular QCD sum rule analysis is impossible. In ref.[44], it was proposed to estimate the soft contributions by using the local quark-hadron duality prescription. It amounts to calculating the amplitude for transitions between the free-quark states produced (or annihilated) by a local current having the hadron's quantum numbers, and then averaging the invariant mass of the quark states over the appropriate duality interval s_0. The latter has the meaning of the effective threshold for the higher hadronic states in the relevant channel and has a specific value for each hadron, $e.g.$, $s_0^\pi \approx 0.7\, GeV^2$ for the pion in the axial current channel.

For the pion form factor, the local quark-hadron duality is supported by the QCD sum rule analysis [34, 35] and agrees well with experimental data. Furthermore, as argued in ref. [44, 45] (see also [46]) the quark-hadron duality prescription has an intuitively appealing interpretation in terms of the light-cone wave functions: it can be treated as a cut-off model $\Psi(\{x_i\}, \{k_{\perp i}\}) \sim \theta(\sum_i (k_{\perp i}^2/x_i) \leq s_0)$ for the soft wave function. The sharp cut-off suggested by the local duality looks like a rough approximation for more smooth wave functions usually adopted in phenomenological quark models. However, the difference is that in the local duality model the width of the $k_\perp$-distribution is directly related to a parameter s_0 characterizing the hadronic spectrum. This parameter s_0 is calculated from the reliable two-point function QCD sum rule and considered as given in the form factor calculations.

The local duality estimate [44] of the soft term for the proton magnetic form factor, based on the standard value $s_0^N \approx 2.3\, GeV^2$ [42] of the nucleon duality interval is very close to available data [47], [48] over a wide region $3\, GeV^2 \lesssim Q^2 \lesssim 20\, GeV^2$. The same calculation [44] also correctly reproduces the observed magnitude of the helicity-nonconservation effects for the proton form factors: $F_2^p(Q^2)/F_1^p(Q^2) \sim \mu^2/Q^2$ with $\mu^2 \sim 1\, GeV^2$ [48]. Within the scenario based on hard scattering dominance, it is rather difficult to understand the origin of such a large scale, since possible sources of helicity nonconservation in pQCD include only small scales like quark masses, intrinsic transverse momenta, $etc.$, and one would rather expect that $\mu^2 \sim 0.1\, GeV^2$.

4 Proton to Delta Transition

Even more drastic difference between predictions of hard and soft scenarios is expected (see, *e.g.*, [49]) in the studies of spin effects in the $\gamma^* p \to \Delta^+$ transition. A renewed attention to this process was raised by the results [50] of the analysis of inclusive SLAC data which indicated that the effective transition form factor drops faster than one would expect from quark counting rules [1, 51, 52]. Within the hard scattering scenario, the DA-sensitivity of this process was originally analyzed in ref.[53]. It was observed there that the hard scattering amplitude in this case has an extra suppression due to cancellation between symmetric and antisymmetric parts of the nucleon distribution amplitude. Hence, from the hard scenario point of view, the faster fall-off found in [50] signalizes the dominance of a non-asymptotic contribution.

In ref.[54], the soft contribution to the $\gamma^* p \to \Delta^+$ transition form factors was estimated within the local quark-hadron duality approach. The duality interval for the Δ-resonance taken there is $s_0^\Delta = 3.5$ GeV^2, which agrees with the results of the two-point function analysis [42]. The results for the effective form factor $G_T(Q^2)$ are close to those obtained from the analysis of inclusive data [55]. This means that the data can be described without a sizable contribution from the hard-scattering mechanism. Furthermore, the $\gamma^* p \to \Delta^+$ transition is described by three independent form factors, and a correct model should not only be able to adjust the absolute magnitude of one of them: it should also be able to explain the relations between different form factors. In particular, the pQCD calculation [53] predicts that the lowest-twist hard contribution always has the property $G_E^{* \, hard}(Q^2) \approx -G_M^{* \, hard}(Q^2)$. This prediction is a specific example of the helicity selection rules [5] inherent in the hard scattering mechanism. Experimentally, the ratio $G_E^*(Q^2)/G_M^*(Q^2)$ is very small [49, 56], which indicates that the leading-twist pQCD term is irrelevant in the region $Q^2 \lesssim 4\,GeV^2$. Small value for $G_E^*(Q^2)/G_M^*(Q^2)$ is also predicted in constituent quark model approaches [57, 58, 59, 60, 61]. However, these approaches usually do not claim applicability in the $Q^2 \gtrsim 2$ GeV^2 region of momentum transfers. The local duality estimates were performed in ref. [54] for several Lorentz structures which appear in the decomposition of the basic γ-odd three-point amplitude. The results obtained from different invariant amplitudes are in satisfactory agreement with each other. All estimates indicate that the transition is dominated by the magnetic form factor $G_M^*(Q^2)$, with electric $G_E^*(Q^2)$ and Coulomb $G_C^*(Q^2)$ form factors being small compared to $G_M^*(Q^2)$ for all experimentally accessible momentum transfers (see Fig. 1).

To summarize, QCD sum rule based results for soft contributions to hadronic form factors are in good quantitative agreement with existing data providing a clear experimental evidence that at available Q^2 hard terms are relatively small.

Acknowledgement. I am grateful to I.G. Aznauryan, V.M. Belyaev, V.D. Burkert, C.E. Carlson, V.V. Frolov, N. Isgur, C. Keppel, N. Mukhopadhyay and P. Stoler for useful discussions. This work was supported by the US Department of Energy under contract DE-AC05-84ER40150.

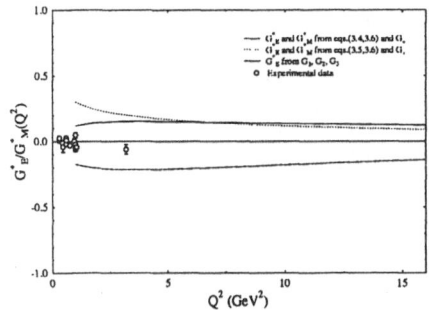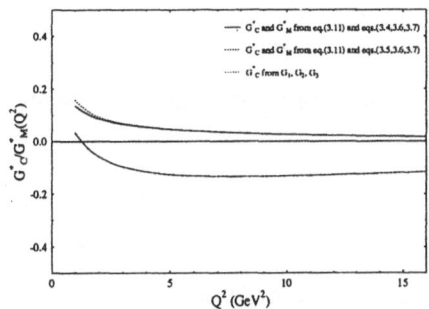

Figure 1. Local duality estimates for the ratio of form factors *a)* $G_E^*(Q^2)$ and $G_M^*(Q^2)$ and *b)* $G_C^*(Q^2)/G_M^*(Q^2)$.

References

1. S.J. Brodsky and G.R. Farrar: Phys. Rev. Lett. **31**, 1153 (1973)

2. R.P. Feynman: *Photon-Hadron Interaction.* Reading: W.A.Benjamin 1972

3. V.L. Chernyak and A.R. Zhitnitsky: JETP Lett. **25**, 510 (1977); Yad. Fiz. **31**, 1053 (1980)

4. A.V. Efremov and A.V. Radyushkin: Phys. Lett. **B94**, 245 (1980); Theor. Mat. Fiz. **42**, 147 (1980)

5. G.P. Lepage and S.J. Brodsky: Phys. Lett. **B87**, 359 (1979); Phys. Rev. **D22**, 2157 (1980)

6. V.L. Chernyak and A.R. Zhitnitsky: Phys. Rep. **112**, 173 (1984)

7. N. Isgur and C.H. Llewellyn-Smith: Nucl. Phys. **B317**, 526 (1989)

8. J. Bolz et al.: Z. Phys. **C66**, 267 (1995)

9. S.J. Brodsky, T.Huang and G.P.Lepage: In: *Particles and Fields 2*, Proceedings of the Banff Summer Institute, Banff, Alberta, 1981, edited by A.Z. Capri and A.N. Kamal, p.143. New York: Plenum 1983.

10. I.V. Musatov and A.V. Radyushkin: Phys. Rev. **D56**, 2713 (1997)

11. V.L. Chernyak, A.R. Zhitnitsky and V.G. Serbo: JETP Lett. **26**, 594 (1977)

12. G.R. Farrar and D.R. Jackson: Phys. Rev. Lett. **43** , 246 (1979)

13. G.P. Lepage and S.J. Brodsky: Phys. Lett. **B87**, 359 (1979)

14. A. Szczepaniak, C.-R. Ji and A. Radyushkin: Phys. Rev. **D57**, 2813 (1998)

15. M.A. Shifman, A.I. Vainshtein and V.I. Zakharov: Nucl. Phys. **B147**, 385, 448 (1979)

16. S.V. Mikhailov and A.V. Radyushkin: Phys. Rev. **D45**, 1754 (1992)

17. A.V. Radyushkin: In: "Continuous advances in QCD", ed. by A.V.Smilga, p. 238. Singapore: World Scientific 1994; hep-ph/9406237

18. A.P. Bakulev and S.V. Mikhailov: Z. Phys. **C65**, 451 (1995)

19. G.'t Hooft: Nucl. Phys. **B75**, 461 (1974)

20. B. Chibisov and A. R. Zhitnitsky: Phys. Lett. **B362** 105 (1995)

21. V.M. Braun and I.Filyanov: Z. Phys. **C44** 157 (1989)

22. D.Daniel, R.Gupta and D.G.Richards: Phys. Rev. **D43** 3715 (1991)

23. V.Yu. Petrov and P.V. Pobylitsa: hep-ph/9712203

24. A.V. Radyushkin and R. Ruskov: Nucl. Phys. **B 481**, 625 (1996)

25. V.M. Belyaev and M. B. Johnson: Phys.Rev. **D56**, 1481 (1997)

26. V.M. Belyaev and M. B. Johnson: Mod. Phys. lett. **A13**, 2909 (1998)

27. V.L. Chernyak and A.R. Zhitnitsky: Nucl. Phys. **B201**, 492 (1984); Erratum: **B 214**, 547 (1984)

28. CLEO collaboration (J. Gronberg et al.): Phys. Rev. **D57**, 33 (1998)

29. E. Braaten, Phys. Rev. **D28**, 524 (1983)

30. S. J. Brodsky, C.-R. Ji, A. Pang and D. G. Robertson: Phys. Rev. **D57**, 245 (1998)

31. O.C. Jacob and L.S. Kisslinger: Phys. Lett. **B243**, 323 (1990)

32. C.R. Ji, P.L. Chung and S.R. Cotanch: Phys. Rev. **D45**, 4214 (1992)

33. B. Chibisov and A. R. Zhitnitsky: Phys.Rev. **D52**, 5273 (1995)

34. B.L. Ioffe and A.V. Smilga: Phys. Lett. **B114**, 353 (1982)

35. V.A. Nesterenko and A.V. Radyushkin: Phys. Lett. **B115**, 410 (1982)

36. V.A. Nesterenko and A.V. Radyushkin: JETP. Lett. **39**, 707 (1984)

37. A.V. Radyushkin: In: "*Baryons '92*", ed. by M.Gai, p.366. Singapore: World Scientific 1993

38. A.P. Bakulev and A.V. Radyushkin: Phys. Lett. **B271**, 223 (1991)

39. V.M. Braun and I. Halperin: Phys. Lett. **B328**, 457 (1994)

40. G. Martinelli and C.T. Sachrajda: Phys. Lett. **B217**, 319 (1989)

41. J. Bolz and P. Kroll: Z. Phys. **A356**, 327 (1996)

42. V.M. Belyaev and B.L. Ioffe: Sov. Phys. JETP **56**, 493 (1982).

43. V.M. Belyaev and I.I. Kogan: Int. J. Mod. Phys. **A8**, 153 (1993)

44. V.A. Nesterenko and A.V. Radyushkin: Phys. Lett. **128B**, 439 (1983); Sov. J. Nucl. Phys. **39**, 811 (1984)

45. A.V. Radyushkin: Acta Phys. Pol. **B15**, 403 (1984)

46. A.V. Radyushkin: Acta Phys. Pol. **B26**, 2067 (1995)

47. A.F. Sill et al.: Phys. Rev. **D48**, 29 (1993)

48. P. Bosted et al.: Phys. Rev. Lett. **68**, 3841 (1992); L. Andivahis et al.: Phys. Rev. **D50**, 5491 (1994)

49. V.D. Burkert and L. Elouadrhiri: Phys. Rev. Lett. **75**, 3614 (1995)

50. P. Stoler: Phys. Rev. Lett. **66**, 1003 (1991); Phys. Rev. **D44**, 73 (1991); Phys. Reports **226**, 103 (1993)

51. V.A. Matveev, R.M. Muradyan and A.N. Tavkhelidze: Lett. Nuovo Cim. **7**, 719 (1973)

52. B.L. Ioffe: Phys. Lett. **B63**, 425 (1976)

53. C.E. Carlson: Phys. Rev. **D34**, 2704 (1986)

54. V.M. Belyaev and A.V. Radyushkin: Phys. Rev. **D53**, 6509 (1996)

55. L.M. Stuart et al.: Phys. Rev. **D58**, 032003 (1998)

56. V.V. Frolov et al.: hep-ex/9808024

57. R.Koniuk and N.Isgur: Phys. Rev. Lett. **44**, 485 (1980); Phys. Rev. **D21**, 1868 (1980)

58. Z.Li and F.E.Close: Phys. Rev. **D42**, 2207 (1990)

59. S. Capstick and G.Karl: Phys. Rev. **D41**, 2767 (1990); S. Capstick: Phys. Rev. **D46**, 2864 (1992)

60. M. Warns, H. Schroeder, W.Pfeil and H. Rollnik: Z. Phys. **C45**, 627 (1990)

61. I.G. Aznaurian: Phys. Lett. **B316**, 391 (1993); Z. Phys. **A346**, 297 (1993)

Few-Body Systems Suppl. 11, 66–73 (1999)

Few-
Body
Systems
© by Springer-Verlag 1999

Light-Baryon Spectroscopy and the Electromagnetic Form Factors in the Quark Model

F. Cardarelli[1], E. Pace[2], G. Salmè[3] and <u>S. Simula</u>[1]

[1]INFN, Sezione Roma III, Via della Vasca Navale 84, I-00146 Roma, Italy
[2]Dipartimento di Fisica, Università di Roma "Tor Vergata", and INFN, Sezione Tor Vergata, Via della Ricerca Scientifica 1, I-00133, Rome, Italy
[3]INFN, Sezione di Roma I, P.le A. Moro 2, I-00185 Rome, Italy

Abstract.

The momentum distributions of the constituent quarks inside the nucleon and the prominent electroproduced nucleon resonances are investigated in the two most sophisticated, available quark potential models, based respectively on the assumption of the valence + gluon dominance and on the exchange of the pseudoscalar Goldstone-bosons arising from the spontaneous breaking of chiral symmetry. It is shown that both models predict a large, similar content of high-momentum components, due to the short-range part of the interquark interaction, which affect the behaviour of both elastic and transition electromagnetic form factors at large values of the momentum transfer. The electromagnetic form factors are calculated within a relativistic approach formulated on the light-front, adopting a one-body current with constituent quark form factors. The results suggest that soft, non-perturbative effects can play a relevant role for explaining the existing data on elastic as well as transition form factors (at least) for $Q^2 \lesssim 10 \div 20 \ (GeV/c)^2$.

1 Introduction

The aim of this contribution is to address few relevant questions concerning the possible consistency of the predictions of the constituent quark (CQ) model with existing data on the electromagnetic (e.m.) properties of the nucleon and the most prominent electroproduced nucleon resonances at large values of the squared four-momentum transfer Q^2. In these kinematical regions ($Q^2 \gtrsim$ few $(GeV/c)^2$) the $pQCD$ hard scattering mechanism appears to be able to explain qualitatively existing data (see, e.g., Ref. [1]), but results from QCD sum rules (see, e.g., Ref. [2]) seem to suggest that also the soft Feynman mechanism can account for the same data as well.

Since the high-Q^2 behaviour of the form factors is correlated with the high-momentum tail of the CQ momentum distribution in the nucleon and its resonances, we first investigate the light-baryon wave functions generated by the two most sophisticated, available quark potential models, based respectively on the assumption of the valence + gluon dominance [3] and on the exchange of the pseudoscalar Goldstone-bosons arising from the spontaneous breaking of chiral symmetry [4]. It will be shown that both models predict a large content of high-momentum components due to the short-range part of the interquark interaction. Moreover, despite the different behaviour of the two models at short distances, the high-momentum tails of the light-baryon wave functions turn out to be quite similar. These high-momentum components are known to affect significantly the large-Q^2 behaviour of both elastic and transition e.m. form factors, which are calculated adopting the light-front quark model of Refs. [5, 6]. We will point out that: i) the introduction of constituent quark form factors in the one-body e.m. current is essential in order to explain the detailed Q^2 behaviour of the nucleon elastic data; ii) the short-range spin-spin interaction generating the $N - \Delta(1232)$ mass splitting is also responsible for the faster-than-dipole fall-off of the $N - \Delta(1232)$ magnetic transition form factor at large Q^2; iii) an approximate dipole fall-off of the $N - S_{11}(1535)$ transition magnetic form factor can be obtained, provided the nucleon elastic data are reproduced. Our results suggest that soft, non-perturbative physics can yield a relevant contribution for explaining the existing data on the high-Q^2 behaviour of elastic and transition e.m. form factors, in accord with the findings of QCD sum rules.

2 Quark potential models and the CQ momentum distribution

The CQ model is known to be a phenomenological model able to explain the basic features of many static hadron properties, like the baryon (and meson) mass spectra. Within this model the CQ's are the only relevant degrees of freedom in baryons, all the other degrees of freedom being frozen in the CQ mass (m_i) and interaction. The baryon wave function Ψ_B is therefore eigenfunction of a Schroedinger-type equation, viz.

$$\hat{H}\Psi_B = [\hat{T} + \hat{V}]\Psi_B = M_B\Psi_B \tag{1}$$

where M_B is the baryon mass, $\hat{T} = \sum_{i=1}^{3} \sqrt{|\boldsymbol{p}_i|^2 + m_i^2}$ is the kinetic term and $\hat{V} = \hat{V}_{conf} + \hat{V}_{s.r.}$ the interaction term, given by a long-range confining part $\hat{V}_{conf}$ and a short-range component $\hat{V}_{s.r.}$ responsible for the hyperfine mass splitting. The confining potential is usually derived from a Lorentz-scalar interaction and, as suggested by the spectroscopy and lattice QCD calculations, it can be taken linearly dependent on the quark-quark distance $r_{ij} \equiv |\boldsymbol{r}_i - \boldsymbol{r}_j|$, namely $\hat{V}_{conf} \to \hat{V}_s = \sum_{i<j} b \cdot r_{ij}$, where b is the string tension. As for the short-range part of the interquark potential, the most sophisticated choices existing in the literature are based on two alternative mechanisms of boson

exchange among CQ's: the one-gluon-exchange (OGE) model of Ref. [3] and the pseudoscalar Goldstone-boson exchange (GBE) model of Ref. [4].

The semi-relativistic Hamiltonian model developed in Ref. [3] is very successful in both meson and baryon sectors: it reproduces a large amount of experimental masses and solves the so-called baryon spin-orbit puzzle. The latter consists in the apparent absence of a significant spin-orbit splitting in the light-baryon mass spectrum at variance with naive expectations. The puzzle was solved by Isgur and co-workers [3] by partially compensating the vector spin-orbit term with the Thomas-Fermi precession spin-orbit term arising from the scalar confining interaction, and by introducing (semi)relativistic corrections to the interquark potential, which yield a significant suppression of the interaction strength in case of light quarks. Nevertheless, a residual problem still remains in the generally good picture given by the OGE model: negative-parity states are below positive-parity ones, in clear contrast to the observation.

In the GBE model of Ref. [4] the short-range part of the CQ interaction is generated by the exchange of the pseudoscalar Goldstone bosons arising from the spontaneous breaking of chiral symmetry. Such a potential model predicts baryon masses in quite good agreement with the experimental data and, in particular, thanks to the flavour dependence of the exchanged mesons, the GBE model is able to yield the correct ordering among positive and negative parity states. However, as pointed out in Ref. [7], the agreement with the mass spectrum is obtained only when the Thomas-Fermi precession spin-orbit term due to the scalar confining interaction is (arbitrarily) neglected. Therefore the baryon spin-orbit puzzle is still to be solved within the GBE model (see for details [7]). Despite the mentioned flaws, we stress that both the OGE and GBE models yield a quite good overall description of the light-baryon spectrum, although they remarkably differ at short interquark distances.

The wave equation (1) has been solved by expanding the wave function Ψ_B onto a (truncated) set of harmonic oscillator basis states and applying to the Hamiltonian the Rayleigh-Ritz variational principle. We have explicitly checked that a sufficiently large number of basis states has been included in order to obtain the full convergence of the quantities considered in this work. The CQ momentum distribution $n(p)$, defined as $n(p) = \int d\Omega_{\mathrm{p}} dp_2 dp_3\, \delta(\boldsymbol{p} + \boldsymbol{p}_2 + \boldsymbol{p}_3)\, |\Psi_B|^2$, calculated using the OGE and GBE models, is shown in Fig. 1 in case of the nucleon and compared with the results obtained adopting only the (linear) confining part of the two interactions. It can be seen that for both models the short-range part of the potential produce a remarkable content of high-momentum components, which turn out to be not very sensitive to the specific interaction model.

The CQ momentum distributions $n(p)$ in the $\Delta(1232)$, $S_{11}(1535)$ and $F_{15}(1680)$ resonances, obtained within the OGE and GBE models, are reported in Fig. 2 and compared with the ones in the nucleon. It can clearly be seen that, although the OGE and GBE models substantially differ at short interquark distances, the high-momentum tails of the baryon wave functions are quite similar in both models with the only (partial) exception of the $S_{11}(1535)$ resonance. Since the high-Q^2 behaviour of the form factors is qualitatively correlated with

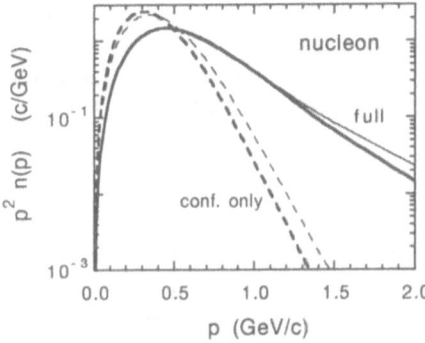

Figure 1. CQ momentum distribution $p^2 n(p)$ in the nucleon versus the internal CQ momentum p. Thick and thin lines correspond to the OGE and GBE models of Refs. [3] and [4], respectively. The solid lines are the results obtained using the full interaction models, whereas the dashed lines correspond to the case in which only their (linear) confining parts are considered.

the high-momentum tail of the CQ momentum distribution, we naively expect the same high-Q^2 behaviour for the elastic nucleon, $N - S_{11}(1535)$ and $N - F_{15}(1680)$ transition form factors, while a faster fall-off is expected in case of the $N - \Delta(1232)$ transition. Such features are indeed present in the existing high-Q^2 data, namely both the elastic nucleon, the $N - S_{11}(1535)$ and the $N - F_{15}(1680)$ transition magnetic form factors exhibit approximately the same dipole fall-off for Q^2 greater than few $(GeV/c)^2$, while the $N - \Delta(1232)$ transition magnetic form factor drops faster than a dipole (see [1]).

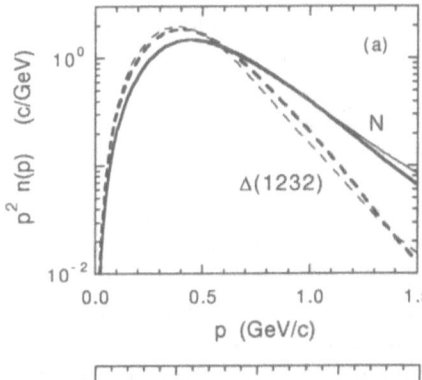

Figure 2. CQ momentum distributions $p^2 n(p)$ in the $\Delta(1232)$ (a), $S_{11}(1535)$ (b) and $F_{15}(1680)$ (c) resonances (dashed lines). Thick and thin lines correspond to the OGE [3] and GBE [4] potential models, respectively. For comparison in each picture the CQ momentum distributions in the nucleon (solid lines) are explicitly shown.

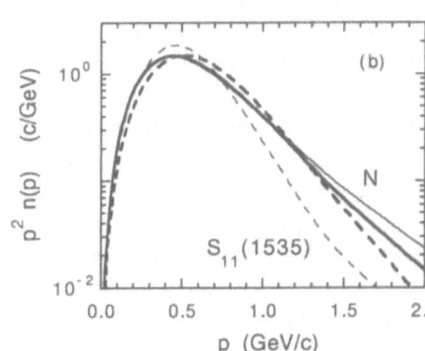

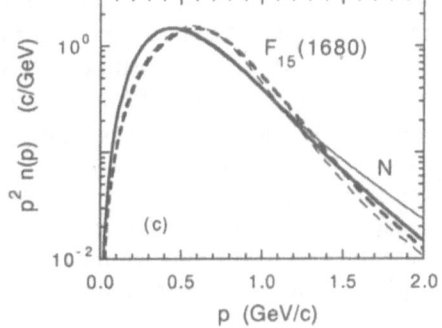

In the next Section the issue of the CQ model predictions for the nucleon elastic and transition e.m. form factors will be addressed and to this end the relativistic quark model of Ref. [5], formulated on the light front, is adopted. Since the high-momentum components generated by the OGE and GBE models turn out to be quite similar, in what follows we will limit ourselves to consider explicitly the light-baryon wave functions of the OGE model only.

3 Elastic and transition e.m. form factors

The effects of the high-momentum tail of the nucleon wave function generated by the OGE model on the elastic nucleon e.m. form factors have been investigated for the first time in Ref. [5], where a relativistic one-body e.m. current was adopted. It was shown that both the relativistic effects and the high-momentum components of the wave function lead to a sizeable overestimation of the proton form factors both at low ($\lesssim 1$ $(GeV/c)^2$) and high ($\gtrsim 1$ $(GeV/c)^2$) Q^2. This result is not surprising, because a pure valence quark model (i.e., without any effect from sea quark pairs) is not expected to describe dynamical properties like the e.m. form factors. One could argue that, in order to keep safe the CQ picture of the hadron structure, the CQ itself can be viewed as a non-elementary object whose structure takes into account in an effective way the presence of non-valence components. Thus, in Ref. [5] a one-body e.m. current with CQ form factors was adopted. The latter ones cannot be derived directly from QCD and therefore one is limited to constrain the CQ form factors by the request of reproducing the nucleon elastic data and to ask if existing data on the transition form factors are consistent with the same one-body e.m. current. This program has been partially carried out in Refs. [5, 6]: adopting the baryon wave functions of the OGE model, the CQ form factors were firstly fixed through the reproduction of the nucleon elastic data and then used to calculate *without free parameters* the $N - \Delta(1232)$ transition form factors, obtaining a good overall description of the existing data[a].

In this contribution we want to point out that, once the elastic data are reproduced, the faster-than-dipole fall-off of the $N - \Delta(1232)$ magnetic transition form factor can be obtained only when the effects from the short-range spin-spin interaction are taken into account in the baryon wave functions. To this end we have calculated the nucleon elastic form factors adopting two different wave functions obtained from the full OGE interaction and from its (linear) confining part only (see solid and dashed lines in Fig. 1, respectively). For each wave functions the CQ form factors have been determined by the request of reproducing the nucleon (and pion) data. The results are reported in Fig. 3 and it can be clearly seen that: i) the introduction of the CQ form factors in the one-body e.m. current is essential in order to explain the detailed Q^2 behaviour of the nucleon elastic data, and ii) once appropriate CQ form factors are introduced, the nucleon data alone cannot distinguish between models with and without short-range interaction effects. It should be mentioned that the

---

[a]Due to the limitations imposed by the violation of the so-called angular condition (see Ref. [6]), we will consider in this work only the predictions of our light-front model for the dominant magnetic transition form factor.

phenomenological CQ form factors associated to the full OGE wave function and to the much softer wave function generated by the linear confining interaction, correspond to quite different values of the CQ size, namely ~ 0.5 and $\sim 0.2\ fm$, respectively.

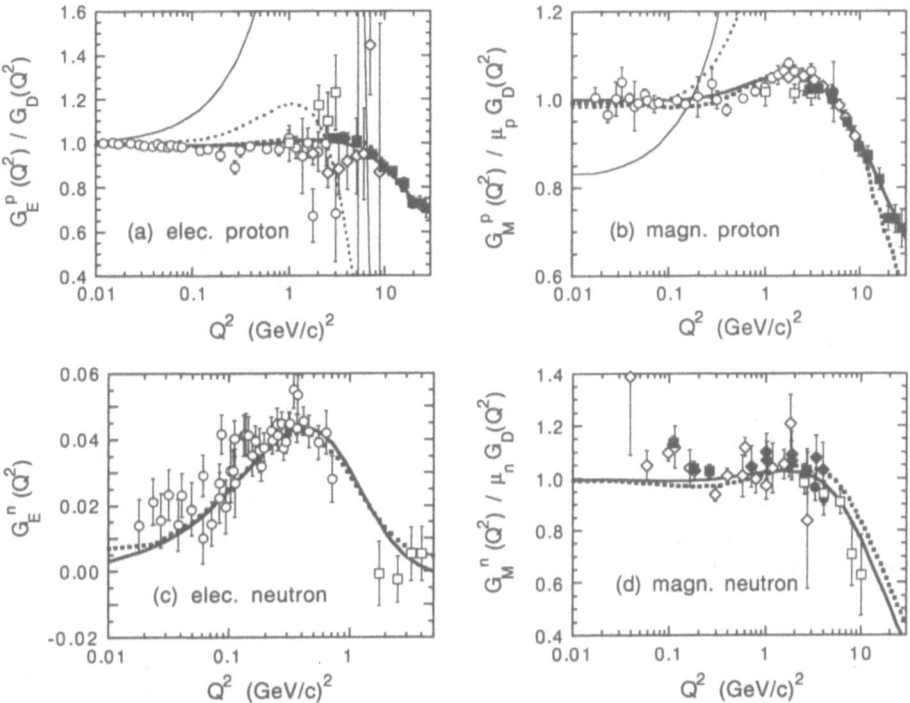

Figure 3. Nucleon form factors $G_E^p(Q^2)/G_D(Q^2)$ (a), $G_M^p(Q^2)/\mu_p G_D(Q^2)$ (b), $G_E^n(Q^2)$ (c) and $G_M^n(Q^2)/\mu_n G_D(Q^2)$ (d) versus Q^2. The solid and dotted thick lines correspond to the results obtained using the nucleon wave functions resulting from the full OGE model [3] and from its (linear) confining part only, including CQ form factors in the one-body e.m. current. In (a) and (b) the dotted and solid thin lines correspond to the case in which point-like CQ's are assumed. Data are quoted in details in Ref. [5]. The dipole form is given by $G_D(Q^2) = 1/(1 + Q^2/0.71)^2$.

In Fig. 4 our parameter-free predictions for the $N - \Delta(1232)$ magnetic transition form factor are compared with existing data. It can clearly be seen that the effects from the short-range part of the CQ interaction, which is responsible for the $N - \Delta(1232)$ mass splitting and also for the different high-momentum tails of the N and $\Delta(1232)$ wave functions (see Fig. 2(a)), are now essential in order to reproduce the faster-than-dipole fall-off of the transition form factor at large Q^2 [b].

Finally, in Fig. 5 the parameter-free predictions for the $N - S_{11}(1535)$ mag-

---

[b]The deviations from the data at low Q^2 are discussed in Ref. [6]. For the photon point see also Ref. [9].

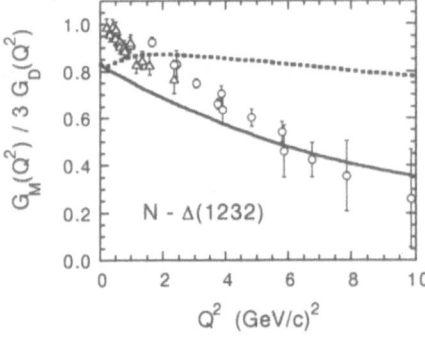

Figure 4. The $N - \Delta(1232)$ transition magnetic form factor $G_M(Q^2)/3G_D(Q^2)$ versus Q^2. The solid and dotted lines correspond to the results obtained using the N and $\Delta(1232)$ wave functions resulting from the full OGE model [3] and from its (linear) confining part only. The CQ form factors are the ones used to reproduce the nucleon form factors (see Fig. 3). Triangles and open dots are from Ref. [8](a) and (b), respectively.

netic transition form factor $G_M^*(Q^2)/G_D(Q^2)$, obtained in Ref. [10] using the full OGE wave functions, are reported and compared with available inclusive electroproduction data [1]. It can be seen that the naive expectation of a dipole fall-off at large Q^2, based on the similar high-momentum behaviours of the N and $S_{11}(1535)$ wave functions (see Fig. 2(b)), is fully confirmed by the explicit calculations. Therefore, we point out that the predictions of our light-front CQ model for the transition form factors to the most prominent electroproduced nucleon resonances are not inconsistent with the data at large Q^2, suggesting that soft, non-perturbative effects can still play a decisive role in the nucleon-resonance transition form factors at least up to $Q^2 \sim 10 \div 20 \ (GeV/c)^2$, in accord with the results of QCD sum rules [2].

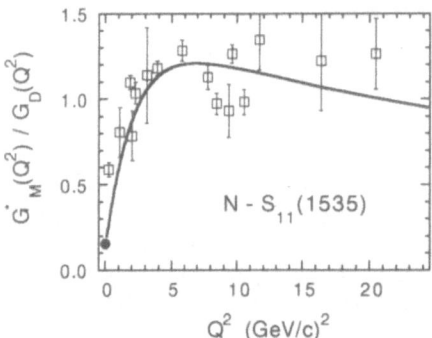

Figure 5. The $N - S_{11}(1535)$ transition magnetic form factor $G_M^*(Q^2)/G_D(Q^2)$ versus Q^2. The solid line corresponds to the results obtained using the N and $S_{11}(1535)$ wave functions resulting from the full OGE model [3]. The CQ form factors are the ones used to reproduce the nucleon form factors (see Fig. 3). Electroproduction data are from Ref. [1], while the photon point is from Ref. [11].

4   Conclusions

The momentum distributions of the constituent quarks inside the nucleon and the prominent electroproduced nucleon resonances have been investigated in the two most sophisticated, available quark potential models, based respectively on the assumption of the valence + gluon dominance and on the exchange of the pseudoscalar Goldstone-bosons arising from the spontaneous breaking of chiral symmetry. It has been shown that both models predict a large, similar content of high-momentum components due to the short-range part of the interquark interaction. Elastic and transition e.m. form factors have been calculated within a relativistic approach formulated on the light-front, adopting a one-body current with constituent quark form factors. The main results are: i) the introduction of constituent quark form factors is essential in order to explain the detailed Q^2 behaviour of the nucleon elastic data; ii) the short-range spin-spin interaction generating the $N - \Delta(1232)$ mass splitting is also responsible for the faster-than-dipole fall off of the $N - \Delta(1232)$ magnetic transition form factor at large Q^2; iii) an approximate dipole fall-off of the $N - S_{11}(1535)$ transition magnetic form factor can be obtained, provided the nucleon elastic data are reproduced. Our results suggest that soft, non-perturbative physics can yield a relevant, decisive contribution for explaining the existing data on the nucleon elastic as well as transition e.m. form factors (at least) up to $Q^2 \sim 10 \div 20 \ (GeV/c)^2$.

References

1. P. Stoler: Phys. Rep. **226**, 103 (1993) and references therein

2. A. Radyushkin: these Proceedings

3. S. Godfrey and N. Isgur: Phys. Rev. **D32**, 189 (1985); S. Capstick and N. Isgur: Phys. Rev. **D34**, 2809 (1986)

4. L. Ya. Glozman et al.: e-print archive hep-ph/9706507; See also L. Ya. Glozman et al.: Phys. Rev. **C57**, 3406 (1998)

5. F. Cardarelli, E. Pace, G. Salmè and S. Simula: Phys. Lett. **B357**, 267 (1995); Few-Body Systems Suppl. **8**, 345 (1995)

6. F. Cardarelli, E. Pace, G. Salmè and S. Simula: Phys. Lett. **B371**, 7 (1996); Nucl. Phys. **A623**, 361c (1997); Phys. Lett. **B 397**, 13 (1997) and in Proc. of the Int. Workshop on *Perspectives in Hadron Physics*, ICTP (Italy), May 1997, eds. S. Boffi, C. Ciofi degli Atti and M.M. Giannini, World Scientific (Singapore, 1998), p. 403

7. F. Cardarelli and S. Simula: preprint INFN-RM3 98/3, e-print archive hep-ph/9809258, to appear in the Proc. of the III Int. Conference on *Quark Confinement and the Hadron Spectrum*, Jefferson Lab (USA), June 1998

8. (a) W. Bartel et al.: Phys. Lett. **B28**, 148 (1968); (b) L.M. Stuart et al.: Phys. Rev. **D58**, 032003 (1998)

9. F. Cardarelli, B. Pasquini and S. Simula: Phys. Lett. **B418**, 237 (1998)

10. E. Pace, G. Salmè and S. Simula: in Proc. of the XVI European Conference on *Few-Body Problems in Physics*, Autrans (France), July 1998, to appear in Few-Body Systems Suppl.

11. Particle Data Group, R.M. Barnett etal.: Phys. Rev. **D54**, 1 (1996)

Few-Body Systems Suppl. 11, 74–77 (1999)

Few-
Body
Systems
© by Springer-Verlag 1999

Charge-Current Operators Consistent with a Semi-relativistic Hamiltonian

P. Demetriou*, S. Boffi, F. Capuzzi and M. Radici

Dipartimento di Fisica Nucleare e Teorica, Università di Pavia, and Istituto Nazionale di Fisica Nucleare, Sezione di Pavia, Italy

Abstract. Charge-current density operators derived from a semi-relativistic Hamiltonian using a functional derivative formalism are presented and applied to electron scattering.

1 Introduction

We present a method of constructing a charge-density operator consistent with a single-particle semi-relativistic Hamiltonian consisting of a kinetic energy

$$T = \sqrt[+]{\boldsymbol{p}^2 + M^2}. \tag{1}$$

and an external energy, and momentum independent potential, avoiding the problem of expanding operators in powers of unsuitable parameters. The derivation makes use of the functional derivative formalism and ensures gauge invariance [1].

2 Electromagnetic interaction in semi-relativistic theories

For a point particle with charge e, momentum $\boldsymbol{p}$ and Hamiltonian $H_0 (= p^0)$ the interaction Hamiltonian with the electromagnetic field is obtained by the minimal substitution $p^\mu \rightarrow p^\mu - eA^\mu$ where $A \equiv A^\mu = (A^0, A^i)$ is the electromagnetic field operator. The resulting Hamiltonian $H(A)$ is gauge invariant and the interaction Hamiltonian with the electromagnetic field,

$$H_{\text{em}}(A) \equiv H(A) - H_0, \tag{2}$$

is expanded in a Taylor series of functional derivatives. From gauge invariance and the law for the derivative of a composed function it turns out that the

---

*E-mail address: vivian@pv.infn.it

quantities defined as the functional derivatives of $H(A)$ with respect to A

$$J_\mu(A, \boldsymbol{x}) \equiv \frac{\delta H(A)}{\delta A^\mu(\boldsymbol{x}, t)}, \tag{3}$$

satisfy the continuity equation

$$\frac{\partial}{\partial x_i} J_i(A, \boldsymbol{x}) = -\mathrm{i}\,[H(A), J_0(A, \boldsymbol{x})], \tag{4}$$

and are interpreted as the components of the charge-current density operator.

Let us now consider the semi-relativistic Hamiltonian

$$H_0 = {}^+\!\sqrt{\boldsymbol{p}^2 + m^2} + V, \tag{5}$$

where the self-adjoint operator V is energy and momentum independent. Under the action of the electromagnetic field $A^\mu(x)$ the minimal substitution gives

$$H(A) = {}^+\!\sqrt{(\boldsymbol{p} - e\boldsymbol{A})^2 + m^2} + eA_0 + V. \tag{6}$$

Since the operators A^μ are bounded, the operator $(\boldsymbol{p} - e\boldsymbol{A})^2$ is self-adjoint. Hence, also the positive square root in Eq. (6) is a well-defined self-adjoint operator [1].

The charge density operator j_0 and the first-order current density operator obtained from Eq. (3) read

$$j_0(\boldsymbol{x}) = e\rho(\boldsymbol{x}), \tag{7}$$

and

$$j_i(\boldsymbol{x}) = \left.\frac{\delta R(A)}{\delta A^i(\boldsymbol{x}, t)}\right|_{A=0}, \tag{8}$$

where

$$R(A) = {}^+\!\sqrt{(\boldsymbol{p} - e\boldsymbol{A})^2 + m^2}. \tag{9}$$

The r.h.s. of Eq. (8) can be obtained from the easily calculable functional derivative of $[R(A)]^2$ [1] and one has finally for the current operator in momentum representation

$$\langle \boldsymbol{k}'|j_\mu(\boldsymbol{x})|\boldsymbol{k}\rangle = e\,\frac{k_\mu + k'_\mu}{k_0 + k'_0}(2\pi)^{-3}\,\mathrm{e}^{\mathrm{i}(\mathrm{k}-\mathrm{k}')\cdot\mathrm{x}}, \tag{10}$$

where

$$k_0 \equiv E_\mathrm{k} = {}^+\!\sqrt{\boldsymbol{k}^2 + m^2}, \qquad k'_0 \equiv E_{\mathrm{k}'} = {}^+\!\sqrt{\boldsymbol{k}'^2 + m^2}. \tag{11}$$

and the continuity equation is satisfied.

To treat a particle with spin $\frac{1}{2}$, one writes the Hamiltonian H_0 of Eq. (5) in the equivalent form

$$H_0 = \sqrt[+]{(\boldsymbol{\sigma} \cdot \boldsymbol{p})^2 + m^2} + V. \tag{12}$$

The minimal substitution gives

$$H(A) = \sqrt[+]{(\boldsymbol{p} - e\boldsymbol{A})^2 - e\boldsymbol{\sigma} \cdot \boldsymbol{B} + m^2} + eA_0 + V, \tag{13}$$

and following the same procedure as before one obtains

$$\langle \boldsymbol{k}', s' | \boldsymbol{j}^S(\boldsymbol{x}) | \boldsymbol{k}, s \rangle = \frac{ie}{k_0 + k_0'} \langle s' | \boldsymbol{\sigma} \times (\boldsymbol{k}' - \boldsymbol{k}) | s \rangle (2\pi)^{-3} e^{i(\mathrm{k}-\mathrm{k}') \cdot \mathrm{x}}, \tag{14}$$

$$\langle \boldsymbol{k}', s' | \boldsymbol{j}_0(\boldsymbol{x}) | \boldsymbol{k}, s \rangle = 0. \tag{15}$$

3 Application to electron scattering

To compare the semi-relativistic charge-current operator obtained in the previous section with other existing operators we consider quasi-elastic electron scattering. Neglecting final-state interactions the coincidence cross section for unpolarized electrons is proportional to the electron-proton cross section σ_{ep} apart from a kinematical factor, and to $S(\boldsymbol{p}, E)$ the spectral function for knocking out a proton of momentum $\boldsymbol{p}$ and energy E. For a coplanar kinematics,

$$\sigma_{\mathrm{ep}} = \sigma_{\mathrm{M}}(\rho_{00}\, g_{00} + \rho_{11}\, g_{11} + \rho_{01}\, g_{01} + \rho_{1-1}\, g_{1-1}), \tag{16}$$

where σ_{M} is the Mott cross section describing elastic Coulomb scattering by a point-like particle. The coefficients $\rho_{\lambda\lambda'}$ only depend on the electron kinematic variables and are well known from quantum electrodynamics [2]. Thus, the four structure functions $g_{\lambda\lambda'}$ (λ the polarization of the virtual photon) are the relevant quantities to study the effects due to the different charge-current density operators. In particular, in order to study the behaviour of the current given by the different approaches one has to focus on the pure transverse structure function, g_{11}.

Assuming the result $g_{11}(\mathrm{cc1})$ obtained within the cross section σ_{ep} most commonly used (σ_{cc1}, [3]) as a reference, the percentage deviation Δ from this result calculated within the other approaches, the non-relativistic reduction corrected to order M^{-2}, M^{-3} and M^{-4} and the semi-relativistic respectively will be considered as an indication of how far one is from the correct relativistic result. In Fig. 1 results are shown in parallel kinematics, i.e. when $\boldsymbol{q}$ and $\boldsymbol{p}'$ are aligned, for three values of $p' = 0.5, 0.75$ and 1 GeV.

From the analysis of [4] it is know that, the non-relativistic result (to order M^{-2}) and that corrected to order M^{-3} are always of opposite sign with respect to the result corrected to order M^{-4}. The latter is within a few percent from the relativistic case for $p' = 0.5$ GeV. However, for larger momenta the deviation rapidly deteriorates, in agreement with [5], where it was stressed that the non-relativistic FW reduction becomes meaningless for values of p/M larger than 0.5, with p a typical momentum. On the contrary, the semi-relativistic result is comparable with the non-relativistic one at low energy and soon merges into the relativistic one for increasing q and p'.

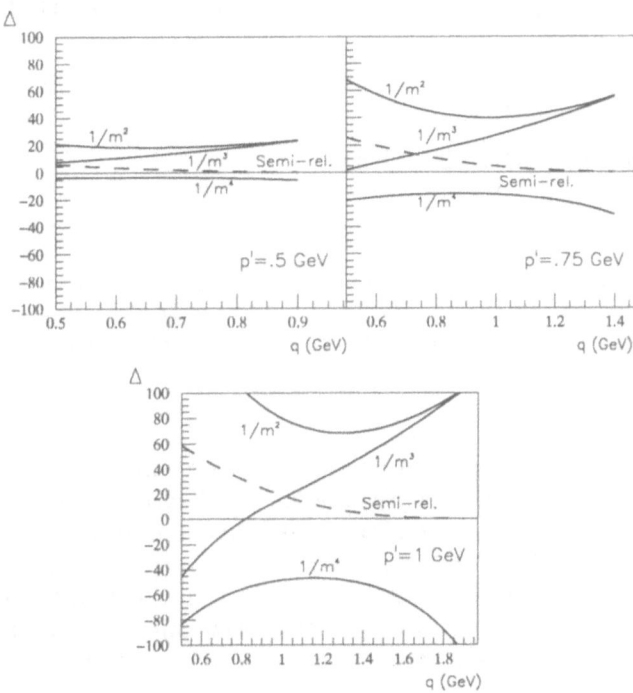

Figure 1. The percentage deviation Δ of the pure transverse structure function g_{11} plotted as a function of q at three values of the momentum of the proton ejected under parallel kinematic conditions.

References

1. S. Boffi, F. Capuzzi, P. Demetriou and M. Radici: Nucl. Phys. **A637**, 585 (1998)

2. S. Boffi, C. Giusti, F.D. Pacati and M. Radici: *Electromagnetic Response of Atomic Nuclei.* Oxford: Oxford University Press 1996

3. T. De Forest Jr.: Nucl. Phys. **A392**, 232 (1983)

4. C. Giusti and F.D. Pacati: Nucl. Phys. **A336**, 427 (1980)

5. K.W. McVoy and L. Van Hove: Phys. Rev. 125, 1034 (1962)

Few-Body Systems Suppl. 11, 78–85 (1999)

Few-
Body
Systems
© by Springer-Verlag 1999

NN and $N\Delta$ Form Factors viewed from ChPT

Thomas R. Hemmert* †

IKP (Th), FZ-Jülich, D-52425 Jülich

Abstract. I discuss recent work on nucleon form factors, magnetic strangeness in the nucleon and the isovector nucleon-delta transition based on the broken chiral symmetry of QCD utilizing recent theoretical developments in ChPT.

1   Introduction

During this workshop we have heard both theoretical and experimental presentations looking for the onset of *perturbative* QCD formulated in explicit (current) quark and gluon degrees of freedom at moderate/high four-momentum transfer, *e.g.* $Q^2 > 1$ GeV^2. At lower momentum transfer one cannot avoid the complications of the strong coupling regime. In this talk I want to discuss some of the constraints resulting from the broken chiral symmetry of QCD for baryon form factors in the *non-perturbative* regime of QCD [1, 2, 3].

2   Nucleon Form Factors and ChPT

The chiral symmetry of the light flavor sector of QCD is spontaneously broken at low energies leading to the existence of Goldstone boson modes. Here we focus on the (u,d)-quark sector only which leads to the identification of the pions as the Goldstone Bosons. All low energy dynamics is governed by these lightest hadronic degrees of freedom and the chiral symmetry puts very strict constraints on their interactions among themselves, with external sources and on their coupling to matter fields (baryons, etc.). This Goldstone-boson dominated regime of *non-perturbative* QCD at low energies can be formulated exactly in an effective lagrangian formalism called Chiral Perturbation Theory (ChPT) [4]. With the pions being the lightest degrees of freedom in the hadron spectrum, ChPT suggests that the long range structure of baryons and its leading momentum dependence is governed by the chiral symmetry of the pion interaction.

---

*E-mail address: th.hemmert@fz-juelich.de
†Invited Lecture

The electro-weak structure of baryons is parameterized via form factors. In the case of the nucleon they have been analyzed in one-loop relativistic baryon ChPT [5] and in a non-relativistic approach called HBChPT [6] in the past. Recently [1], we have repeated this analysis utilizing a phenomenological extension of ChPT called the "small scale expansion" [7]. In this approach one includes the first nucleon resonance $\Delta(1232)$ as an explicit degree of freedom in a phenomenologically resummed chiral expansion. In [1] all 6 form factors of the nucleon are discussed, here I will only address the isovector Pauli form factor $F_2^v(q^2)$.

Consider the nucleon matrix element of the isovector component of the quark vector current $V_\mu^i = \bar{q}\gamma_\mu(\tau^i/2)q$, which involves a vector (Dirac) and a tensor (Pauli) form factor,

$$\langle N(p_2)|V_\mu^i(0)|N(p_1)\rangle = \bar{u}(p_2)\left[F_1^v(q^2)\,\gamma_\mu + \frac{i}{2M_N}F_2^v(q^2)\,\sigma_{\mu\nu}q^\nu\right]u(p_1)\,, \quad (2.1)$$

where $u(p)$ is a Dirac spinor and $q^2 = (p_2 - p_1)^2$ is the invariant momentum transfer squared. The radii of these form factors should be determined by the extension of the pion cloud. E.g. for the radius of $F_2^v(q^2)$ one finds to leading order in HBChPT [6], [SSE [1]]

$$\begin{aligned}
(r_2^v)^2 &= \frac{g_A^2 M_N}{8F_\pi^2 \kappa_v \pi m_\pi} + \left[\frac{g_{\pi N\Delta}^2 M_N}{9F_\pi^2 \kappa_v \pi^2 \sqrt{\Delta^2 - m_\pi^2}} \log\left[\frac{\Delta}{m_\pi} + \sqrt{\frac{\Delta^2}{m_\pi^2} - 1}\right]\right] \\
&= 0.52\text{fm}^2 \left[+\,0.09\text{fm}^2\right],
\end{aligned} \quad (2.2)$$

compared with the empirical value, $(r_2^v)^2 = 0.80\text{fm}^2$ [8]. The only parameters are the pion decay constant (mass) F_π (m_π), the πNN $(\pi\Delta N)$ couplings g_A $(g_{\pi N\Delta})$, the nucleon mass M_N, the mass-splitting $\Delta = M_\Delta - M_N$ and the anomalous isovector magnetic moment κ_v. One can see that already the *leading* HBChPT result for the extension of the pion cloud provides a good estimate for the size of the nucleon in this channel. Inclusion of explicit delta components in the nucleon wavefunction around which the pions can fluctuate provides a 17% correction in the right direction [1]. In the chiral limit, we recover the well known $1/m_\pi$ singularity, which is not touched by the resonance contribution in accord with general decoupling requirements. Other form factor results are discussed in [1].

3 Magnetic Strangeness in the Nucleon

So far we have focused on baryon ChPT involving two light flavors. The analysis presented in the previous section can be generalized in a straightforward fashion to a SU(3) chiral symmetry of QCD, *i.e.* the inclusion of explicit strange degrees of freedom like kaons, lambdas etc. Repeating the analysis of the isovector form factors of the nucleon in SU(3) HBChPT one finds an extra contribution from kaon loops to $(r_2^v)^2$ (Eq.(2.2)) of the order of a few percent. This contribution vanishes if the strange quark mass becomes very heavy. For the case of the isovector nucleon form factors SU(3) HBChPT therefore reproduces our physical expectation that the kaon contributions are much less important than their pion counterparts due to the much larger mass. Here the pions clearly dominate the long-range physics and control the size of the nucleon. This is also seen in the spectral function [9].

However, this is not always the case. For example in the isoscalar form factors of the nucleon the leading chiral contribution to the radius is given by the kaon cloud, as the pionic contribution only begins via 3 pion intermediate states at the 2 loop level. The $\mathcal{O}(p^3)$ analysis suggests that roughly 30% of the isoscalar radius of the nucleon comes from structure related to the kaons in the nucleon! For details regarding these issues we refer to [1, 2, 9].

Another sector where explicit strange degrees of freedom figure prominently concerns–quasi by definition–the recent interest in the so-called "strangeness content of the nucleon", e.g. see ref.[10]. In the following we focus on the strangeness vector current of the nucleon defined as

$$\langle N | \, \bar{s} \, \gamma_\mu \, s \, | N \rangle = \langle N | \, \bar{q} \, \gamma_\mu \, (\lambda^0/3 - \lambda^8/\sqrt{3}) \, q \, | N \rangle = (1/3) J_\mu^0 - (1/\sqrt{3}) J_\mu^8 \, , \quad (3.1)$$

with $q = (u, d, s)$ denoting the triplet of the light quark fields and $\lambda^0 = I \, (\lambda^a)$ the unit (the $a = 8$ Gell–Mann) SU(3) matrix. Assuming conservation of all vector currents, the corresponding singlet and octet vector current for a nucleon can then be written as

$$J_\mu^{0,8} = \bar{u}_N(p') \left[F_1^{(0,8)}(q^2) \gamma_\mu + F_2^{(0,8)}(q^2) \frac{i \sigma_{\mu\nu} q^\nu}{2 M_N} \right] u_N(p) \, . \quad (3.2)$$

Here, $q_\mu = p'_\mu - p_\mu$ corresponds to the four–momentum transfer to the nucleon by the external singlet $(v_\mu^{(0)} = v_\mu \lambda^0)$ and the octet $(v_\mu^{(8)} = v_\mu \lambda^8)$ vector source v_μ, respectively. The strangeness Dirac and Pauli form factors are defined via

$$F_{1,2}^{(s)}(q^2) = \frac{1}{3} F_{1,2}^{(0)}(q^2) - \frac{1}{\sqrt{3}} F_{1,2}^{(8)}(q^2) \, , \quad (3.3)$$

subject to the normalization $F_1^{(s)}(0) = 0$, $F_2^{(s)}(0) = \kappa_B^{(s)}$, with $\kappa_B^{(s)}$ the (anomalous) strangeness moment. In the following we concentrate our analysis on the "magnetic" strangeness form factor $G_M^{(s)}(q^2)$, which in analogy to the (electro)magnetic Sachs form factor is defined as

$$G_M^{(s)}(q^2) = F_1^{(s)}(q^2) + F_2^{(s)}(q^2) \, . \quad (3.4)$$

In the case of a nucleon $G_M^{(s)}(0) \equiv \mu_N^{(s)}$ defines the so called "strange magnetic moment" of the nucleon whose sign/size is heavily contested in theoretical analyses. Furthermore, it is precisely this form factor at $q^2 = -0.1 \text{GeV}^2$ which has been analyzed in the recent Bates measurement [12].

One expects that for low q^2 ChPT can give a prediction for this (as of 1998!) unknown quantity. For the case of $\mu_N^{(s)}$ this is only partially correct as one needs additional information about an unknown isosinglet counterterm [2, 11].

However, even if one cannot calculate the overall normalization of $G_M^{(s)}(q^2)$ at $q^2 = 0$, the evolution of this form factor with q^2 can be predicted in terms of well-known low energy quantities! To $\mathcal{O}(p^3)$ in SU(3) HBChPT one finds [2]

$$G_M^{(s)}(Q^2) = \mu_N^{(s)} + \frac{\pi M_N m_K}{(4\pi F_\pi)^2} \frac{2}{3} \left(5D^2 - 6DF + 9F^2\right) f(Q^2) , \quad (3.5)$$

with $Q^2 = -q^2$, $D \simeq 3/4$, $F \simeq 1/2$, $F_\pi \equiv (F_\pi + F_K)/2 \simeq 102\,\text{MeV}$ the average pseudoscalar decay constant and m_K being the kaon mass. The momentum dependence is given entirely in terms of the function

$$f(Q^2) = -\frac{1}{2} + \frac{4 + Q^2/m_K^2}{4\sqrt{Q^2/m_K^2}} \arctan\left(\frac{\sqrt{Q^2}}{2m_K}\right) . \quad (3.6)$$

For small and moderate Q^2 it rises almost linearly with increasing Q^2.

I emphasize that Eq.(3.5) only contains the leading order chiral contribution which stems exclusively from the kaon-cloud of the nucleon. It will be interesting to calculate the next-to-leading order (*i.e.* $\mathcal{O}(p^4)$) correction to this result in order to check possible contributions from vector mesons which are usually assumed to dominate this form factor [13]. However, already at $\mathcal{O}(p^3)$ one can implicitly include some of the higher order corrections if one analyzes the magnetic isoscalar form factor $G_M^{I=0}(Q^2)$ and the strange magnetic form factor $G_M^{(s)}(Q^2)$ simultaneously [2]. One obtains the *model-independent* connection

$$G_M^{(s)}(Q^2) = \mu_N^{(s)} + \mu_N^{I=0} - G_M^{I=0}(Q^2) + \mathcal{O}(p^4) , \quad (3.7)$$

where $\mu_N^{I=0}$ denotes the isoscalar magnetic moment of the nucleon. To $\mathcal{O}(p^3)$ one therefore predicts that the low Q^2 behavior of the strange magnetic form factor of the nucleon is *exactly* controlled by the well-known isoscalar form factor of the nucleon! For details I refer to [2].

Eqs.(3.5,3.7) can be considered as a lower, upper bound on the q^2 evolution of the strange magnetic form factor at low momentum transfer [2]. Both relations can be used to extrapolate from the experimentally determined values for $G_M^{(s)}(Q^2)$ at $Q^2 > 0$ to the sought after strange magnetic moment $\mu_N^{(s)}$ of the nucleon at $Q^2 = 0$. Clearly, with improving experimental accuracy on $G_M^{(s)}(Q^2)$ one also needs to calculate the $\mathcal{O}(p^4)$ corrections to both relations. Furthermore, comparing Eqs.(3.5,3.7) we are also looking forward to the mapping of the low Q^2 dependence of $G_M^{(s)}(Q^2)$ by the G0 collaboration at J-Lab [14].

4 The Isovector $N\Delta$ Transition

Finally, I want to give a brief update on the ongoing calculations [3, 15] regarding the isovector nucleon-delta transition multipoles and form factors. Recent interest is mainly triggered by three observations:

1. In a multipole analysis one finds that in the photoexcitation of $\Delta(1232)$ $[\gamma N \rightarrow \Delta]$ one can only have magnetic dipole (M1) or electric quadrupole

(E2) transitions from the nucleon to the delta. Simple constituent quark models of the nucleon however generally assume all quarks to be in an s-wave state and therefore predict zero strength for the E2 transition. Several fits to pion photoproduction data in the delta region however show a non-zero ratio of E2/M1 strength of about -1% to -3% (*e.g.* [16]), indicating non-radial/many-body components in the ground-state wave function of the nucleon.

2. For electroproduction of $\Delta(1232)$ $[\gamma^* N \to \Delta]$ the transition multipoles M1,E2 do not only develop a dependence on the four-momentum transfer (squared) Q^2 but one can now have additional contributions from a Coulomb quadrupole transition C2. Our knowledge of the Q^2 dependence of these three multipoles for $0 < Q^2 < 1\text{GeV}^2$ mainly stems from experiments of the 1970s [17]. Recently, new measurements have started at Bonn which seem to validate the old analyses showing interesting differences in the Q^2 behavior among these multipoles for $Q^2 < 0.3\text{GeV}^2$ [18]. Furthermore, one would also like to compare the Q^2-falloff of the $N\Delta$ transition form factors with the well-known dipole behavior of the electric/magnetic Sachs form factors of the nucleon, *e.g.* see [17, 19].

3. Perturbative QCD predicts that for very large four-momentum transfer the ratio of E2/M1 for the case of delta electroproduction should tend to unity. At which finite Q^2 the crossover from a negative to a positive ratio should happen and whether this point is kinematically accessible at present/future electron scattering machines is an issue of current theoretical debate, *e.g.* [20].

We have started two collaborations [3, 15] to look into these topics from the viewpoint of ChPT. In particular, we are using the recently developed SSE formalism [7] in order to treat the delta resonance in a systematic fashion. What kind of results can one expect from these efforts?

1. There exist already 2 calculations regarding the ratio of E2/M1 at the real photon point utilizing ChPT [22]. Our present understanding is that one needs to take into account non-zero contributions from *three* different ingredients—namely pion loops, 1/M corrections and counterterms. While the loop contribution is relatively easy to calculate and agreed upon, the contributions from the 1/M corrections and counterterms have not been handled with the same accuracy so far. SSE offers a systematic formalism to address both aspects. At this point we can say that the actual number for E2/M1 in ChPT is quite sensitive to the treatment/size of several unknown counterterms. In order to settle this issue one needs a full calculation of pion-photoproduction in the delta resonance region [15] in order to fix these unknowns with the accuracy required for E2/M1. Only then one can expect a new systematic prediction for E2/M1 from ChPT. We also note that in the past only the leading delta contribution to the s-wave multipole E_{0+} had been calculated explicitly in SSE

[7]. The p-wave multipoles are known to receive large contributions from $\Delta(1232)$, but so far these effects have only been included via "resonance saturation" in higher order couplings [21]. Utilizing SSE [7], we are now analyzing explicit $\Delta(1232)$ components in the three p-wave multipoles. It will be interesting to see how far in energy the inclusion of explicit delta degrees of freedom can extend the applicability of ChPT to pion-photoproduction off nucleons into the delta resonance region. [15].

2. Surprisingly, the determination of the Q^2-evolution of the three $N\Delta$ transition form factors and of the corresponding three transition multipoles is a much simpler problem in ChPT, but has not been addressed so far. The important point to realize is that most of the unknown couplings/counterterms only concern the $Q^2 = 0$ values. Once one fixes the form factors/multipoles at the measured real photon values [16] one obtains their Q^2-dependence in terms of very few parameters which are under control. It is then straightforward to extract radii for the transition form factors and compare with the form factors of the nucleon. This project [3] is close to being finished once the problem of the scaling in the radii (discussed below) is fixed.

3. Concerning the third issue, ChPT can certainly not answer the problem of the onset of perturbative QCD in the E2/M1 ratio, probably even the zero-crossing point is at too high a momentum transfer for this approach. However, it should be possible to say whether E2/M1 first drops even more negative for low momentum transfer and whether there is a *turning point* after which the curve moves towards a positive value.

Finally, I want to address a problem that we encountered during the calculation of the isovector $N\Delta$ transition form factors. Assuming conservation of the vector current as well as invariance under P,C,T symmetry operations one concludes that in general there exist 3 independent structures for such a transition. To be more specific, let's assume that we are talking about the process $\Delta \to N\gamma^*$. The matrix-element is then typically written as [23][1]

$$
i\mathcal{M}_{\Delta \to N\gamma}^{full} = \frac{e}{2M_N}\bar{u}(p_N)\gamma_5\left[g_1(q^2)(\not{\epsilon}\epsilon_\mu - \not{q}q_\mu) + \frac{g_2(q^2)}{2M_N}(p_N \cdot \epsilon\, q_\mu\right.
$$
$$
\left. -p_N \cdot q\,\epsilon_\mu) + \frac{g_3(q^2)}{2M_N}(q \cdot \epsilon\, q_\mu - q^2\epsilon_\mu)\right] u_\Delta^\mu(p_\Delta) . \qquad (4.1)
$$

Here M_N is the nucleon mass, $p_{N,\Delta}$ denotes the momentum of the nucleon, delta and q, ϵ are the photon momentum and polarization vectors, respectively. The delta is described in the Rarita-Schwinger formalism, i.e. as an axial-vector spinor u_Δ^μ. Now one proceeds to calculate this matrix element in a *non-relativistic microscopic approach*, in our case SSE [7]. Calculating to third order in the expansion scheme the calculation can produce up to two inverse

---

[1]Minor differences to the form of Eq.(4.1) arise via field-redefinitions utilizing the equations of motion for the baryons. However, this does not change the thrust of the above argument.

powers of the expansion scale, *i.e.* one is sensitive to structures up to $1/M_N^2$. In order to match the calculation with the most general matrix element Eq.(4.1) one also needs to expand it up to the same power in $1/M_N^2$. One finds

$$
\begin{aligned}
i\mathcal{M}_{\Delta->N\gamma}^{(3)} =\ & e\,\bar{u}_v(r_N)\left\{(S\cdot\epsilon)q_\mu\left[\frac{g_1(q^2)}{M_N}+\mathcal{O}(1/M_N^3)\right]\right.\\
& +(S\cdot q)\epsilon_\mu\left[-\frac{g_1(q^2)}{M_N}-\frac{\Delta}{2M_N^2}g_1(0)+\frac{\Delta}{4M_N^2}g_2(0)+\mathcal{O}(1/M_N^3)\right]\\
& +(S\cdot q)(v\cdot\epsilon)q_\mu\left[\frac{(g_1(0)-\frac{1}{2}g_2(0))}{2M_N^2}+\mathcal{O}(1/M_N^3)\right]\\
& \left.+(S\cdot q)(q\cdot\epsilon)q_\mu\left[0+\mathcal{O}(1/M_N^3)\right]\right\}u_{v,\Delta}^\mu(0)\ .
\end{aligned}
\tag{4.2}
$$

Here S_μ denotes the Pauli-Lubanski vector, v_μ corresponds to the velocity vector of the delta reference frame and $\Delta = M_\Delta - M_N$. As one can see from Eq.(4.2) the $1/M_N^2$-expansion demands that there are no explicit structures proportional to $\epsilon\cdot q$ to this order. Nevertheless the SSE calculations yield such terms! We therefore have to conclude that the often-used form for the isovector $N\Delta$ transition Eq.(4.1) is *not compatible* with (non-relativistic) microscopic calculations of this transition that rely on a systematic $1/M$ expansion. It is therefore mandatory to rescale $g_2(q^2)$, $[g_3(q^2)]$ by M_N/Δ, $[M_N^2/\Delta^2]$ in order to achieve a systematic matching between the microscopic calculations and the most general amplitude. Furthermore, without this rescaling of the form factors their transition radii would scale as $r_i^2 \sim M_N^n$; $i = 2, 3; n \geq 1$, *i.e.* one would see no $1/M$ suppression[2] in the radii compared to the $q^2 = 0$ point! Finally, we note that phenomenological analyses of data utilizing Eq.(4.1) are not affected by this problem, as long as all amplitudes are treated in a fully relativistic form. A detailed publication describing all these aspects is in preparation [3].

Acknowledgement. I would like to thank my collaborators V. Bernard, H.W. Fearing, G.C. Gellas, C.N. Ktorides, U.-G. Meißner, G.I. Poulis and S. Steininger and acknowledge helpful financial support from ECT*.

References

1. V. Bernard, H.W. Fearing, T.R. Hemmert and U.-G. Meißner: Nucl. Phys. **A635**, 121 (1998)

2. T.R. Hemmert, U.-G. Meißner and S. Steininger: Phys. Lett. **B437**, 184 (1998)

3. G.C. Gellas, T.R. Hemmert, C.N. Ktorides and G.I. Poulis: hep-ph/9810426

4. S. Weinberg, Physica **A96**, 327 (1979); J. Gasser and H. Leutwyler: Ann. Phys. **158**, 142 (1984); Nucl. Phys. **B250**, 465 (1985)

---

[2]The only known case of such an anomalous scaling concerns the pseudoscalar form factor of the nucleon which contains the (light) pion-pole, see *e.g.* [1].

5. J. Gasser, M.E. Sainio and A. Švarc: Nucl. Phys. **B307**, 779 (1988)

6. V. Bernard, N. Kaiser, J. Kambor and U.-G. Meißner: Nucl. Phys. **B388**, 315 (1992); V. Bernard, N. Kaiser and U.-G. Meißner: Int. J. Mod. Phys. **E4**, 193 (1995)

7. T.R. Hemmert, B.R. Holstein and J. Kambor: Phys. Lett. **B395**, 89 (1997); J. Phys. **G24**, 1831 (1998)

8. P. Mergell, U.-G. Meißner and D. Drechsel: Nucl. Phys. **A596**, 367 (1996)

9. V. Bernard, N. Kaiser and U.-G. Meißner: Nucl. Phys. **A611**, 429 (1996)

10. M.J. Musolf et al.: Phys. Rep. **239**, 1 (1994)

11. M.J. Ramsey-Musolf and H. Ito: Phys. Rev. **C55**, 3066 (1997)

12. B. Mueller et al.: Phys. Rev. Lett. **78**, 3824 (1997)

13. R.L. Jaffe: Phys. Lett. **B229**, 275 (1989); H.-W. Hammer, U.-G. Meißner and D. Drechsel: Phys. Lett. **B367**, 323 (1996)

14. TJNAF experiment E91-017 (D. Beck, spokesperson)

15. V. Bernard, T.R. Hemmert and U.-G. Meißner: "The Role of Delta(1232) in Pion Photoproduction.", forthcoming.

16. R.M. Davidson, N.C. Mukhopadhyay and R.S. Wittman: Phys. Rev. **D43**, 71 (1991); O. Hanstein, D. Drechsel and L. Tiator: Phys. Lett. **B385**, 45 (1996); R.A. Arndt, I.I. Strakovsky and R.L. Workman: Phys. Rev. **C56**, 577 (1997); R. Beck et al.: Phys. Rev. Lett. **78**, 606 (1997); G. Blanpied et al.: Phys. Rev. Lett. **79**, 4337 (1997)

17. Particle Data Group: Rev. Mod. Phys. **48**, S163 (1976)

18. F. Kalleicher et al.: Z. Phys. **A359**, 201 (1997)

19. G. Sterman and P. Stoler: Ann. Rev. Nucl. Part. Sci. **47**, 193 (1997)

20. C.E. Carlson and N.C. Mukhopadhyay: Phys. Rev. Lett. **81**, 2646 (1998)

21. V. Bernard, N. Kaiser and U.-G. Meißner: Z. Phys. **C70**, 483 (1996)

22. M.N. Butler, M.J. Savage and R.P. Springer, Phys. Lett. **B304**, 353 (1993); *ibide* **B314**, 122 (1993); M. Napsuciale and J.L. Lucio: Nucl. Phys. **B494**, 260 (1997)

23. H.F. Jones and M.D. Scadron: Ann. Phys. **81**, 1 (1973); R.C.E. Devensish, T.S. Eisenschitz and J.G. Körner: Phys. Rev. **D14**, 3063 (1976)

Few-Body Systems Suppl. 11, 86–93 (1999)

Missing Baryons

Simon Capstick*

Supercomputer Computations Research Institute and Department of Physics,
Florida State University, Tallahassee, Florida 32306, USA

Abstract. Many excited baryon states which are predicted to exist in models of the nucleon which treat three constituent quarks symmetrically are not present in analyses of pion-nucleon elastic scattering. The properties and quantum numbers of the nonstrange baryons accessible at new electron and photon facilities are described in a model of the baryon spectrum and strong decays based one-gluon exchange. A model which is able to describe almost all of the channels open for strong decays of the established and missing states is outlined. This model gives a reasonable description of the strong couplings of known states, and is used to determine the best decay channels for discovering missing baryon states.

1 Introduction

Models of baryon structure exist which describe the nucleon and its excitations in terms of a diquark and a quark [1]. If the diquark is tightly bound its internal excitation is costly in energy, so that the low-lying excitations of the nucleon will not include excitations of the diquark. In SU(6) language this means that, in addition to the lightest negative-parity excitations in the [70,1$^-$] multiplet which have all been seen in $N\pi$ elastic scattering, the low-lying positive-parity excitations of the nucleon lie in symmetric 56-plet and mixed-symmetry 70-plet representations. Models which treat all three quarks symmetrically have more low-lying excitations. Light positive-parity excited states are present in the [56$'$,0$^+$], [70,0$^+$], [56,2$^+$], [70,2$^+$], and [20,1$^+$] SU(6) multiplets, with J^P = 1/2$^+$, 3/2$^+$, 5/2$^+$, and 7/2$^+$, for a total of twenty-one positive-parity excited states. A collective model based on a spectrum-generating algebra developed by Bijker, Iachello and Leviatan [2] also predicts many missing baryons, although with different quantum numbers and masses. Note that some diquark models also allow for the quark and diquark to interact and exchange a quark and so their identity in order to maintain overall antisymmetry [1], and may have

---

E-mail address: capstick@scri.fsu.edu

similar numbers of positive-parity excited states, although the [20,1$^+$] multiplet is still excluded.

Of the twenty-one low-lying positive-parity excited nonstrange states predicted by symmetric quark models, nine are considered well established by the Particle Data Group (PDG) [3], there are three tentative (one or two stars) nucleon states, and two tentative Δ states, for a total of fourteen, of which five need confirmation. The remaining predicted states are defined as missing by Koniuk and Isgur [4]. This definition can be expanded to include any baryon predicted by symmetric quark models but noticeably absent in the analyses, such as the many higher mass negative-parity states above 1900 MeV. Although not discussed in detail here, the situation is similar for the strange baryons Λ and Σ, except there are fewer excited states present in the analyses, with a few missing low-lying negative-parity states and more missing positive-parity states.

These states may be missing because of strong diquark clustering in the light-quark baryons. There is some evidence from the lattice [5] and from an analysis of baryon strong decays [6] that such strong diquark clustering may not be present. The explanation adopted here, and in the work of others [4, 7], is that such states have either small $N\pi$ couplings in a partial wave which includes other light strongly coupled states, or are close in mass to a more strongly coupled state, both of which make extraction of a signal from $N\pi$ elastic scattering difficult. This explanation can be understood using a quark model of the spectrum and wavefunctions of these states and their strong decay, which are described briefly below. This model can then be used to show how such states can be found.

2 Model of the Baryon Spectrum

The model of the baryon spectrum described here is based on the assumption that baryons can be described by constituent quarks (with $m_{u,d} = 220$ MeV, $m_s = 420$ MeV, and finite size) which interact at short distances *via* one-gluon exchange, and are confined by a potential which is given by the meson string tension times the minimum length of a Y-shaped string which connects them. The short distance potential consists of a color-Coulomb interaction along with spin-dependent interactions. Alternative explanations are given by Bijker, Iachello and Leviatan in their algebraic model [2], and by Glozman and Riska [8] and others [9], who assume that the contact splittings in the spectrum are generated by one-boson exchange (OBE) between the quarks. The latter approach to the spectrum is described in these proceedings by Wagenbrunn [11]; some results for baryon pionic decays in this model using pointlike pions are available [10].

One-gluon exchange (OGE) models have the advantage that they give a unified description of the meson and baryon spectra, using the same quark masses, confining string tension, and short-distance potential [12, 13]. In QCD quarks exchange gluons at short distances, and overwhelming evidence for one-gluon exchange potentials exists in the $c\bar{c}$ and $b\bar{b}$ heavy-quark mesons. There is some evidence from the regularity of the splittings between the tensor ($^3P_2$), vector ($^3S_1$) and pseudoscalar ($^1S_0$) mesons as the quark masses are lightened [14] that suggests that the same mechanism operates in light-quark systems. OGE models also give a reasonable description of the spectrum *and* decays of baryons and mesons. On the other hand, OGE and OBE are at the same order in $1/N_c$ and some evidence from the lattice exists [15] which suggests that OBE is responsible for a substantial part of the $\Delta - N$ and $\rho - \pi$ splittings. It is likely that both mechanisms are in operation in the light-quark baryons.

This OGE model can only describe the soft physics of conventional excitations below the gluonic excitation energy. It also ignores possibly large mass shifts from coupling to $qqq(q\bar{q})$ states through baryon-meson loops. It is not necessarily nonrelativistic; the quark Hamiltonian used has the form

$$H = \sum_i (p_i^2 + m_i^2)^{\frac{1}{2}} + V, \tag{1}$$

where in the nonrelativistic limit (which is *not* taken in the model) the potential V tends to a confining potential made up of the string potential plus Thomas-precession spin-orbit terms, plus a one-gluon-exchange term made up of a color-Coulomb term, hyperfine contact and tensor terms, and spin-orbit terms. Away from the nonrelativistic limit the momentum dependence of the potentials is parametrized, and the potentials undergo relativistic smearing. The strong coupling constant is also allowed to run from its value of 0.6 at $Q^2 = 0$. The dynamical problem is solved for the masses and wavefunctions of the states by expanding the wavefunctions in a large coupled oscillator basis, forming a matrix of the Hamiltonian in this basis, and diagonalizing. For details see Ref. [13].

3 Strong Decay Models

Before the resulting spectrum can be compared to the known states it is necessary to use the wavefunctions that result to calculate their strong decays, in order to find out which of the many predicted states are likely to be seen in the analyses. Strong decay models can be based on an algebraic approach, where the invariance of the strong interaction causing the decay under isospin, its approximate invariance under SU(3)$_f$, along with its conservation of angular momentum, are used (with some additional assumptions) to write down the tensor structure of the decay operator in the flavor, quark spin, and orbital spaces. Hadron wavefunctions are built in these spaces, the Wigner-Eckart theorem is used to give relations between the amplitudes, and several reduced matrix elements are fit to well known decays. The most successful formulation of this approach is called SU(6)$_W \times$O(3), and is reviewed by Hey and Litchfield [16].

Such algebraic models are not as predictive as those which assume a definite structure for the hadron states and the decay operator. An example of the latter is the elementary-meson emission model used by Koniuk and Isgur [4], and used within the algebraic approach of Bijker, Iachello and Leviatan [17]. Here pointlike mesons couple to the quarks; if the meson is a pseudoscalar the decay operator is proportional in the nonrelativistic limit to

$$g\boldsymbol{\sigma}_3 \cdot \mathbf{K} + h\boldsymbol{\sigma}_3 \cdot \mathbf{p}_3, \tag{2}$$

where $\boldsymbol{\sigma}_3$ and $\mathbf{p}_3$ are the spin and momentum of the third quark, and $\mathbf{K}$ is the momentum of the emitted meson. The other quarks are spectators, although the final baryon must be allowed to recoil, which ultimately alters their momentum. Note that the identity of the interacting quark is summed over, and the decay operator is taken to be an isoscalar. In the work of Koniuk and Isgur [4] configuration-mixed wavefunctions were used, and although in principle all decay rates could be given in terms of the couplings g and h, better results were found by fitting certain linear combinations to the measured amplitudes.

In the $^3P_0$ model of Micu [18] and LeYaouanc and collaborators [19] the emitted meson is given a quark substructure, which requires that a model be given for the created quark pair. Here the created quarks have the quantum numbers of the vacuum ($^3P_0$). Another possibility is that the quark pair is created with gluon ($^3S_1$) quantum numbers. This structure results if the QCD interaction Hamiltonian is applied to an oriented flux tube in meson decays. However, although for certain meson decays $^3S_1$ and $^3P_0$ produce similar results, it has been shown by Geiger and Swanson [20] that the experimental S/D ratios in meson decays involving vector mesons strongly favor the $^3P_0$ model. This assumed form is shown to be favored by a microscopic model of decays by Ackleh, Barnes and Swanson [21].

4 Baryon Strong Decays

A variant of the $^3P_0$ model has been applied extensively to meson decays by Kokoski and Isgur[22], and several final states in the decays of many baryon states have been examined in the $^3P_0$ model by Forsyth and Cutkosky [23], who also consider other created pair structures, and by Stancu and Stassart [7].

The wavefunctions resulting from the OGE spectral calculation of Ref. [13] described above (with single Gaussian wavefunctions for the emitted mesons) have been used in a $^3P_0$ model to make a broad survey of the strong decays of baryons into $N\pi$ [24], and various quasi-two-body channels in $N\pi\pi$ [25] such as $N\rho$, $N\omega$, $N\eta$, $N\eta'$, $\Delta\pi$, $N(1440)\pi$, and $\Delta(1600)\pi$. The isospin selective final states $\Delta\eta$ and $\Delta\omega$ have also been examined recently [26], as the relatively high thresholds for these reactions (and the complementary isospin selective reactions $N\eta$ and $N\omega$) will simplify the discovery of new states in an analysis, and such experiments are now feasible using the Crystal Ball in the AGS at Brookhaven [27].

With the completion of a recent calculation of the couplings to strange final states of the nonstrange baryons, this model can be used to give upper limits on branching fractions and total widths into various final states for the nonstrange baryons. Note these are upper limits because in some cases the above channels will not include decays into all possible available final states. Examples of other two-pion channels are $N(\pi\pi)'_S$ where the two pions are in a correlated S-wave, which can contribute [28] to the decays of several known baryons, and decays into negative-parity baryons and a pion.

Tables 4.1 and 4.2 show upper limits for the predicted total widths and branching fractions for nonstrange positive-parity baryons below about 2000 MeV in partial waves with predicted missing states. Also shown are our assignments to states in the analyses, based largely on the $N\pi$ decays and masses. It is clear that the states seen in the analyses correspond, in number and in mass (with some model error) to those predicted to have the largest $N\pi$ branches in their partial wave. There are some exceptions; for example the two-star state $N(2000)F_{15}$ is predicted to have a small $N\pi$ branch and is seen predominantly in the $N\rho$ and $\Delta\pi$ channels, as predicted for the model state $N[5/2^+]_3(1995)$.

State	Γ_{tot}	$N\pi$	$\Delta\pi$	$N\rho$	$N\eta$	$N\omega$
$[N\frac{1}{2}^+](1540)$	425	.97	.03	.00	.00	.00
$N(1440)P_{11}$****	250-450	0.6-0.7				
$[N\frac{1}{2}^+](1770)$	305	.06	.63	.04	.22	.00
$N(1710)P_{11}$***	50-250	0.1-0.2				
$[N\frac{1}{2}^+](1880)$	155	.05	.49	.03	.18	.14
$[N\frac{1}{2}^+](1975)$	45	.08	.47	.14	.00	.22
$[N\frac{3}{2}^+](1795)$	300	.66	.01	.04	.22	.00
$N(1720)P_{13}$****	100-200	0.1-0.2				
$[N\frac{3}{2}^+](1870)$	190	.20	.12	.02	.26	.11
$N(1900)P_{13}$**	500 ± 80	0.26 ± 0.06				
$[N\frac{3}{2}^+](1910)$	390	.00	.75	.03	.00	.17
$[N\frac{3}{2}^+](1950)$	140	.12	.43	.11	.00	.28
$[N\frac{3}{2}^+](2030)$	85	.04	.57	.15	.00	.16
$[N\frac{5}{2}^+](1770)$	55	.77	.05	.16	.01	.00
$N(1680)F_{15}$****	120-140	0.6-0.7				
$[N\frac{5}{2}^+](1980)$	270	.01	.89	.02	.00	.03
$[N\frac{5}{2}^+](1995)$	185	.00	.51	.33	.04	.08
$N(2000)F_{15}$**	300 ± 200	0.06 ± 0.02				
$[N\frac{7}{2}^+](2000)$	45	.13	.53	.03	.21	.06
$N(1990)F_{17}$**	350 ± 150	0.06 ± 0.02				

Table 4.1. Predicted branching fractions of positive-parity N^* baryons below roughly 2000 MeV. The $N\eta'$, $N(1440)\pi$ and $\Delta(1600)\pi$ branching fractions (< 0.02 for all cases) and strange hadron branches have been excluded for brevity. States from the analyses [3] are listed under the predicted state to which they are assigned.

From the branching fractions listed in the tables it is obvious that several missing positive-parity nonstrange baryon states should be discovered in the

State	Γ_{tot}	$N\pi$	$\Delta\pi$	$N\rho$	$N(1440)\pi$	$\Delta\eta$
$[\Delta\frac{1}{2}^{+}](1835)$	315	.05	.63	.20	.01	.07
$\Delta(1750)P_{31}$*	300 ± 120	0.08 ± 0.03				
$[\Delta\frac{1}{2}^{+}](1875)$	265	.33	.27	.14	.01	.07
$\Delta(1910)P_{31}$****	190-270	0.15-0.3				
$[\Delta\frac{3}{2}^{+}](1795)$	165	.48	.43	.01	.06	.00
$\Delta(1600)P_{33}$***	250-450	0.10-0.25				
$[\Delta\frac{3}{2}^{+}](1915)$	225	.08	.44	.32	.00	.10
$\Delta(1920)P_{33}$***	150-300	0.05-0.2				
$[\Delta\frac{3}{2}^{+}](1985)$	215	.05	.44	.25	.00	.17
$[\Delta\frac{5}{2}^{+}](1910)$	80	.15	.30	.52	.00	.02
$\Delta(1905)F_{35}$****	280-440	0.05-0.15				
$[\Delta\frac{5}{2}^{+}](1990)$	350	.00	.56	.10	.00	.28
$\Delta(2000)F_{35}$**	250-400	0.02-0.07				
$[\Delta\frac{7}{2}^{+}](1940)$	85	.58	.27	.08	.02	.02
$\Delta(1950)F_{37}$****	290-350	0.35-0.4				

Table 4.2. Predicted branching fractions of positive-parity nonstrange Δ baryons below roughly 2000 MeV. $\Delta(1600)\pi$, $\Delta\omega$ branching fractions (< 0.03 for all cases) and strange hadron branches have been excluded for brevity. States from the analyses [3] are listed under the predicted state to which they are assigned.

$N\pi\pi$ channels $\Delta\pi$ and $N\rho$, for example in the $p\pi^{+}\pi^{-}$ TJNAF experiment of Napolitano *et al.* [29]. It is also clear that the isospin-selective channels $N\eta$ and $N\omega$, which will also be studied at TJNAF in this energy range [30], should show good signals for missing baryons, and require a somewhat simpler analysis. The same is true of the more difficult $\Delta\eta$ and $\Delta\omega$ channels [27]. Although not detailed here, recent work [31] within this model on decays into strange branches has shown that, along with information from the channels mentioned above, strangeness production experiments [32] should also discover many new excited negative-parity baryons.

5 Conclusions

Models of the baryon excited states must simultaneously describe their spectrum, electromagnetic couplings, and strong decays. Algebraic approaches, and those based on potential models with one-gluon exchange are the most well-developed, although the consequences of models which ascribe the spin-spin interactions in baryons to one-boson-exchange are currently being worked out. The one-gluon exchange model described here is able to explain why there are many missing positive-parity nonstrange baryons, and gives definite predictions for the channels, mass regions, and partial waves within each channel where these missing states should be found. This calculation could be improved by taking into account the effects of decay-channel mixings and the effects of meson loops in a coupled-channels approach. As baryons with excited glue are likely to first appear in this mass range it will also be necessary to know their masses, quantum numbers, strong decays and mixings with conventional baryon states to fully understand the physics operating here. Work on these topics is currently underway.

Acknowledgement. This work was supported in part by the U.S. Department of Energy through Contract No. DE-FG05-86ER40273, and by the Florida State University Supercomputer Computations Research Institute which is partially funded by the Department of Energy through Contract No. DE-FC05-85ER250000.

References

1. G.R. Goldstein, *Workshop on diquarks*, Eds. M. Anselmino and E. Predazzi (World Scientific, Singapore, 1989), p. 159; M. Anselmino et al.: Rev. Mod. Phys. **65**, 1199 (1993).

2. R. Bijker, F. Iachello and A. Leviatan: Ann. Phys. **236**, 69 (1994)

3. C. Caso et al.: Eur. Phys. J. **C3**, 1 (1998)

4. R. Koniuk and N. Isgur: Phys. Rev. **D21**, 1868 (1980)

5. D.B Leinweber: Phys. Rev. **D47**, 5096 (1993)

6. C.P. Forsyth and R.E. Cutkosky: Nucl. Phys. **B178**, 35 (1981)

7. See P. Stassart and Fl. Stancu: Z. Phys. **A351**, 77 (1995), and references therein

8. L.Ya. Glozman and D.O. Riska: Phys. Rep. **268**, 263 (1996)

9. L.Ya. Glozman, W. Plessas, K. Varga, and R.F. Wagenbrunn: Nucl. Phys. **A631**, 469c (1998)

10. A. Krassnig et al.: talk presented at the *16th European Conference on Few-Body Problems in Physics (FB98)* (June 1998)

11. R.F. Wagenbrunn: these proceedings

12. S. Godfrey and N. Isgur: Phys. Rev. **D32**, 189 (1985)

13. S. Capstick and N. Isgur: Phys. Rev. **D34**, 2809 (1986)

14. N. Isgur and G. Karl: Physics Today **36**, No. 11, 36 (1983)

15. K.-F. Liu et al.: hep-ph/9806491

16. A.J.G. Hey and Litchfield: Nucl. Phys. **B95**, 516 (1975)

17. R. Bijker, F. Iachello and A. Leviatan: Phys. Rev. **D55**, 2862 (1997)

18. L. Micu: Nucl. Phys. **B10**, 521 (1969)

19. See A. Le Yaouanc, L. Oliver, O. Pène and J. C. Raynal: *Hadron Transitions In The Quark Model*, Gordon and Breach, 1988, and references therein

20. P. Geiger and E.S. Swanson: Phys. Rev. **D50**, 6855 (1994)

21. E.S. Ackleh, T. Barnes, and E.S. Swanson: Phys. Rev. **D54**, 6811 (1996)

22. R. Kokoski and N. Isgur: Phys. Rev. **D35**, 907 (1987)

23. C.P. Forsyth and R.E. Cutkosky: Z. Phys. **C18**, 219 (1983)

24. S. Capstick and W. Roberts: Phys. Rev. **D47**, 1994 (1993)

25. S. Capstick and W. Roberts: Phys. Rev. **D49**, 4570 (1994)

26. S. Capstick and W. Roberts: Phys. Rev. **D57**, 4301 (1998)

27. B. Nefkens, private communication; D.M Manley, private communication

28. D. M. Manley and E. M. Saleski: Phys. Rev. **D45**, 4002 (1992)

29. J. Napolitano et al.: TJNAF experiment E-93-033

30. S. Dytman et al.: TJNAF experiment E-89-039; B.G. Ritchie et al.: TJNAF experiment E-91-008; V. Burkert, H. Funsten, D.M. Manley, B. Mecking, et al.: CEBAF experiment E-91-024

31. S. Capstick and W. Roberts: Phys. Rev. **D58**, 074011 (1998)

32. M. Mestayer, K.H. Hicks et al.: TJNAF experiment E-93-030; R. Schumacher et al.: TJNAF experiment E-89-004; G. S. Mutchler et al.: TJNAF experiment E-89-024; L. Dennis, H. Funsten et al.: TJNAF experiment E-89-043

Few-Body Systems Suppl. 11, 94–99 (1999)

Amplitude Analysis

Ron Workman

Department of Physics, Virginia Tech, Blacksburg, VA 24061, USA

Abstract. Even if a 'complete set' of experimental observables were measured for the elastic scattering or photo/electroproduction of pseudoscalar mesons, ambiguities would remain in the extracted partial-wave and isospin decomposed amplitudes. As these problems are not widely understood, the present work outlines the way model-dependence enters into analyses of data from both hadronic and electromagnetic facilities.

1 Introduction

In order to understand the significance of amplitude analysis, one must first recognize that there is generally a gap between the quantities 'predicted' through models and the observables which are actually 'measured' in experiments. The experimental quantities are generally cross sections, often with polarizations fixed or measured for the beam, target and recoiling particles. In contrast, the 'predicted' quantities (for example, the coupling constants, sigma terms, and resonance properties) typically require not only the underlying amplitudes, but also extrapolations of these into unphysical regions and/or separations into background and resonant contributions.

It is often assumed that at least the first step (data→amplitudes) in comparisons between theory and experiment could be carried out in a 'model-independent' way, if sufficient observables were available. This leads to a working definition of amplitude analysis:

"Amplitude analysis is a model-independent determination of the (helicity/transversity) amplitudes using only experimental data and the relations between amplitudes and observables, resulting in a set of amplitudes at discrete energies and angles."

The usual goal in this type of analysis is the smallest set of observables removing all ambiguities in the extracted amplitudes, apart from an overall phase. Unfortunately, one rarely has the required set of observables. As a result, further theoretical input is generally required. In the following we will first review some of the most important (formal) results of these studies. We will then consider how relevant these results are for the N* program. Finally, since most analyses depart from the ideal defined above, we will consider how model-dependence enters into and effects a range of more standard analyses.

2   Some Formal Results

Most of the formal work on amplitude (and partial-wave) analysis has been confined to spinless and spin-0−spin-1/2 cases. A good source of references is the Introduction of Höhler's 'bible' on the subject [1]. Another very readable source is the review of Bowcock and Burkhard [2]. More mathematical discussions are contained, for example, in the papers of Sabba-Stefanescu [3] and the book of Chadan and Sabatier [4].

The basic problem is most easily demonstrated in spinless scattering. Here we have only the cross section

$$\frac{d\sigma}{d\Omega} = |f|^2, \tag{1}$$

which gives the scattering amplitude,

$$f = |f|e^{i\phi(E,\theta)}, \tag{2}$$

up to an overall phase. The difficulty comes in determining $\phi(E,\theta)$ with the addition of isospin, analyticity and unitarity constraints. It appears [5] that this is possible in spinless and spin-0−spin-1/2 scattering. Clearly the task is easier in the forward direction (optical theorem) and for elastic scattering.

When the scattered particles have spin, there are more amplitudes to determine and more observables to choose from. If we once again ignore the overall phase, the object is to pick a set of observables which fixes all the *relative* phases between the amplitudes. The allowable sets [6] have been determined by Dean and Lee [7] (for πN scattering), Chiang and Tabakin [8] (for pseudoscalar meson photoproduction), and Dmitrasinovic, Donnelly and Gross [9] (for scalar and pseudoscalar electroproduction).

3   Are we asking the right question?

Suppose now that we have been given a complete set of observables for a particular reaction, and have constructed the helicity amplitudes up to an overall phase. Is this sufficient to determine all of the contributing N* states? For two important reasons, the answer is negative. What we really want are the partial-wave amplitudes for each isospin state. Once the (model-dependent) separation of resonant and background contributions has been carried out, we can compare with model predictions. Unfortunately, the unknown overall phase is an important element in both the partial-wave and isospin decompositions.

To illustrate the first point, we can return to the simple spinless case, where the full amplitude

$$f = \sum_l (2l+1)f_l(E)P_l(x), \tag{3}$$

may be written as an infinite sum of partial-wave amplitudes. Here x denotes $\cos(\theta)$. Clearly, if f has a phase which is an unknown function of θ, this relation cannot be inverted to give unique values of $f_l(x)$.

The effect of this overall phase on isospin decompositions is also serious, but has been discussed less often. This may be due to the fact that, in πN scattering, the potential ambiguities can be easily removed. Suppose, in the πN case, that we have constructed amplitudes for elastic $\pi^\pm p$ scattering and charge-exchange ($\pi^- p \to \pi^0 n$) scattering. In principle, the amplitudes for each reaction have a *different* overall phase. Therefore we cannot simply combine these quantities, using the usual relations, in order to obtain isospin amplitudes. However, we know one more piece of information. The amplitudes for the charge states satisfy a triangle relation

$$A^- + \sqrt{2}A^0 = A^+, \qquad (4)$$

which, since we know the lengths of all 3 sides, fixes the relative angles.

Unfortunately, this method doesn't work in all cases. As an example, consider pion photoproduction. Here there are 4 different amplitudes for the production of $\pi^0 p$, $\pi^+ n$, $\pi^- p$, and $\pi^0 n$. These can be constructed from 3 isospin amplitudes. In this case, we also have a relation

$$\sqrt{2}A^{\pi^0 n} + A^{\pi^- p} + A^{\pi^+ n} = \sqrt{2}A^{\pi^0 p}, \qquad (5)$$

between the charge states. However, knowing the lengths of the 4 sides is now *not* sufficient to determine the relative angles. This problem remains even if the overall phases have been fixed for 2 of the reactions.

Some readers may at this point be wondering how an 'unmeasureable' phase could be so important. It should first be noted that this phase is not equivalent to the unmeasureable phase of a wavefunction in quantum mechanics. Instead, this is the relative phase between the scattered and unscattered waves in the familiar relation

$$\psi \sim e^{i\mathbf{k}\cdot\mathbf{x}} + f(\theta)\frac{e^{ikr}}{r} \quad \text{as} \quad r \to \infty. \qquad (6)$$

It is actually possible to see effects of such phases via multiple scattering [1, 2] and to (in principle) measure some of them using the Hanbury-Brown–Twiss method [10]. In practice, however, some theoretical input is necessary to fix this phase.

4   What is generally done

Given the problems with 'pure' amplitude analysis, it is not surprising that alternate approaches have been employed in the study of N* physics. These include restricted multipole analyses, the use of dispersion relations, direct model fits to data, and fits based on Breit-Wigner plus background contributions.

Before discussing these, however, we should mention an approach to the amplitude analysis problem which has been applied to elastic πd scattering.

In this work [11] there were insufficient observables for a model-independent amplitudes analysis. As a result, amplitudes were given with 3 different levels of model-dependence. First, it was shown that certain combinations of amplitudes could be extracted directly from the existing data. Models then provided the additional information necessary to complete a full amplitude analysis. Finally, a model was used to fix the overall phase, thus allowing a partial-wave analysis. An analysis of this type in the N* arena would also be interesting, particularly as more polarization data become available.

One way of avoiding the overall phase problem is to fit multipoles directly. In the simplest variant of this method, the partial-wave series is cut off after a few terms. Examples include analyses near threshold and over the first resonance region. Clearly this method won't work if the neglected terms sum up to a sizable contribution [12]. Often the higher waves are assumed to be well approximated by the Born terms. A particularly interesting study of this kind is the Grushin [13] fit. This analysis was carried out over the first resonance without the help of pion nucleon phases (which give the multipole phases via Watson's theorem). Here the overall $\pi^+ n$ phase was determined through interference with the (real) Born terms. A relation between the $\pi^0 p$ and $\pi^+ n$ multipoles was then used to fix the overall $\pi^0 p$ phase. Such an analysis, free from πN input, can form the basis for a relatively unbiased determination of the Δ^+ resonance position [14].

Dispersion relations, applied either to the invariant or partial-wave amplitudes give the least model-dependent way to supplement experimental data. Analyticity, particularly for fixed-t, has been an important element in studies of uniqueness for spinless and spin-0−spin-1/2 scattering. These constraints are particularly useful in elastic scattering as they give the real parts of forward amplitudes, once the imaginary parts have been determined via the optical theorem and total cross sections. To the author's knowledge, no formal studies (analogous to those carried out by Sabba Stefanescu for πN scattering), have determined the minimal theoretical input required to fix all the phases for pseudoscalar meson photo and electroproduction. Here, even the appropriate question is less obvious, as there are many more observables which could be measured [15].

If the analytic properties of the amplitudes are assumed in advance (for example, in Breit-Wigner plus background or model-based fits), we are able to obtain results from minimal sets of data. A complete set of experiments is not required, and the results are obviously model dependent. Here the ambiguity lies in the choice of model, which must necessarily give only an approximation to the true analytic structure. This approach can also yield useful information, particularly when it fails. Failure within a model (hopefully one which builds in the most important constraints from analyticity and unitarity) indicates a missing element.

5 Implications for CLAS data analysis

At present, our knowledge of the N* spectrum comes from analyses of data from decades of measurements at numerous laboratories, carried out using a variety of techniques (each having different inherent systematic errors). The promise of CLAS data has been a set of measurements with high precision and linked systematic errors. This would greatly reduce one of the most difficult problems in amplitude and partial-wave analyses. It is therefore interesting to consider how much information we can obtain from CLAS data alone.

If one is able to control the beam and target polarization, while measuring in both θ and ϕ, it has been shown [16] that one can obtain the type-S (cross section and single-polarization) and BT (beam-target double-polarization) data, including the recoil polarization P. In pion photo- or electroproduction, for example, this leaves only one relative phase undetermined, apart from the overall phase. How one could obtain P without recoil-polarization measurements is also evident in the relation

$$FG - EH = P - T\Sigma, \tag{7}$$

between the type-S and BT observables [16].

The problem, from an 'amplitude analysis' point of view, is that this remaining relative phase requires further double-polarization measurement involving recoil polarization. As CLAS was not designed for such measurements, it might seem that non-CLAS data would be required to complete a 'model-independent' analysis. However, as we have argued above, in order to extract partial-wave amplitudes, the overall phase must be fixed. By fixing this overall phase, relative phases are also restricted. It seems reasonable to 'conjecture' that any physical input sufficient to fix this overall phase will simultaneously remove (at least) the one relative phase undetermined from type-S and BT experiments. The remaining question then has a more theoretical nature. What (minimal) theory input is required to fix the overall phase in photo- electroproduction reactions?

Acknowledgement. The author thanks B.M.K. Nefkens for sparking his interest in these and related questions. This work was supported by U.S. Department of Energy grant No. DE-FG02-97ER41038.

References

1. G. Höhler: *Pion-Nucleon Scattering*. Edited by H. Schopper, Landolt-Börnstein, Vol. **I/9b2** (Springer-Verlag, Berlin, 1983)

2. J.E. Bowcock and H. Burkhardt: Rep. Prog. Phys. **38**, 1099 (1975)

3. I. Sabba Stefanescu: J. Math. Phys. **23**, 1190 (1982)

4. K. Chadan and P.C. Sabatier: *Inverse problems in quantum scattering theory*. New York: Springer-Verlag 1989

5. While the work of Ref. [3] claims a unique solution *in principle*, whether current solutions are truly unique has recently been questioned by I.G. Alekseev *et al.*, Phys. Rev. **C55**, 2049 (1997)

6. Note that forward and backward scattering are special cases for the amplitude reconstruction problem. For example, in πN scattering and photoproduction, the spin observables are either zero or ± 1 at these angles. Therefore we need only a cross section measurement to determine these (essentially degenerate) transversity amplitudes up to an overall phase.

7. N.W. Dean and P. Lee: Phys. Rev. **D5**, 2741 (1972)

8. W.-T. Chiang and F. Tabakin: Phys. Rev. **C55**, 2054 (1997)

9. V. Dmitrasinovic, T.W. Donnelly and F. Gross: *Complete measurements of scalar and pseudoscalar electroproduction*, in *Research Program at CEBAF (III), RPAC III*, edited by F. Gross (CEBAF, Newport News, 1988), p. 547

10. This is briefly discussed in Refs. [1] and [2]. A modern perspective is given by G. Baym: *The physics of Hanbury Brown–Twiss interferometry: from stars to nuclear collisions*, lectures at the XXXVII Zakopane School, June 1997; nucl-th/9804026

11. H. Garcilazo, E.T. Boschitz, W. Gyles, W. List, C.R. Ottermann, R. Tacik, and M. Wessler: Phys. Rev. **C39**, 942 (1989)

12. The neglect of high partial waves can alter the resulting low partial waves, as has been discussed by A. Donnachie: Rep. Prog. Phys. **36**, 695 (1973)

13. V.F. Grushin, A.A. Shikanyan, E.M. Leiken, and A. Ya. Rotvain: Yad. Fiz. **38**, 1448 (1983) [Sov. J. Nucl. Phys. **38**, 881 (1983)]

14. R.L. Workman: Phys. Rev. **C56**, 1645 (1997)

15. In Ref. [3], arguments were based on the assumed existence of cross section and P data, but not R and A data.

16. I.S. Barker, A. Donnachie and J.K. Storrow: Nucl. Phys. **B95**, 347 (1975)

Few-Body Systems Suppl. 11, 100–103 (1999)

Few-
Body
Systems
© by Springer-Verlag 1999

Baryon Resonance Spectrum Error Analysis, Background Effects

S. A. Dytman *

Department of Physics and Astronomy, University of Pittsburgh, Pittsburgh, PA 15260

Abstract. Our analysis of the baryon spectrum and many results have been reported in previous proceedings of this workshop and elsewhere. The main purpose of this note is to give recent progress, in particular some interesting features that may not get as much treatment in a full article. We present a new treatment of errors along with a discussion of resonant vs. nonresonant processes.

1 Introduction

The Carnegie-Mellon-Berkeley (CMB) model [1] was first developed by Cutkosky and collaborators in the 1970's to analyze the baryon resonance spectrum derived from πN elastic and inelastic data. At that time, the available inelastic data was fairly meager; however, the model has many important and valuable features that make it perhaps the best model in existence today for determining the baryon spectrum. It has multichannel unitarity, and can easily accommodate every reaction channel for which data exists. The amplitudes are analytic as a result of a dispersion relation at the core of the calculation. Thus, resonances are found as poles in the T-matrix rather than as model parameters in a Breit-Wigner shape. In addition, the analytic phase space factors allow for the full complexity of inelastic channel openings, e.g. allowing for on-shell and off-shell rescattering. This is particularly important for describing the $S_{11}(1535)$ resonance which is very close to the very strong inelastic opening of the ηN channel. Finally, the model has a careful treatment of the interference of overlapping resonances.

Although no Lagrangian is solved, this model has the same properties that naturally result from use of one because it is fully consistent with an underlying Lagrangian. Most Lagrangian models are only valid below two pion threshold

---

E-mail address: dytman+@pitt.edu

because of the difficulty in maintaining unitarity where Watson's theorem is no longer applicable.

We have reproduced the original features of the CMB model and extended it to handle photoproduction as well [2, 3]. Born terms are included and the photon-nucleon channel is added at the same level as the hadron-nucleon channels. Thus, the photon-nucleon interaction is included to all orders. The cost of maintaining multichannel unitarity is that the amplitudes are no longer Gauge invariant. Full details can be found in the thesis of Tom Vrana [2]. The paper recently submitted [3] includes a pedagogical discussion. In the work completed to date, all πN elastic and $\pi N \to \pi\pi N$ inelastic T-matrices [4, 5], pion photoproduction [6], $\pi N \to \eta N$ [2], and the total cross sections for $\gamma N \to \eta N$ reactions are fit simultaneously, one partial wave at a time. A new baryon spectrum with masses, partial decay widths, and photocoupling amplitudes is available.

2 Error Analysis

The fitting of many parameters makes error determination difficult. Many models estimate error bars, hence often ignoring systematic effects. To do a fuller error analysis, we fit parts of the data set with a simpler model. Specifically, we use a $2x2$ real K-matrix to fit each reaction channel under study along with the remaining channel with the largest branching fraction for determination of each partial decay width. The two largest channels are fit to determine the error on the mass and width of each resonance, often $\pi N \to \pi\Delta$, the mass, width, branching fraction are varied along with a small number of background parameters in 2 separate fits. In the first fit, 3 background parameters describing a flat background are varied. In the second fit, a fourth parameter which describes a linear background shape is added. The fractional error in the parameter under study is a linear combination of the largest relative error and the absolute difference in the results of the two fits. For the photocoupling amplitude, we add the difference in the amplitude derived from separate fits with 2 different Born term ranges.

Thus, the error bars have both random and systematic components. In some cases, the error bar is larger than that previously reported. For example, $P_{11}(1440)$ is reported by Manley and Saleski [7] with a mass and width of 1462 ± 10 MeV and 391 ± 34 MeV, respectively. We find a significant sensitivity to the shape of the background (see following section for more details) and our final results for the same quantities are 1479 ± 80 MeV and 490 ± 120 MeV. Our fractional error bars for the higher mass state ($P_{11}(1710)$) are also larger than those of Ref. [7] and we find a much smaller width.

3 Role of Nonresonant Processes

Although nonresonant background to the resonances can be important, few articles on baryon resonance extraction discuss its role. In the CMB model, back-

102

ground used is empirical, but has the same analytic phase space that is applied to the resonances. (We plan to substitute a theoretically based background in the future.) With the Carnegie-Mellon-Berkeley form of the full T matrix, nonresonant and resonant couplings are intermixed in the dressed propagator. Thus, it is impossible to write the amplitudes as a simple sum of resonance

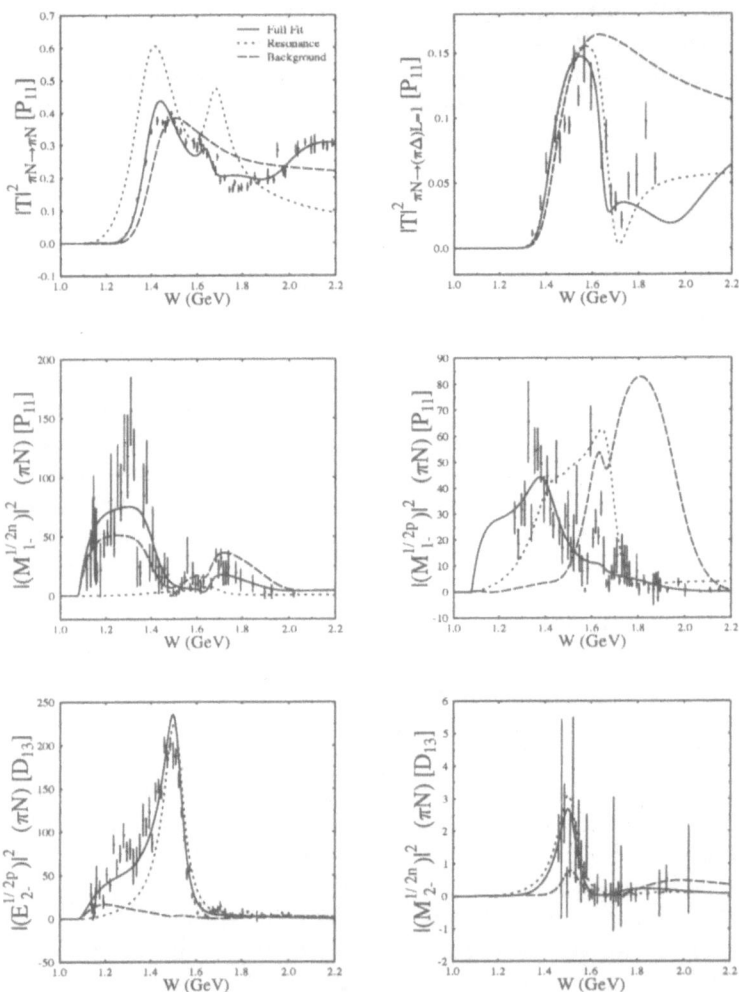

Figure 1. Upper four figures show P_{11} partial wave T matrices and mulitpoles for various reactions. Lower two figures show two of the multipoles in the D_{13} partial wave. See text for details.

and nonresonant pieces in either a T-matrix or a K-matrix. The best way to show the roles of each is then unclear. In the figures, we calculate the 'resonance only' amplitude with all background couplings turned off. For the πN reactions, the 'nonresonant' amplitude is determined by turning off all resonant couplings. For the photon multipoles 'nonresonant' amplitudes, the πN resonant couplings are left on so that the 'nonresonant' amplitude is unitary.

In the figure we show the most important distributions of $|T(W)|^2$ for πN interactions and the inner product of the multipole for $\gamma N \to \pi N$ reactions. We show the distributions for the P_{11} and D_{13} partial waves. These are near the extremes; the former has very strong nonresonant terms while the latter has inconsequential nonresonant terms in the final fits. As a result, the resonance parameters for $D_{13}(1520)$ are much better determined than $P_{11}(1440)$. The higher mass resonances ($P_{11}(1710)$ and $D_{13}(1700)$) are more poorly determined for a number of reasons- interference with a strong state at lower energy and poor data quality.

The P_{13} partial wave has the strongest nonresonant contribution. Since the Born diagrams are sufficient to fit the $\gamma N \to \pi N$ data in this partial wave, the photocoupling amplitudes must be poorly determined. Another interesting example is the S_{11} partial wave at $W \sim 1.5$ GeV. It has moderate nonresonant contribution in the dominant elastic and $\pi N \to \eta N$ reactions. Nevertheless, parameters for the $S_{11}(1535)$ are difficult to determine because these reactions strongly couple to each other. These are difficulties with model dependence in the resonant shape rather than the nonresonant shape.

References

1. R.E. Cutkosky, C.P. Forsyth, R.E. Hendrick, and R.L. Kelly: Phys. Rev. **D20**, 2839 (1979)

2. T.P. Vrana: University of Pittsburgh Ph. D. thesis, unpublished

3. T.P. Vrana, S.A. Dytman, and T.S.-H. Lee: submitted to Phys. Rev. C; to be published

4. R.A. Arndt, I.I. Strakovsky, R.L. Workman, and M.M. Pavan: Phys. Rev. **C52**, 2120 (1995)

5. D.M. Manley, R.A. Arndt, Y. Goradia, and V.L. Teplitz: Phys. Rev. **D30**, 904 (1984)

6. R.A. Arndt, I.I. Strakovsky, R.L. Workman: Phys. Rev. **C53**, 430 (1996)

7. D.M. Manley and E.M. Saleski, Phys. Rev.**D45**, 4002 (1992)

Few-Body Systems Suppl. 11, 104–111 (1999)

Coupled-Channel Analysis of N^* Photoproduction

D. Mark Manley*

Department of Physics and Center for Nuclear Research, Kent State University, Kent, OH 44242 USA

Abstract. Experiments involving πN and γN scattering provide data that may be represented in terms of energy-dependent partial-wave amplitudes. A determination of these amplitudes requires a sufficiently varied and high-quality data base, supplemented by various theoretical constraints. It is further necessary to parametrize the energy dependence of these amplitudes in order to extract pole positions and decay amplitudes for nucleon resonances. For consistency, such parametrizations require a coupled-channels framework. This paper describes a novel multichannel, multiresonance method for parametrizing πN and γN amplitudes, and presents some preliminary results of fits. In addition, quark-model predictions are presented as a guide to suggest promising future experimental studies of nucleon resonances.

1 Introduction

Many methods exist in the literature for parametrizing a unitary, multichannel S-matrix, although all have drawbacks as well as advantages. Early attempts to parametrize the global energy dependence of πN and $\pi\pi N$ amplitudes were based on a simple K-matrix formalism [1, 2]. The determination of T-matrix parameters required additional fits that were restricted in energy. More recently, the sophisticated model developed by Cutkosky [3] has been revived [4, 5]; this model, while stressing the underlying analytic structure of the amplitudes, is hampered by the nontrivial effort required to estimate uncertainties in resonance parameters. For these reasons, a new multichannel, multiresonance parametrization was developed (see the next section). The new parametrization is a simple multiresonance generalization of a multichannel Breit-Wigner representation. It has the advantage that its fitting parameters often correspond exactly to the parameters predicted by quark models [6].

The work described here is an extension of an earlier project to extract resonance parameters from a consistent description of purely hadronic $\pi N \rightarrow$

---

*E-mail address: manley@zeus.kent.edu

πN and $\pi N \rightarrow \pi\pi N$ amplitudes [7]. The present work currently uses πN elastic amplitudes from the WI94 solution of Arndt et al. [8], $\pi N \rightarrow \pi\pi N$ amplitudes from the 1984 isobar-model analysis of Manley et al. [9], and $\gamma N \rightarrow \pi N$ amplitudes from the SP93 solution of Arndt et al. [10]. A discussion of how the $\gamma N \rightarrow \pi N$ multipole amplitudes from Ref. [10] were converted into dimensionless Argand amplitudes can be found in Ref. [11].

The following section presents the essential ingredients of the new parametrization. A fitted photoproduction amplitude is presented as an example. Next we briefly discuss the quality of existing data for hadronic reactions and the impact this will have on future fits involving electromagnetic channels. The possibility of improving and extending the hadronic data base by performing new measurements using the Crystal Ball spectrometer at Brookhaven National Laboratory's AGS is briefly discussed. A figure based on quark-model predictions is shown to illustrate the important resonances that may contribute for two selected reactions involving all-neutral final states with pure isospin.

2 Multichannel Parametrization

The amplitudes determined from partial-wave analyses of $\pi N \rightarrow \pi N$, $\pi N \rightarrow \pi\pi N$, etc. form elements of a dimensionless partial-wave transition matrix (or T-matrix). This T-matrix corresponds to a unitary scattering matrix (or S-matrix) according to the relation $\mathbf{T} = (\mathbf{S} - \mathbf{I})/2\mathrm{i}$, where $\mathbf{I}$ is an identity matrix. In the present work, the S-matrix was parametrized as

$$\mathbf{S} = \mathbf{S}_B^T \, \mathbf{S}_R \, \mathbf{S}_B, \tag{1}$$

where $\mathbf{S}_R$ is symmetric and unitary, and describes contributions from all s-channel resonances, and $\mathbf{S}_B$ is unitary (but not generally symmetric) and describes background contributions (arising, for example, from t- and u-channel exchanges). In Eq. (1), $\mathbf{S}_B^T$ denotes the transpose of $\mathbf{S}_B$. Hence, by construction, $\mathbf{S}$ is both unitary and symmetric (as required by time-reversal invariance).

Let us first consider the simple case of a single resonance. We may then write

$$\mathbf{S}_R = e^{(2\mathrm{i}\delta)\,\mathbf{XX}}, \tag{2}$$

where δ is an energy-dependent resonant phase and $\mathbf{XX}$ is a matrix having unit trace and the property, $(\mathbf{XX})^2 = \mathbf{XX}$. It follows that the exponential may be expanded to rewrite $\mathbf{S}_R$ as

$$\mathbf{S}_R = \mathbf{I} + (e^{2\mathrm{i}\delta} - 1)\,\mathbf{XX}. \tag{3}$$

If we define the resonant T-matrix as $\mathbf{T}_R = (\mathbf{S}_R - \mathbf{I})/2\mathrm{i}$, then for a single resonance, we obtain

$$\mathbf{T}_R = (e^{\mathrm{i}\delta} \sin \delta)\,\mathbf{XX}. \tag{4}$$

We may also define a real, symmetric resonant K-matrix such that

$$\mathbf{S}_R = (\mathbf{I} + \mathrm{i}\mathbf{K}_R)(\mathbf{I} - \mathrm{i}\mathbf{K}_R)^{-1}, \tag{5}$$

$$\mathbf{T}_R = \mathbf{K}_R(\mathbf{I} - \mathrm{i}\mathbf{K}_R)^{-1} \; . \tag{6}$$

For a single resonance, we obtain

$$\mathbf{K}_R = (\tan \delta) \; \mathbf{XX} \; . \tag{7}$$

These results accommodate various forms for the energy dependence of the resonant T-matrix amplitudes. In the present work, we chose the matrix elements to be given by

$$T_{ij} = \frac{(1/2) \; g_i \; g_j}{M - W - \mathrm{i}\Gamma/2} \; . \tag{8}$$

In this case, we have

$$\tan \delta = \frac{\Gamma/2}{M - W} \; , \qquad \mathbf{XX}_{ij} = \frac{g_i g_j}{\Gamma} \; , \tag{9}$$

where M is the conventional Breit-Wigner mass, W is the total c.m. energy, and $\Gamma = \sum_k g_k^2$ is the energy-dependent total width. The quantity g_k evaluated at $W = M$ is the "decay amplitude". Its magnitude equals the square root of the partial width for decay of the resonance into the kth channel, and its phase is determined relative to that for the πN channel. An alternate choice would be to take

$$T_{ij} = \frac{W \; g_i \; g_j}{M^2 - W^2 - \mathrm{i}\Gamma W} \; . \tag{10}$$

Then

$$\tan \delta = \frac{\Gamma W}{M^2 - W^2} \; , \qquad \mathbf{XX}_{ij} = \frac{g_i g_j}{\Gamma} \; . \tag{11}$$

The energy dependence of the decay amplitudes was taken into account by writing $g_k(W) = \gamma_k \Phi(W)$, where γ_k is a fitting parameter and $\Phi(W)$ is an energy- and angular-momentum-dependent phase-space factor. If $\Phi(M) > 0$, then we had $g_k(W) = g_k(M)(\Phi(W)/\Phi(M))$. A strong advantage of the present procedure is that the fitting parameters were M and $g_k(M)$, which are exactly the parameters predicted by some quark-model calculations [6].

The goal of developing a new parametrization for $\mathbf{T}_R$ was to extend the single-resonance parametrization discussed above to multiple resonances while maintaining the simple physical significance of the fitting parameters. Our procedure may be illustrated by considering the extension to two resonances. In that case, the resonant K-matrix has the form

$$\mathbf{K}_R = (\tan \alpha) \; \mathbf{XX} + (\tan \beta) \; \mathbf{YY} \; . \tag{12}$$

The corresponding T-matrix may be shown to be

$$\mathbf{T}_R = \frac{T_\alpha \; (\mathbf{XX}) + T_\beta \; (\mathbf{YY}) + \mathrm{i}(\hat{x} \cdot \hat{y}) \; T_\alpha T_\beta (\mathbf{XY} + \mathbf{YX})}{1 + (\hat{x} \cdot \hat{y})^2 \; T_\alpha T_\beta} \; , \tag{13}$$

where $T_\alpha = e^{\mathrm{i}\alpha} \sin \alpha$ and $T_\beta = e^{\mathrm{i}\beta} \sin \beta$. We may write $\mathbf{XX}_{ij} = x_i x_j$, $\mathbf{YY}_{ij} = y_i y_j$, $\mathbf{XY}_{ij} = x_i y_j$, and $\mathbf{YX}_{ij} = y_i x_j$. Here $\hat{x} \cdot \hat{y} = \sum_k x_k y_k$ is the trace of $\mathbf{XY}$ and $\mathbf{YX}$. Poles of the resonant T-matrix occur when

$$1 + (\hat{x} \cdot \hat{y})^2 \; T_\alpha T_\beta = 0 \; . \tag{14}$$

If we assume that

$$\tan \alpha = \frac{G_\alpha}{M_\alpha - W} \ , \qquad \tan \beta = \frac{G_\beta}{M_\beta - W} \ , \tag{15}$$

where $M_\alpha = M_\alpha(W)$ and $M_\beta = M_\beta(W)$, then the pole condition Eq. (14) may be factorized as $(W - W_x)(W - W_y) = 0$, with $W_x = M_x - iG_x$ and $W_y = M_y - iG_y$. Given the Breit-Wigner T-matrix parameters as input, it is straightforward to find the energy-dependent K-matrix parameters from the relations

$$M_\alpha + M_\beta = M_x + M_y \ , \tag{16}$$

$$G_\alpha + G_\beta = G_x + G_y \ , \tag{17}$$

$$M_\alpha G_\beta + M_\beta G_\alpha = M_x G_y + M_y G_x \ , \tag{18}$$

$$M_\alpha M_\beta - [1 - (\hat{x} \cdot \hat{y})^2] \, G_\alpha G_\beta = M_x M_y - G_x G_y \ . \tag{19}$$

Note that the poles m_1 and m_2 in $\mathbf{K}_R$ are given by $m_1 = M_\alpha(m_1)$ and $m_2 = M_\beta(m_2)$. The generalization to more than two resonances is straightforward.

The space available here greatly limits what results can be discussed using this new multichannel parametrization. As one example, we show in Fig. 1 the magnetic P_{11} amplitude for $\gamma p \to \pi N$. The curve shows the results of a preliminary three-resonance fit. As a second example, we successfully fitted the F_{15} amplitudes, including all γp and γn amplitudes, using two resonances and a consistent set of hadronic couplings to describe both the γp and γn amplitudes. The lower-mass resonance, $F_{15}(1680)$, is well established as the first Regge recurrence of the nucleon. Our fitted values for its mass, width, and elasticity are 1688(2) MeV, 137(5) MeV, and 0.67(1), respectively. In good agreement with prior analyses, our helicity amplitudes for the $F_{15}(1680)$ are $A^p_{1/2} = -0.010(2)$, $A^p_{3/2} = 0.144(3)$, $A^n_{1/2} = 0.025(3)$, and $A^n_{3/2} = -0.033(3)$ in units of GeV$^{-1/2}$. Our hadronic decay amplitudes also agree well with prior analyses. Based on unitarity, we can deduce that the $F_{15}(1680)$ must have very small branching fractions to ηN, $K\Lambda$, and ωN, and, indeed, to all hadronic channels other than πN and $\pi\pi N$. The higher-mass F_{15} resonance is not established, and our fitted values for its mass, width, and elasticity are 1809(6) MeV, 179(33) MeV, and 0.10(2), respectively.

3 Hadronic Reactions: An Opportunity to Improve the Data Base

Studies of baryon spectroscopy are concerned with trying to uncover possible symmetries in the observed pattern of excited states, and with trying to understand the basic mechanism of quark confinement. High-quality hadronic data will be necessary to capitalize on the great wealth of meson photoproduction and electroproduction data that will result from the N^* program at Jefferson Lab (JLab). For example, one needs to know hadronic decay amplitudes in order to determine γN helicity couplings from photoproduction data. For self-consistency, multichannel analyses are required; at present, only

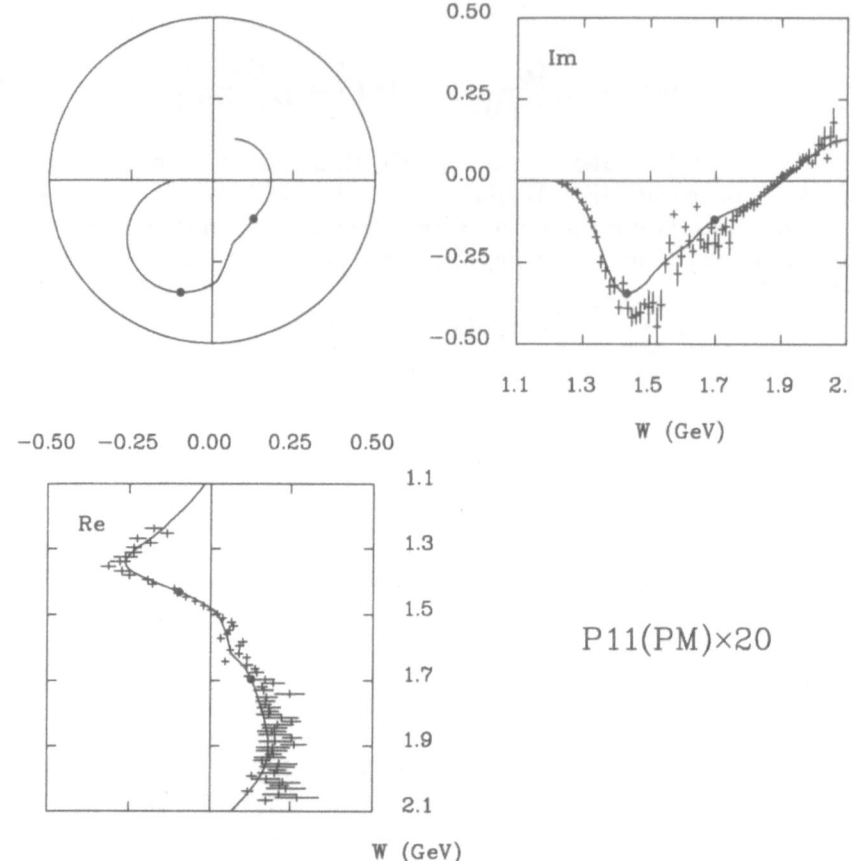

Figure 1. P_{11} Argand magnetic multipole amplitude for $\gamma p \to \pi N$. The data and curve have been scaled by a factor of 20 to facilitate viewing details. The curve shows the result of a multichannel fit with three resonances.

data for πN elastic scattering have the quality and variety needed from the hadronic sector. High-statistics data are desperately needed for several important hadronic reactions, including $\pi N \to \eta N$ above the threshold region, as well as for $\pi N \to \pi\Delta$, $\pi N \to \rho N$, $\pi N \to \omega N$, $\pi N \to K\Lambda$, and $\pi N \to \eta' N$. For several of these reactions, quark-model predictions are our only guide as to what might be discovered in a partial-wave analysis. As an example, Fig. 2 shows predictions based on the work of Capstick and Roberts [6] for the $I = \frac{1}{2}$ reactions $\pi^- p \to \eta' n$ and $\pi^- p \to \omega n$. Nothing is known experimentally about the nucleon resonances that couple to either final state.

The quantity shown in Fig. 2 is the effective cross section, defined by

$$\sigma_{\text{eff}} = \frac{4\pi}{q^2} \left(J + \frac{1}{2}\right) C_I^2 \, |T_{\text{eff}}|^2 \, . \tag{20}$$

Here q is the relative c.m. momentum of the incident pion and proton evaluated at $W = M$, where M is the predicted resonance mass. Also, J is the spin of the resonance, C_I is an isospin Clebsch-Gordon coefficient, and $T_{\text{eff}} = \sqrt{\Gamma_{\pi N}\Gamma_f}/\Gamma$ is the effective resonant T-matrix element, which is determined entirely from quark-model predictions. The total width Γ is obtained by summing predicted partial widths for all available hadronic channels (that is, for πN, $\pi\Delta$, ρN, ηN, $\eta' N$, ωN, $\pi N(1440)$, and $\pi\Delta(1600)$). Both $\pi^- p \to \omega n$ and $\pi^- p \to \eta' n$ should be valuable in searching for "missing resonances," those states predicted to exist by quark-model calculations but which have not been observed experimentally. In particular, the $\pi^- p \to \eta' n$ reaction may excite a few $N = 3$ negative-parity resonances rather strongly (none established), while the $\pi^- p \to \omega n$ reaction at fairly low energy may excite two missing P_{13} resonances strongly.

The opportunity exists to measure differential cross sections for several of the important hadronic reactions using the Crystal Ball spectrometer in the D-line at Brookhaven National Laboratory's AGS. The Crystal Ball spectrometer is a large solid-angle gamma-ray detector, which is nearly ideal for investigating reactions of the type $\pi^- p \to \gamma n$. For example, $\pi^- p \to \omega n$ events would be identified by detecting the ω meson through its decay $\omega \to \pi^\circ\gamma \to 3\gamma$, and by using missing mass to detect the final-state neutron. Similarly, $\pi^- p \to \eta' n$ events would be identified by detecting the η' meson through its decay $\eta' \to \pi^\circ\pi^\circ\eta$, and again using missing mass to detect the final-state neutron. The neutral pion and η meson would be identified through their decays $\pi^\circ \to 2\gamma$ and $\eta \to 2\gamma$. Events for many other resonance reactions, including $\pi^- p \to \pi^\circ n$, $\pi^- p \to \pi^\circ\pi^\circ n$, $\pi^- p \to \eta n$, and $\pi^- p \to \gamma n$, could be measured simultaneously.

4 Summary

In order to describe consistently the many different reactions that may occur through formation of intermediate N^* resonances, it is necessary to implement some type of coupled-channel description. As data (amplitudes) for more channels are included in a multichannel fit, one gains confidence that the resulting fit provides a unique set of resonance parameters; however, as one includes additional channels, it becomes increasingly difficult to obtain a satisfactory fit of all available data. This problem stems at least in part from inherent inconsistencies in the data. Another contributing factor is that adding decay channels increases the number of fitting parameters, so one has a larger parameter space to search. Nevertheless, the benefits of the coupled-channel method warrant the difficulties that inevitably arise when attempting to describe a large body of data in a self-consistent manner.

Studies of baryon spectroscopy may help answer some very important and basic questions. A better knowledge about the existence and decay properties of high-spin baryon resonances could help us understand quark confinement at relatively large quark separations. High-quality scattering data may help us distinguish between purely dynamical resonances and those that are true quark excitations. Electromagnetic reactions that do not involve the πN channel will

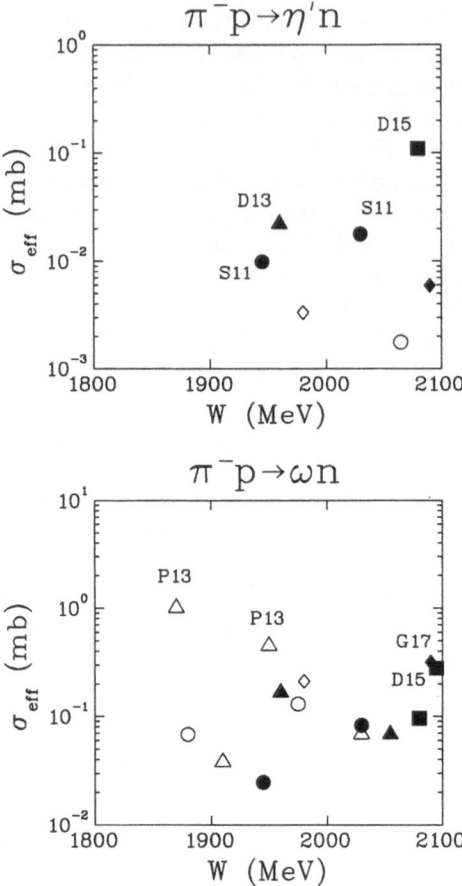

Figure 2. Quark-model predictions for the relative strength of resonances in the pure $I = \frac{1}{2}$ reactions $\pi^- p \to \eta' n$ (top panel) and $\pi^- p \to \omega n$ (bottom panel). Resonances with $J^P = \frac{1}{2}^-$, $\frac{3}{2}^-$, $\frac{5}{2}^-$, and $\frac{7}{2}^-$ are indicated by filled circles, triangles, squares, and diamonds, respectively, while resonances with $J^P = \frac{1}{2}^+$, $\frac{3}{2}^+$, $\frac{5}{2}^+$, and $\frac{7}{2}^+$ are indicated by open circles, triangles, squares, and diamonds, respectively.

be studied at JLab to search for missing resonances. New hadronic reactions may also provide evidence for missing resonances, and the new data may also be used to test models of baryon decay. The unambiguous observation of extra resonances (those not explained as bound states of three constituent quarks) would provide evidence for new forms of hadronic matter, such as hybrids. A more complete description of the baryon spectrum may reveal or confirm hints of new symmetries (such as parity doublets).

The N^* program at JLab will provide a wealth of new electromagnetic data on meson photoproduction and electroproduction. Complementary high-quality hadronic data will be crucial to maximize the physics potential of the electromagnetic data. For example, data for $\gamma N \to \eta N$, $\gamma N \to \pi N$, $\pi N \to \pi N$, and $\pi N \to \eta N$ could be analyzed together to determine γN helicity couplings and the πN and ηN decay amplitudes more precisely. Currently, the data base for $\pi N \to \eta N$ is poor except in the threshold region where $S_{11}(1535)$ dominates. Data for $\pi N \to \pi \Delta$ are poor above $W = 1700$ MeV. Data at all energies for $\pi N \to \rho N$ are worse than those for $\pi N \to \pi \Delta$. For other important reactions, such as $\pi N \to \omega N$, almost nothing is known from experiment about the contribution from N^* resonances. The experimental situation for the hadronic reactions is unlikely to improve any time soon unless the Crystal Ball program at Brookhaven National Laboratory is able to continue into the high-momentum AGS D-line, which can furnish high-intensity pions covering the resonance region up to $W = 2100$ MeV.

Acknowledgement. I am grateful to my friend and former doctoral student, Martin Niboh, for his contributions to this work. I am indebted to Richard Arndt and Ron Workman for providing me with their amplitudes for $\pi N \to \pi N$ and $\gamma N \to \pi N$. I would also like to thank the organizers of the Joint ECT*/JLab Workshop on N^* Physics and Nonperturbative QCD for their support and hospitality during my visit to Trento, Italy. This work was supported in part by a grant from the U.S. National Science Foundation.

References

1. R.S. Longacre and J. Dolbeau: Nucl. Phys. **B122**, 493 (1977)

2. R.S. Longacre et al.: Phys. Rev. **D17**, 1795 (1978)

3. R. E. Cutkosky et al.: Phys. Rev. **D20**, 2839 (1979)

4. M. Batinić et al.: Phys. Rev. **C51**, 2310 (1995)

5. S. A. Dytman, T.P. Vrana, and T.-S.H. Lee: πN Newsletter **13**, 90 (1997)

6. S. Capstick and W. Roberts: Phys. Rev. **D49**, 4570 (1994)

7. D.M. Manley and E. M. Saleski: Phys. Rev. **D45**, 4002 (1992)

8. R.A. Arndt and R.L. Workman: Private Communication

9. D.M. Manley et al.: Phys. Rev. **D30**, 904 (1984)

10. Z. Li, R.A. Arndt, L.D. Roper, and R.L. Workman: Phys. Rev. **C47**, 2759 (1993)

11. D.M. Manley: πN Newsletter (to appear)

Few-Body Systems Suppl. 11, 112–115 (1999)

Few-
Body
Systems
© by Springer-Verlag 1999

Parity-Violating $\Delta(1232)$ Electroweak Production: Axial Structure and New Physics

S.J. Pollock[1], N. Mukhopadhyay[2], M. Ramsey-Musolf[3], J. Líu[4], H. Hammer[4]

[1] University of Colorado, Boulder CO 80303 USA
[2] Rensselaer Polytechnic Institute, Troy, NY 12180-3590 USA
[3] University of Connecticut, Storrs, CT 06260 USA
[4] TRIUMF, 4004 Wesbrook Mall, Vancouver, B.C., Canada V6T 2A3

Abstract.
We examine the theoretical interpretability and feasibility of a measurement of parity-violating electroweak excitation of the $\Delta(1232)$ via polarized electron scattering [1]. We examine the implications for hadron structure theory, and the extent to which we could constrain various extensions of the Standard Model.

1 Introduction

If one measures a parity-*violating* (PV) asymmetry in electroproduction of the Δ using longitudinally polarized electrons on unpolarized nucleons, the resonant multipoles cancel, yielding a clean and simple structure. The $N \to \Delta$ transition is appealing for a PV experiment, since the Δ resonance is isolated from other nucleon resonances, the cross section is relatively large, and the transition is isovector. This isovector nature eliminates uncertainties associated with the unknown strangeness content of the nucleon, and yields a unique sensitivity to possible contributions from additional heavy particles not appearing in the Standard Model (SM). $N \to \Delta$ offers the additional advantage that it only couples strongly to one channel, $N\pi$, allowing one to treat unitarity issues quite rigorously. In a recent paper [1], we examined this observable in some detail. The asymmetry has the following structure [1, 2]:

$$A_{LR} = \frac{N_+ - N_-}{N_+ + N_-} = \frac{-G_\mu}{\sqrt{2}} \frac{|q^2|}{4\pi\alpha} \left[\Delta^\pi_{(1)} + \Delta^\pi_{(2)} + \Delta^\pi_{(3)} \right], \qquad (1)$$

where N_+ (N_-) are the number of scattered electrons for a beam of positive (negative) helicity electrons, q^2 is the square of the four-momentum transfer to the target; α and G_μ are the fine structure constant and the Fermi constant.

The quantities $\Delta^\pi_{(i)}$ denote the three primary contributions to A_{LR}. $\Delta^\pi_{(1)} = 2(1 - 2\sin^2\theta_W)$ includes the entire resonant hadronic vector-current contribution to the asymmetry, at tree level in the SM. This has no dependence on hadronic form factors, due to a cancellation between terms in the helicity-dependent and -independent cross sections. $\Delta^\pi_{(2)}$ contains residual contributions from non-resonant vector-current isoscalar backgrounds. The third term involves the axial-vector $N \to \Delta$ coupling: $\Delta^\pi_{(3)} \approx 2(1 - 4\sin^2\theta_W)F(q^2, s)$ (plus axial vector background contributions). $F(q^2, s)$ involves a ratio of electroweak response functions, s is the square of the total energy in the center of mass.

$\Delta^\pi_{(1)}$ provides a window on physics beyond the SM. We have found that a 1% knowledge of $\Delta^\pi_{(1)}$ provides constraints comparable to those presently obtained from atomic PV. A demanding experimental setup might be able to achieve this level, if non-resonant backgrounds and axial contributions can both be understood at 25-30% levels or better. The third term, $\Delta^\pi_{(3)}$, is interesting from the standpoint of hadron structure. To a good approximation, $F(q^2, s)$ is proportional to the ratio of two transition form factors: C_5^A/C_3^V, where V (A) correspond to hadronic vector (axial vector) currents. This ratio is the off-diagonal analog of G_A/G_V. $\Delta^\pi_{(3)}$ could thus allow tests of low-energy consequences of chiral symmetry, such as the off-diagonal Goldberger-Treiman relation and its (small) chiral corrections. Also, a determination of C_5^A could provide tests of lattice and quark model calculations and of recipes proposed for correcting vector $N \to \Delta$ transition form factor discrepancies.

The most serious uncertainties appear in two guises: (i) background contributions in $\Delta^\pi_{(2)}$, and (ii) radiative corrections including two-boson exchange "dispersion corrections" and corrections induced by PV quark-quark interactions in the hadronic vertex. (The former enter the analysis of all of the $\Delta^\pi_{(i)}$, while the latter contribute to $\Delta^\pi_{(3)}$ only) An estimate of such hadronic PV corrections goes beyond the scope of this work, but we emphasize the importance of performing such an estimate when extracting C_5^A/C_3^V from $\Delta^\pi_{(3)}$.

2 Results

For numerical estimates we assume [3] 1000 hours of 100% polarized beam with a luminosity of 2×10^{38} cm^{-2} s^{-1}, solid angle 20 msr, energy spread for outgoing electrons 0.2 GeV. The figure of merit can easily be scaled for other assumptions. (E.g. this solid angle is grossly conservative for backward angles.) Fig. 1 shows the Q^2-dependence of the axial term for several electron energies. We included a variety of models [1]; the theory spread is commensurate with the precision with which one might expect to determine the axial term. A measurement of A_{LR} may thus only be marginally useful as a discriminator among models. Still, it will afford a determination of C_5^A/C_3^V at the level of discrepancies arising in the vector sector. More detailed numbers can be found in Table 2.1 and Fig. 2.

We have used a model [4] to estimate the background contributions to A_{LR}. E.g., at $\theta = 10°$ and $\epsilon = 1$ GeV, we find $\Delta^\pi_{(2)}/\Delta^\pi_{(3)} \approx 6\%$, so even a large

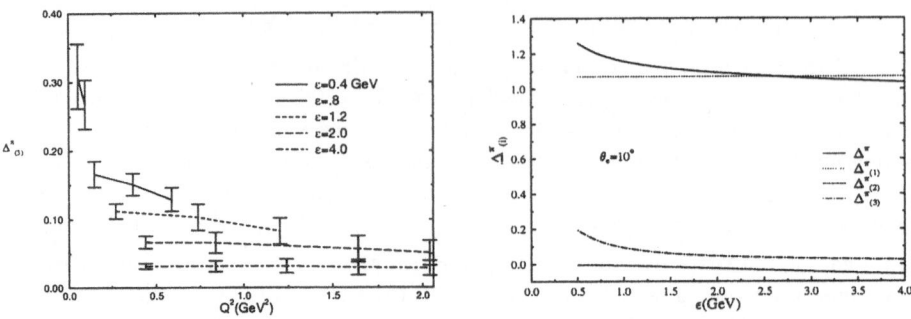

Figure 1. a) $\Delta^{\pi}_{(3)}(Q^2)$ for different incident electron energies, ϵ. Error bars show a rough spread of theoretical predictions. b) Contributions to the asymmetry from the $\Delta^{\pi}_{(i)}$ as a function of incident energy for $\theta = 10°$. $\Delta^{\pi} = \Delta^{\pi}_{(1)} + \Delta^{\pi}_{(2)} + \Delta^{\pi}_{(3)}$.

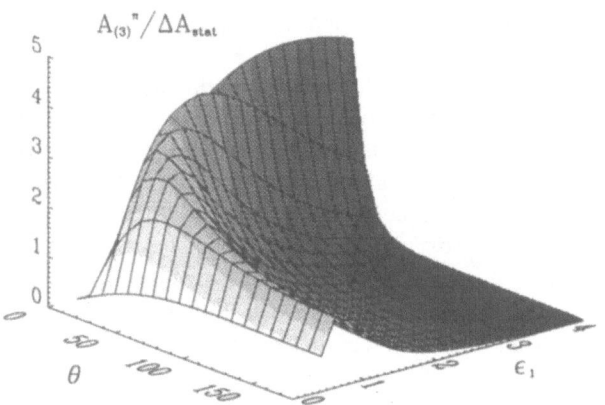

Figure 2. A 3-D plot of $\frac{A_{(3)}/A_{\text{tot}}}{\delta A_{\text{stat}}/A_{\text{tot}}}$, versus both incident energy and electron scattering angle. The shading is determined by the value of $\Delta^{\pi}_{(3)}$, smaller values are shaded darker.

uncertainty in $\Delta^{\pi}_{(2)}$ has negligible effect on an extraction of the axial term. At kinematics suited for a determination of $\sin^2 \theta_W$, the vector-current background contributes 4-6% of the total asymmetry. A probe for new physics at the 1% level would require a theoretical uncertainty in $\Delta^{\pi}_{(2)}$ to be below 15-25%, which may be difficult to achieve. It appears that a model-independent approach with an *experimental* isospin decomposition of EM pion production offers the best hope for eliminating vector-current background uncertainties.

In summary, we have analyzed the PV $N \to \Delta$ asymmetry. We considered sensitivities to scenarios for physics beyond the SM – leptoquarks, extra neutral gauge bosons, and fermion compositeness – as well as to transition form

Table 2.1. Our estimates of A_{LR}, experimental statistical uncertainty for A_{LR} (given our experimental assumptions), vector-current backgrounds, contribution of axial multipoles, and ratio of axial contribution to statistical uncertainty, respectively. $\Delta^{\pi} = \Delta^{\pi}_{(1)} + \Delta^{\pi}_{(2)} + \Delta^{\pi}_{(3)}$, $A_{(3)}$ is the contribution to the A_{LR} arising from $\Delta^{\pi}_{(3)}$.

E (GeV)	θ_{lab} (Deg.)	Q^2 (GeV^2)	$10^5 A_{tot}$	$\frac{\delta A_{stat}}{A_{tot}}$ (%)	$\frac{\Delta^{\pi}_{(2)}}{\Delta^{\pi}}$ (%)	$\frac{\Delta^{\pi}_{(3)}}{\Delta^{\pi}}$ (%)	$\frac{A_{(3)}}{\delta A_{stat}}$
.5	10.	.002	-.03	45.9	-.24	20.4	.4
1.0	10.	.020	-.22	5.9	-.7	11.0	1.9
4.0	10.	.418	-4.17	.7	-5.5	2.6	4.0
.5	180.	.157	-1.81	11.6	-3.70	15.5	1.3
1.0	180.	.846	-8.85	7.5	-7.0	7.1	.9
4.0	180.	6.150	-60.08	43.8	-10.4	.6	.0

factors of interest to hadron structure theory. We estimated the precision with which A_{LR} might be determined in a realistic experiment, and the scale of background effects. The use of $A_{LR}(N \rightarrow \Delta)$ as a probe of hadron structure appears to be a feasible prospect. A $\sim 25\%$ determination of the hadronic axial vector response could be carried out with realistic running times. At reasonable kinematics for such a measurement, backgrounds appear to be sufficiently under control. While a 25% determination of C_5^A/C_3^V would not allow a detailed discrimination among model predictions, it would significantly improve upon knowledge from CC neutrino reactions, and test model predictions at the level of discrepancies arising in the vector current sector. A complete theoretical analysis of $\Delta^{\pi}_{(3)}$, including effects of potentially large and uncertain radiative corrections associated with hadronic PV, awaits future study.

References

1. N.C. Mukhopadhyay, M.J. Ramsey-Musolf, S. Pollock, J. Liu, and H.-W. Hammer: Nucl. Phys. **A633**, 481 (1998)

2. M. J. Musolf et al.: Phys. Rep. **239**, 1 (1994)

3. S.P. Wells, N. Simicevic, and K. Johnston: JLab proposal E97-104

4. H.-W. Hammer and D. Drechsel, Z. Phys. **A353**, 321 (1995)

Few-Body Systems Suppl. 11, 116–123 (1999)

Few-
Body
Systems
© by Springer-Verlag 1999

Nucleon Axial Matrix Elements

Barry R. Holstein

Department of Physics and Astronomy, University of Massachusetts
Amherst, MA 01002 USA
and
Institut für Kernphysik, Forschungszentrum Jülich, D-52425 Jülich, Germany

Abstract. Current issues associated with nucleon axial matrix elements are studied, including the Goldberger-Treiman discrepancy, the induced pseudoscalar, and SU(3) chiral perturbation theory.

1 Introduction

I have been given the task of speaking about the nucleon axial matrix elements. In comparison with the many exciting things being discussed at this meeting this may seem rather prosaic. However, I will try to convince you otherwise by discussing issues associated with the Goldberger-Treiman discrepancy, recent and future measurements of the induced pseudoscalar, and the renormalizations of the axial couplings within SU(3) chiral perturbation theory.

2 The Goldberger-Treimen Discrepancy OR Time Dependence of Fundamental Constants

Many years ago Dirac noticed that the ratio of electrical to gravitational forces between a pair of electrons was equal to $\alpha/(Gm_e^2) \sim 10^{40}$. In asking himself how such an enormous dimensionless number could arise he noticed that 10^{40} is also the age of the universe measured in fundamental units of time—*i.e.* the time it takes light to traverse an elementary particle—10^{17} sec./ 10^{-23} sec.! He then asked if it were possible that the electrical to gravitational ratio might change as the universe evolved. It turns out on further analysis that this is extremely unlikely—the consequences of even relatively small changes to either the fine structure or gravitational constants turn out to be significant (the anthropic principle) [1], but I am prepared today to point out an arena where the changes in a fundamental coupling have been major, and they have occured within a generation—-the nucleon axial coupling. Below I give a list of values which I have gathered from various sources, and it is clear that there has been a seven percent increase in g_A within a decade!

$$1959 : g_A/g_V = 1.17 \pm 0.02 \ [2] \quad 1965 : g_A/g_V = 1.18 \pm 0.02 \ [3] \quad \quad (1)$$
$$1967 : g_A/g_V = 1.24 \pm 0.01 \ [4] \quad 1969 : g_A/g_V = 1.26 \pm 0.02 \ [5]$$

In fact this number continues to increase—the latest published experiments from Grenoble give $g_A/g_V = 1.266 \pm 0.004$ [6], which is the value I shall employ in this note.

Now my facetious discussion in the previous paragraph is at one level amusing, but at another has some important ramifications when considered in terms of the Goldberger-Treiman relation [15]

$$M_N g_A(0) = F_\pi g_{\pi NN}(0) \quad (2)$$

which is required by chiral invariance and hence by QCD. Now I have indicated in Eq. 2 that the axial coupling g_A and the pion-nucleon coupling constant $g_{\pi NN}$ are *both* to be evaluated at zero momentum transfer. However while the former is the ("time-dependent") number quoted above, the latter is *not* a physical quantity. What *is* directly measurable is the pion-nucleon coupling evaluated at the pion mass-squared—$g_{\pi NN}(m_\pi^2)$, so it is useful to examine the Goldberger-Treiman *discrepancy*

$$\Delta_\pi = 1 - \frac{M g_A(0)}{F_\pi g_{\pi NN}(m_\pi^2)} = 1 - \frac{g_{\pi NN}(0)}{g_{\pi NN}(m_\pi^2)} \quad (3)$$

In the venerable text Bjorken and Drell this number is described as being less than 0.1, but I want to argue that it must be *much* less. Indeed, while $g_{\pi NN}(0)$ is not an observable, one can show that in reasonable models such as the linear sigma model or via correlated $\pi - \rho$ exchange one should expect $\Delta_\pi \simeq 0.02$. However, there is another interesting approach via the so-called Dashen-Weinstein theorem [7], which uses the fact that while there does not exist a prediction for Δ_π in chiral SU(2), since it is given in terms of an a priori unknown counterterm, in SU(3) this quantity is given in terms of a sum of quark masses times an SU(3) octet operator. Thus a relation exists—the Dashen-Weinstein theorem—*between* corresponding kaon and pion quantities [8, 9]

$$\Delta_\pi = \frac{\sqrt{3}}{2} \frac{F_K}{F_\pi} \frac{m_u + m_d}{m_u + m_s} \left(\frac{g_{\Lambda KN}}{g_{\pi NN}} \Delta_K^\Lambda - \frac{1}{\sqrt{6}} \frac{g_{\Sigma KN}}{g_{\pi NN}} \Delta_K^\Sigma \right) \quad (4)$$

where

$$\Delta_K = 1 - \frac{(M_N + M_{\Sigma,\Lambda}) g_A(0)}{\sqrt{2} F_K g(m_K^2)} = \begin{cases} 0.32 & \Lambda \\ -0.05 & \Sigma \end{cases} \quad (5)$$

where I have used the values [11, 12]

$$\begin{array}{ccc} & g(m_K^2) & g_A(0) \\ \Lambda & -13.5 & -0.72 \\ \Sigma & 4.3 & 0.34 \end{array} \quad (6)$$

Then using $(m_u + m_d)/(m_u + m_s) = m_\pi^2/m_K^2$ we find $\Delta_\pi^{theo} \simeq 0.028$. Besides changes in the size of the axial coupling, however, the size of F_π decreased by 1% in 1990 when it was realized that previous evaluations had not included the running of the weak coupling constant [10], and there has been continuous debate about the size of $g_{\pi NN}(m_\pi^2)$, with current analyses favoring either the Karsruhe value 13.4 [13] or the VPI number 13.05 [14]. Thus we find

$$g_{\pi NN} = 13.4 \longrightarrow \Delta_\pi = 4.1\%, \quad \text{or} \quad m_s/\hat{m} \approx 48$$
$$g_{\pi NN} = 13.05 \longrightarrow \Delta_\pi = 1.5\%, \quad \text{or} \quad m_s/\hat{m} \approx 17 \qquad (7)$$

so that if the low value of $g_{\pi NN}$ is confirmed, the Goldberger-Treiman discrepancy would strongly favor the conventional χpt picture $(m_s/\hat{m} = 25)$ over its generalized version, which predicts $m_s/\hat{m} << 25$ [9].

3   The Induced Pseudoscalar

The axial matrix element of the nucleon consists in general of two pieces, the usual axial coupling and the induced pseudoscalar[1]

$$< p(p')|A_\mu|n(p) >= \bar{u}(p')(g_A(q^2)\gamma_\mu\gamma_5 + g_P(q^2)\frac{q_\mu}{2M}\gamma_5)u(p) \qquad (8)$$

and chiral considerations require that this new piece is dominated by its pion pole contribution

$$g_P(q^2) = \frac{4MF_\pi}{m_\pi^2 - q^2}g_{\pi NN}(q^2) \simeq \frac{4MF_\pi}{m_\pi^2 - q^2}g_{\pi NN}(m_\pi^2) - \frac{2M^2}{3}g_A(0)r_A^2 \qquad (9)$$

where r_A is the axial radius. This result is generally used in the combination

$$r_P = \frac{m_\mu}{2Mg_A(0)}g_P(q^2 = -0.9m_\mu^2) = 6.7 \qquad (10)$$

relevant for muon capture. This is the standard approach and is used because the contraction of the four-vector q_μ with the lepton tensor results in a factor of the lepton mass accompanied by the nucleon matrix element of γ_5, which brings in an additional suppression $|q|/2M$, meaning that despite the extraordinary precision of modern nuclear beta decay experiments, any effects from g_P arise only at $\mathcal{O}[r_P m_e^2/(2Mm_\mu)] \sim 10^{-5}$!. On the other hand in muon capture this factor becomes $r_P m_\mu/2M$, which means that the pseudosclar contributes at the same order as weak magnetism and becomes in principle measurable. The one problem here is that typically one has only a single number—the capture rate—to work with so that in order to extract the desired value of r_P one must make three reasonable, but still model-dependent, assumptions—i) the validity of CVC in order to extract $f_V(q^2)$, $f_M(q^2)$ from electron; scattering data; ii) the validity of the impulse approximation to evaluate $g_A(q^2)$; and iii)

---

[1]In general one could also allow an axial tensor coupling, but this "second class current" is disallowed by G-invariance.

the assumption of G-invariance to rule out the presence of second class currents. Using these assumptions one finds the experimental values

$$r_P = \begin{cases} 6.5 \pm 2.4 & H \ [17] \\ 6.9 \pm 0.2 & {}^3He \ [18] \\ 9.0 \pm 1.7 & {}^{12}C \ [19] \end{cases} \qquad (11)$$

which are in agreement with the chiral expectations. It should be noted that the extraordinary precision associated with the ${}^3He$ number is allowed because of a spectacular new PSI experiment which measured the capture rate to 3%

$$\Gamma_\mu({}^3He) = 1496 \pm 4 sec.^{-1} \qquad (12)$$

In order to eliminate some of this model dependence, there are additional approaches which have been and which are being pursued

i) Radiative muon capture on hydrogen: This is the approach which has received the most recent attention, because the result [20]

$$r_P = 9.8 \pm 0.7 \pm 0.3 \qquad (13)$$

is at variance with the chiral prediction at the 3σ level. Now this TRI-UMF measurement is extraordinarily difficult both because of the tiny 10^{-8} branching fraction compared to ordinary capture and because of the presence of many possible experimental backgrounds. However, it has the advantage that at the maximum photon energy $k_{max} = 100 MeV$ the momentum transfer is $q_{max} = m_\mu^2$ compared to the value $q_\mu^2 = -0.9 m_\mu^2$ which obtains in the ordinary muon capture case. Integrated over the photon spectrum this leads to an enhancement of about a factor of three for pseudoscalar effects in RMC over those in OMC [21]. Clearly this is an experiment that should be repeated.

ii) Threshold pion photoproduction on hydrogen: This might seem a strange place to study nucleon *axial* matrix elements, but this is possible because of the PCAC relation [22]

$$< \pi^+ n_{p'} | V_\mu^{em} | p_p > \xrightarrow{q \to 0} \frac{-i}{\sqrt{2} F_\pi} < n_{p'} | A_\mu^- | p_p > \qquad (14)$$

The variation in q^2 which is allowed by the use of electroproduction rather than photoproduction permits a check of the q^2-variation of both axial matrix elements. A recent Saclay experiment produced in this way a measurement of the axial radius r_A which was in good agreement with parallel neutrino scattering measurements, when a small chiral symmetry offset is included, but more importantly for our case for the first time a measurement was made of the *shape* of $g_P(q^2)$ which was in good agreement with the pion pole dominance assumption [23].

iii) Correlations in polarized muon capture on ^3He: The final method which is being pursued at present goes back to an old idea to measure the correlation of the final neutrino direction with initial state polarization in the case that the muon and target are polarized [24]. Of course, the muon is almost completely longitudinally polarized at the time of its capture, but unless the target too is polarized this polarization for the most part lost as the muon cascades down through the various atomic levels before finally reaching the ground state–1S–level from which it is captured. In the general case, when one has muon polarization $P'\hat{n}$ and (spin 1/2) target polarization $P\hat{n}$ the decay distribution is found to be of the form

$$\frac{d^2\Gamma_\mu}{d\Omega_{\hat{k}}} = A - 2PP'B - \frac{1}{2}(P + P')C\hat{n} \cdot \hat{k} + 2PP'D\left[(\hat{n} \cdot \hat{k})^2 - \frac{1}{3}\right] \quad (15)$$

where the structure functions A, B, C, D are functions of the weak form-factors g_v, g_M, g_A, g_P whose specific forms can be found in the literature. The important feature for our case is that D has a strong dependence on g_P, and thus measurement of the angular correlations allows one to pick out the induced pseudoscalar. This measurement was attempted unsuccessfully many years ago at LAMPF, but only recently has the ability to polarize ^3He at high levels given hope that this experiment can actually be carried out. Preliminary results at TRIUMF are encouraging, but the precision is not yet at a level where anything definitive can be said [25].

4   Axial Couplings and SU(3)

The existence of semileptonic hyperon decays such as $\Lambda, n \rightarrow pe^-\bar{\nu}_e, \Sigma^-, \Xi^- \rightarrow \Lambda e^-\bar{\nu}_e, \Xi^- \rightarrow \Sigma^0 e^-\bar{\nu}_e$, etc. allows the probing of axial matrix elements in SU(3). Indeed it has long been known that to leading order in chiral symmetry one can describe such decays in terms of simple f,d parameters, *e.g.*

$$\begin{aligned}
g_V^{pn} &= f_V & g_A^{np} &= f_A + d_A \\
g_V^{p\Lambda} &= -\sqrt{\frac{3}{2}}f_V & g_A^{p\Lambda} &= -\sqrt{\frac{1}{6}}(3f_A + d_A) \\
g_V^{n\Sigma^-} &= -f_V & g_A^{n\Sigma^-} &= -f_A + d_A
\end{aligned} \quad (16)$$

This type of fit yields remarkably good results—$\chi^2_{d.o.f} \approx 8.5$ for ten degrees of freedom, when small $\leq 5\%$ quark model symmetry breaking effects are added [26]. One can try to do even better by including chiral loops using heavy baryon chiral perturbation theory. At one loop one finds results [27]

$$g_A^{ij} = (f_A, d_A)^{ij} + \sum_m \beta_m^{ij} m_m^2 \ln \frac{m_m^2}{\mu^2} \quad (17)$$

However, this inclusion of supposedly *model-independent* corrections brings in modifications to the axial couplings at the level of 30-50% which results in a vastly increased χ^2. Of course, one can restore experimental agreement by the addition of appropriately chosen higher order counterterms, but then one worries about the convergence of the chiral expansion and is certainly justified in asking what is going on. Our answer is that this simple chiral picture omits an important piece of physics, which is finite hadronic size [28]. The simple chiral expansion assumes (at lowest order) propagation of mesons between *point* baryons, while in the real world any such propagation takes place between objects about a fermi or so in size. That means that only the *long*-distance component of the meson loop is really model-independent and to be trusted. One can eliminate such short distance components by use of a cutoff regularization with scale $\sim$300 MeV $\leq \Lambda \leq\sim$ 600 MeV of order inverse baryon size rather than the usual dimensional regularization which mixes both long and short distance effects [29]. The specific form of the cutoff function is unimportant, so for calculational purposes it is useful to use a simple dipole. The result is that the heavy baryon integral responsible for loop corrections to axial couplings

$$I_{ij}(m^2) = \int \frac{d^4k}{(2\pi)^4} \frac{k_i k_j}{(k_0 - i\epsilon)^2 (k^2 - m^2 + i\epsilon)} = \frac{-i\delta_{ij}}{16\pi^2} m^2 \ln \frac{m^2}{\mu^2} \qquad (18)$$

is replaced by

$$\tilde{I}_{ij} = \int \frac{d^4k}{(2\pi)^4} \frac{k_i k_j}{(k_0 - i\epsilon)^2 (k^2 - m^2 + i\epsilon)} \left(\frac{\Lambda^2}{\Lambda^2 - k^2} \right)^2 = \frac{-i\delta_{ij}}{16\pi^2} J(m^2) \qquad (19)$$

where

$$J(m^2) = \frac{\Lambda^4}{(\Lambda^2 - m^2)^2} m^2 \ln \frac{m^2}{\Lambda^2} + \frac{\Lambda^4}{\Lambda^2 - m^2} \qquad (20)$$

We see then that unlike Eq. 18 which emphasizes heavy meson (short-distance) propagation over that of light mesons, in the large mass limit

$$J(m^2) \stackrel{m^2 \gg \Lambda^2}{\longrightarrow} \frac{\Lambda^4}{m^2} \ln \frac{m^2}{\Lambda^2} \to 0 \qquad (21)$$

On the other hand in the large cutoff limit we have

$$J(m^2) \stackrel{\Lambda^2 \gg m^2}{\longrightarrow} \Lambda^2 + m^2 \ln \frac{m^2}{\Lambda^2} \qquad (22)$$

which reproduces the usual dimensional regularization result plus a quadratic term in Λ. That this latter piece does not destroy the chiral invariance can be seen from the feature that it can be absorbed in a renormalization of the basic couplings [30]

$$
\begin{aligned}
d_A^r &= d_A^{(0)} - \frac{3}{2} d_A (3d_A^2 + 5f_A^2 + 1) \frac{\Lambda^2}{16\pi^2 F_\pi^2} \\
f_A^r &= f_A^{(0)} - \frac{1}{6} f_A (25d_A^2 + 63f_A^2 + 9) \frac{\Lambda^2}{16\pi^2 F_\pi^2}
\end{aligned}
\qquad (23)
$$

	dim.	$\Lambda=300$	$\Lambda=400$	$\Lambda=500$	$\Lambda=600$
$g_A(\bar{p}n)$	1.72	0.37	0.53	0.69	0.84
$g_A(\bar{p}\Lambda)$	-1.78	-0.34	-0.51	-0.67	-0.84
$g_A(\bar{\Lambda}\Sigma^-)$	1.17	0.23	0.34	0.45	0.56
$g_A(\bar{n}\Sigma^-)$	0.36	0.07	0.10	0.14	0.17
$g_A(\bar{\Lambda}\Xi^-)$	0.83	0.15	0.23	0.31	0.39
$g_A(\bar{\Sigma}^0\Xi^-)$	2.46	0.45	0.68	0.91	1.15

Table 4.1. Given are the nonanalytic contribtions to g_A for various transitions in dimensional regularization and for various values of the cutoff parameter Λ in MeV.

However, in this procedure with reasonable values of the cutoff, the SU(3) chiral expansion is now under control, as can be seen in Table 1.

This brings such results into agreement with typical chiral bag calculations, such as the cloudy bag [31], and there is no longer any need to append large counterterm contributions in higher orders.

5 Conclusion

We have above considered an old subject—that of nucleon axial matrix elements—from the point of view of modern experiments. I hope that I have convinced you that despite the age of the field, the new results in the areas of Goldberger-Treiman discrepancies, induced pseudoscalar measurements, and SU(3) chiral perturbative studies promise continued interest even as we approach the millenium.

Acknowledgement. It is a pleasure to acknowledge support from the Alexander von Humboldt Foundation and the hospitality of Forschungszentrum Jülich. This work was also supported by the National Science Foundation.

References

1. For a recent slant and for further references see V. Agrawal, S.M. Barr, J.F. Donoghue, D. Seckel: Phys. Rev. **D57**, 5480 (1998)

2. A.N. Sosnovskii et al.: Sov. Phys. JETP **8**, 739 (1959)

3. S.A. Adler: Phys. Rev. Lett. **14**, 1051 (1965)

4. C.J. Christensen et al.: Phys. Lett. **B26**, 11 (1967); Phys. Rev. **D5**, 1628 (1972)

5. R.J. Blin-Stoyle, *Fundamental Interactions and the Nucleus*, North-Holland, New York (1969)

6. K. Schreckenbach et al.: Phys. Lett. **B259**, 353 (1991)

7. R. Dashen and M. Weinstein, Phys. Rev. **188**, 2330 (1969)

8. C.A. Dominguez: Riv. del Nuovo Cimento **8**, 1 (1985)

9. N.H. Fuchs, H. Sazdjian, and J. Stern, Phys. Lett. **B238**, 380 (1990)

10. B.R. Holstein, Phys. Lett. **B244**, 83 (1990)

11. Particle Data Group, Phys Rev. **D50**, 1173 (1996)

12. H. Haberzettl et al.: Phys Rev. **D58**, 40 (1998)

13. R. Koch and E. Pieterinin: Nucl. Phys. **A336**, 331 (1980)

14. R. Arndt et al.: Phys. Rev. Lett. **65**, 157 (1990)

15. M.L. Goldberger and S.B. Treiman: Phys. Rev. **110**, 1478 (1958)

16. J.D. Bjorken and S.D. Drell, *Relativistic Quantum Mechanics*, McGraw-Hill, New York (1964)

17. G. Bardin et al.: Phys. Lett. **B104**, 320 (1981)

18. P. Ackerbauer et al.: Phys. Lett. **B417**, 224 (1998)

19. V. Roesch et al.: Phys. Rev. Lett. **46**, 1507 (1981)

20. G. Jonkmans et al.: Phys. Rev. Lett. **77**, 4512 (1996)

21. H.W. Fearing: Phys. Rev. **C21**, 1951 (1980); D.S. Beder and H.W. Fearing: Phys. Rev. **D39**, 3493 (1989)

22. A.I Vainshtein and V.I. Zakharov: Nucl. Phys. **B36** (1972); V. Bernard, N. Kaiser, and U.-G. Meissner: Nucl. Phys, **A607**, 379 (1996)

23. S. Choi et al.: Phys. Rev. Lett. **71**, 3927 (1993)

24. See, *e.g.* B.R. Holstein: Phys. Rev. **C3**, 1964 (1972)

25. W.J. Cummings et al.: Proc. WEIN '95, ed. H. Ejiri, T. Kishimoto, and T. Sato, World Scientific, Singapore (1995), p. 381; G. Cates: private communication

26. J.F. Donoghue, B.R. Holstein, and S.W. Klimt: Phys. Rev. **D35**, 934 (1987)

27. J. Bijnens, H. Sonoda, and M.B. Wise: Nucl. Phys. **B261**, 185 (1985)

28. J.F. Donoghue, B.R. Holstein, and B. Borasoy: hep-ph/9804281

29. J.F. Donoghue and B.R. Holstein: hep-ph/9803312

30. M.A. Luty and M. White: Berkeley prepring LBL33993 (1993)

31. See, it e.g. T. Yamaguchi et al.: Nucl. Phys. **A500**, 129 (1989); K. Kubodera et al.: Nucl. Phys. **A439**, 695 (1985); S. Theberge et al.: Phys. Rev. **D22**, 2838 (1980); A.W. Thomas, J. Phys. **G7**, L283 (1981); R.E. Stuckey and M.C. Birse, J. Phys. **G23**, 29 (1997)

Few-Body Systems Suppl. 11, 124–129 (1999)

Few-
Body
Systems
© by Springer-Verlag 1999

Electroproduction of the $\Delta(1232)$ and $S_{11}(1535)$ Resonances at High Momentum Transfer.

Paul Stoler

Physics Department, RPI, Troy, NY 12180, USA

Abstract. The evolution of baryon resonance transition amplitudes with increasing Q^2, in order to learn more about the transition from the low Q^2 domain of the quark model to the the high Q^2 domain, where hard QCD processes play an important role, is discussed. In particular, the results of an experiment performed at Jefferson Lab. which studies the $\Delta(1232)$ and $S_{11}(1535)$ resonances in the π^0 and η dacay channels respectively, up to $Q^2 \sim 4 GeV^2/c^2$, and future plans to extend these measurements to higher Q^2 are presented.

Fundamental issues in hadron physics are which degrees of freedom are appropriate to describe exclusive reactions at experimentally accessible momentum transfers, and how are models at low, medium and high Q^2 related? Constituent quark models (CQM) appear to work well at the low Q^2 limit, and it is widely believed that valance pQCD will be valid in the limit of high Q^2. How far the validity of these models extend in Q^2 is still an open question.

Many exclusive reactions exhibit constituent scaling behavior at moderate Q^2 (few GeV^2/c^2) which is interpreted by some as the onset of perturbative QCD. Others [2, 3] argue that the observed scaling is not a manifestation of pQCD, and that at the experimentally accessible range of Q^2 exclusive reactions are explained primarily by soft Feynman mechanism. Recently[4, 3], there have been promising developments in bridging the high and low Q^2 extremes bridging with a QCD quark-parton description of exclusive reactions. In this approach the perturbative hard part of the reaction, which is calculable is isolated from the non-perturbative soft, physics which is parameterized in terms of *non-forward parton distributions* (NFPD). In the limit of forward scattering it is shown that the NFPD become the usual inclusive parton distribution functions.

An attractive aspect of NFPD's, is that the same NFPD are common to different exclusive reactions which involve the same set of hadrons. I believe an important advance of the new formalism will be to explicitly show the connection between these NFPD's and valance pQCD at the high Q^2 limit, and the CQM at the low Q^2.

Baryon resonance transition amplitudes, which we plan to measure at JLAB to the highest possible Q^2, are appropriate testing grounds for this and other theoretical approaches to exclusive reactions. The various models for exclusive reactions make very definite predictions. For example, valance pQCD predicts $1/Q^4$ scaling for the leading helicity conserving form factors, and $1/Q^5$ and $1/Q^6$ dependence for next to leading helicity *non*-conserving form factors, which asymptotically vanish relative to the helicity conserving form factors. The various versions of the CQM, and NFPD approach make very different predictions about the Q^2 evolution of these amplitudes.

Here, I would like to report on the first completed JLAB experiments on resonance production at high Q^2, the measurement of the the the reactions $e + p \to e + p + \pi^0$ and $e + p \to e + p + \eta$ for a range in W from pion threshold to about 1.7 GeV^2/c^2. In this range of W the π^0 final state is dominated by the $\Delta(1232)$ below the 2-pion threshold and the eta final state is dominated by the $S_{11}(1535)$. The experiments were carried out in Hall C, at high luminosity ($\sim 1 \times 10^{38}$ cm^{-1}s^{-1}). Scattered electrons were detected in the Short Orbit Spectrometer (SOS), and recoil protons were detected in the High Momentum Spectrometer (HMS). The incident beam energies were 3.2 and 4.0 GeV which resulted in $Q^2 = 2.8$ and 4.0 GeV^2/c^2 for the $\Delta(1232)$ measurement and $Q^2 = 2.4$ and 3.6 GeV^2/c^2 for the $S_{11}(1535)$.

Since the experiment is kinematically complete, the undetected single π^0 and η final states were selected by missing mass. An example of the experimental missing mass distribution is shown in Figure 1.

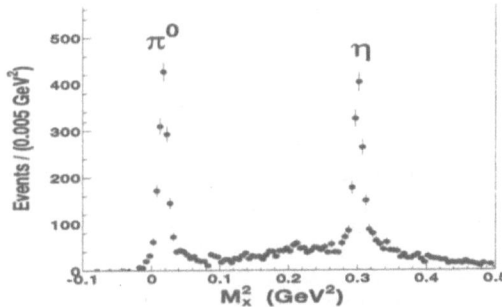

Figure 1. Example of a raw missing mass distribution for the reaction $p(e, e'p)X$. The data are from one experimental proton arm setting at $Q^2 = 2.8$ GeV^2.

At higher Q^2 the protons emerge in a narrow cone around the q vector so that a significant fraction of 4π c.m. solid angle was obtained in the decay of the $\Delta \to p\pi^0$, and somewhat less in the decay of the $S_{11}(1535)$ to $p\eta$.

For each reaction we evaluated the invariant mass W and the center of mass decay angles $\cos\theta_{\pi,\eta}^{cm}$, and $\phi_{\pi,\eta}^{cm}$. For the π^0 channel there is a significant background from the elastic radiative tail, most of which was removed by accepting events only for $| \phi_\pi^{cm} | > 8°$. For the η the 2 and 3 pion continuum had to be subtracted. The data were binned with $\Delta W = 30$ MeV, $\Delta\cos\theta_{\pi,\eta}^{cm} = 0.2$, and $\Delta\phi_{\pi,\eta}^{cm} = 30°$. The bin size in Q^2 (≈ 0.5 GeV^2) was determined by the apparatus acceptance. Details of the data analysis for the Δ and S_{11} experiments can be found in [11] and [10], respectively.

Assuming the one photon exchange approximation, the differential cross section of single meson electroproduction is related to the center of mass differential cross section for meson production by virtual photons, $d\sigma/d\Omega_{\pi,\eta}^{cm}$, as follows :

$$\frac{d\sigma}{dW\,dQ^2 d\Omega_{\pi,\eta}} = \Gamma\frac{d\sigma}{d\Omega_{\pi,\eta}} \qquad (1)$$

where Γ is the virtual photon flux (Hand convention). $d\sigma/d\Omega_{\pi,\eta}^{cm}$ was obtained at more than 2000 intervals in $\cos\theta_{\pi,\eta}^{cm}$, $\phi_{\pi,\eta}^{cm}$, and W. Examples of the pion angular distributions at $W = 1.235$ GeV are shown in Figure 1. The errors shown in the figures are statistical only. They do not include an estimated systematic uncertainty in overall normalization of about 5 to 6%.

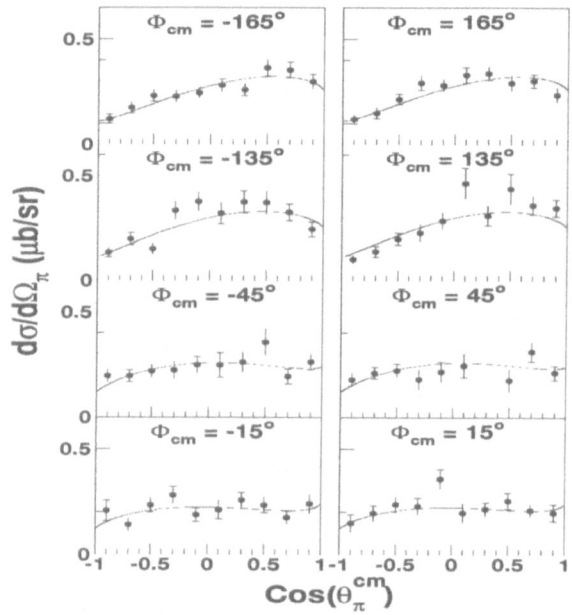

Figure 2. Examples of the angular distributions for π^0 production at $W = 1.235$ GeV and $Q^2 = 2.8 \text{GeV}^2/c^2$ [11]. The curves are due to a model independent analysis (see text).

First we consider the $\Delta(1232)$ [11]. The models make predictions about the Q^2 dependence of the transition form factor, G_M^*, [21] as well as the relative contributions of M_{1+}, E_{1+}, and S_{1+} multipole amplitudes. At low Q^2, according to the CQM the $N-\Delta$ transition is predominantly M_{1+}, and $R_{EM} \equiv E_{1+}/M_{1+}$ and $R_{SM} \equiv S_{1+}/M_{1+}$ are small, which is experimentally confirmed near $Q^2 = 0$ [5, 6]. In the high Q^2 limit, according to valence pQCD only helicity conserving amplitudes should contribute, leading to the prediction $R_{EM} = +1$ and $R_{SM} =$

0. The earlier data from DESY [7] at $Q^2 = 3.2$ GeV^2 have been evaluated , suggesting [8, 9] that R_{EM} is small, but with large errors.

The differential cross sections are not sufficient to extract the amplitudes without further assumptions. In order to obtain a model independent (MI) idea of the magnitude of the form factor G_M^*, and the relative contributions of the amplitudes R_{EM}, and R_{SM} , for each W bin the angular distributions were fit assuming S and P wave contributions and M_{1+} dominance. The overall fit to the data was quite good (at W = 1.235 GeV $\chi_{pdf}^2 \sim 1.36$ and 1.22 for Q^2 = 2.8 and 4.0 GeV^2/c^2 respectively) and verifies the M_{1+} dominance at the resonance pole.

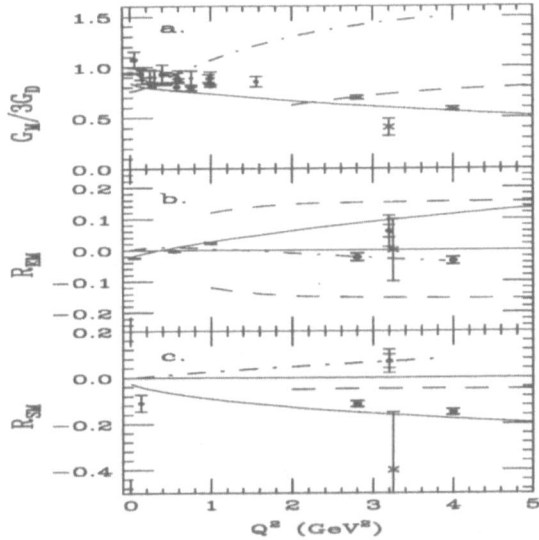

Figure 3. The $N \to \Delta(1232)$ transition form factor $G_{M_{1+}}$, and the ratios R_{EM} and R_{SM}, compared to the predictions. **a.)** G_M^*: The data for $Q^2 < 2$ GeV^2 are evaluations of previous data [9, 16]. **b.)** R_{EM}: The data at $Q^2 = 0$ are from [5]. **c.)** R_{SM}: The datum at $Q^2 = 0.13$ GeV^2 is from [20]. For R_{EM} and R_{SM} the points denoted by small circles (o) at $Q^2 =0.55, 1.0$, and 3.2 GeV^2 are evaluations of DESY data[7, 8], and those denoted by $\times$ at $Q^2 = 3.2$ GeV^2 are the evaluation of [9]. The data points at $Q^2 = 2.8$ and 4.0 GeV^2, denoted by • are the effective Lagrangian [9] extracted values of R_{EM} and R_{SM} from the present JLAB experimental data. The dash-dot curve is due to a light-front calculation [16] in the framework of the relativistic CQM. The solid curve is due to a light-front CQM calculation [17] including constituent quark form-factors. The dashed curve is the result of a pQCD sum rule calculation [3]. An independent effective Lagrangian calculation in the framework of [19] confirms this result.

To extract more reliable quantitative values of the amplitudes an effective Lagrangian approach [14] was employed and is in good agreement with the above MI approach. The resulting resonance part of G_M^*, R_{EM} and R_{SM}, shown in Figure 2., are nearly identical with the combined resonant plus non-resonant values, and agree with the MI analysis very well considering the approximations of that procedure. Also shown in Figure 2. are the results of earlier experiments and selected theoretical calculations.

The following features are noteworthy. The Δ form factor $G_{M_{1+}}$ exhibits a greater decrease with Q^2 than the dipole form as suggested by previous analysis of inclusive data [12, 13]. The ratio R_{EM} remains very small, and consistent within errors with its value at $Q^2 = 0$; while the ratio R_{SM} come out to be non-zero and negative. All these results show that the hadron helicity is not conserved in this exclusive reaction and the valence pQCD description is not applicable for this reaction at these momentum transfers.

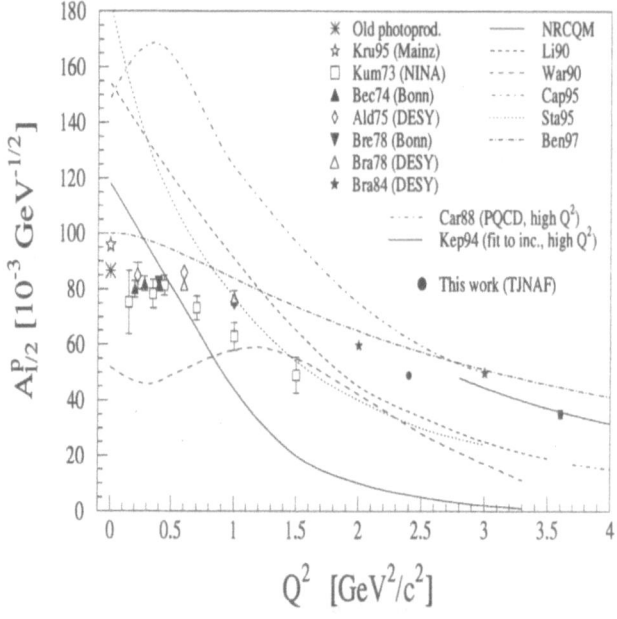

Figure 4. The amplitude $A_{1/2}$ for $N \rightarrow S_{11}(1535)$ assuming a resonance width $\Gamma = 150$ MeV, and η branching ratio 0.55. The error bars on the preliminary JLAB results at $Q^2 = 2.4$ and 3.6 GeV^2/c^2 are statistical only, and do not include systematic errors in the η branching ratio, or in our resonance width measurement. The other data points and theoretical curves are referenced in [10]

For the $S_{11}(1535)$ data [10], a partial wave fit was made to the data, assuming S and P wave contributions, and E_{0+} dominance. In addition a smooth non-resonant background with realistic threshold dependence was included in the fit. To extract the helicity conserving electromagnetic amplitude $A_{1/2}$ from the differential cross section requires the input of the η branching ratio, as well as the resonance width Γ, neither of which have previously been unambiguously measured. We found that the best fit value of Γ (154 ± 20 MeV) is very close to the PDG [15] preferred value of 150 MeV. For the branching ratio we chose 0.55, which is consistent with analysis of inclusive data [13]. The obtained $A_{1/2}$, along with previously published data, and the results of a number of different theoretical quark model based curves are shown in Figure 3.

Acknowledgement. The author thanks Volker Burkert, Nimai Mukhopadhyay, Bijan Saghai and Silvano Simula for their excellent organization of a fruitful workshop.

References

1. S.J. Brodsky and G.P. Lepage: Phys. Rev. **D23**, 1152 (1981); C. E. Carlson and J.L. Poor: *ibid* **D38**, 2758 (1988)

2. N. Isgur and C.H. Llewellyn-Smith: Phys. Lett. **52**, 1080 (1984)

3. A.V. Radyushkin: Nucl. Phys. **A527**, 53 (1991)

4. A. Radyushkin: Phys. Lett. **B385**, 333 (1996); X. Ji: *ibid* **78**, 610 (1997)

5. R.Beck et al.: Phys. Rev. Lett **78**, 609 (1997); G.Blanpied et al.: *ibid* **79**, 4337 (1997)

6. F. Kalleicher et al.: Z. Phys. **A359**, 201 (1997)

7. R. Haidan: PhD Thesis, DESY report F21-79/03 (1979)

8. V. Burkert and L. Elouadrhiri: Phys. Rev. Lett. **75**, 3614 (1995)

9. R. Davidson and N. Mukhopadhyay: private communication

10. C.S. Armstrong: PhD thesis, College of William and Mary; C.S. Armstrong et al.: to be published

11. V.V. Frolov: PhD thesis, Rensselaer Polytechnic Institute; V.V. Frolov et al.: to be published

12. P. Stoler: Phys. Rep. **226**, 103 (1993)

13. C. Keppel: "*Workshop on CEBAF at Higher Energies*, CEBAF pp. 237 (1994)

14. R. Davidson, N. Mukhopadhyay and R. Wittman: Phys. Rev. Lett. **56**, 804 (1986)

15. Review of Particle Physics: Phys. Rev. **54**, 1 (1996)

16. S. Capstick and B.D. Keister: Phys. Rev. **D51**, 3598 (1995)

17. F. Cardarelli et al.: Phys. Lett **B371**, 7 (1996)

18. V.M. Belyaev and A.V. Radyushkin: Phys. Rev. **D53**, 6509 (1997)

19. T. Sato and T.-S. H. Lee, Phys. Rev. **C54**, 2606 (1996)

20. F. Kalleicher et al.: Z. Phys. **A359**, 201 (1997)

21. H.F. Jones and M.D. Scadron, Ann. Phys. **81**, 1 (1979)

Few-Body Systems Suppl. 11, 130–133 (1999)

Few-
Body
Systems
© by Springer-Verlag 1999

Axial Vector Form Factor $G_A^{N\Delta}$ for the $N \to \Delta(1232)$ Transition

Latifa Elouadrhiri*

Christopher Newport University, 1 University Place, NN, VA 23606, USA and Jefferson Laboratory, 12000 Jefferson Avenue, Newport News, VA 23606, USA

and the

CLAS collaboration

Abstract.
Using current algebra and the Partially Conserved Axial Vector Current hypothesis (PCAC) for soft pions, the $ep \to e'\Delta^{++}\pi^-_{soft}$ cross section near threshold can be related to the axial vector form factor $G_A^{N\Delta}$ for the $N \to \Delta(1232)$ transition. The first priliminary data from this reaction from the CLAS detector at Jefferson Lab is presented.

1 INTRODUCTION

When the theory of electro- and photoproduction of pions was developed during the decade of the 1950's, Kroll and Ruderman[1] derived model-independent predictions for the threshold region, the so-called Low Energy Theorems (LET), by asserting gauge and Lorentz invariance. Following the presentation of a general pion production formalism by Chew, Goldberger, Low and Nambu[2], Fubini et al.[3] extended earlier work by including the hypothesis of Partially Conserved Axial Current (PCAC).

The relationship between low-energy pion electroproduction and the axial vector nucleon vertex $< N|A_\mu|N >$ was widely studied during the seventies. In particular, by means of current algebra techniques and PCAC it was shown how $G_A^N(Q^2)$, the axial vector form factor of the nucleon, might be deduced from the threshold cross section for the $\gamma_v p \to \pi^+ n$ reaction.[4] This offered a practical alternative to the more direct but experimentally difficult means of determining $G_A^N(Q^2)$ from the $\nu N \to \mu N$ reaction. This conjecture was verified in an experiment at Saclay [5].

This theoretical work has been revised by V. Bernard et al. in the framework of Chiral Perturbation Theory (ChPT).[6]

*e-mail: latifa@jlab.org

It has also been suggested[7, 8, 9, 10] that two-pion production $\gamma_v p \rightarrow e'\Delta^{++}\pi^-$ could be exploited to obtain information on the $< \Delta|A_\mu|N >$ vertex. This reaction has the advantage that the axial transition form factor is expected to provide the main contribution at threshold. This dominance facilitates a meaningful comparison between theoretical predictions and experimental data.

Bartl *et al.*[10] have utilized current algebra in the soft pion limit to derive a relationship between the $< \Delta|A_\mu|N >$ vertex and threshold $\Delta^{++}\pi^-$ photo- and electroproduction.

A more recent theoretical development is the extension of Chiral Perturbation Theory (ChPT) to baryons. Chirally invariant effective Lagrangians have been now constructed for baryons fields, and there has been much recent work on both one-and two-pion photo-and electroproduction. Calculations of the $\Delta^{++}\pi^-$ electroproduction cross section using the Heavy Baryon Chiral Perturbation Theory (HBChPT) with explcit deltas are under way.

2 PREVIOUS EXPERIMENTAL RESULTS

The only existing near-threshold data on the $ep \rightarrow e'\pi^-\Delta^{++}$ reaction come from an experiment at DESY[11, 12, 13]. In this case, a beam of 7.2 GeV electrons from the DESY synchrotron was directed onto a 9 cm long liquid hydrogen target situated inside a 1 m long streamer chamber that was in a 1.8 T magnetic field. Scintillators and shower counters served to detect the scattered electron and trigger the streamer chamber. Produced hadrons were detected with full angular acceptance. The total flux incident on the target was 3.3×10^{12} electrons, corresponding to an integrated luminosity of 1.26×10^6 μb^{-1}.

Figure 1(a) shows results from DESY data, the $ep \rightarrow e'\pi^-\Delta^{++}$ cross section manifests a rapid rise from threshold and near-isotropic production of the Δ^{++} with respect to the virtual photon direction in the $\gamma_v p$ center-of-momentum system. These features were also seen in the photoproduction results[14], and are consistent with the dominance near threshold of the Born contact, or seagull term. The dominance of the contact term near threshold is crucial for the determination of the properties of the $< \Delta|A_\mu|N >$ vertex.

To facilitate the comparison with theory one should seek to examine the angular distribution closer to threshold, selecting a small W-range for which Δ^{++} production remains isotropic. Inadequate counting statistics precluded such an analysis of the DESY data.

3 A FIRST LOOK AT $ep \rightarrow e'\Delta^{++}\pi^-_{soft}$ WITH CLAS

Figure 1(b) shows the invariant mass distribution of $ep \rightarrow e'\Delta^{++}\pi^-_{soft}$ as measured with the CLAS detector. [15] In this case, we require full detection of the scattered electron and p and π^+ products of the Δ^{++} decay in each of the three drift chamber regions. Time-of-flight information was also required, as was the detection of the electron in the Cerenkov counter and electromagnetic calorimeter.

The detection of the low energy π^- in this case is unnecessary because the good resolution of CLAS permits unambiguous reconstruction of soft pions using missing mass techniques. The kinematic region of interest in this experiment, $W < 1.5$ GeV, lies below the $W \approx 1.6$ GeV threshold for ρ-meson production. There then remain two main background processes to the $ep \to \pi^- \Delta^{++}$ reaction we seek to study; these lead to the $\Delta^0 \pi^+$ and non-resonant 3-body $p\pi^+\pi^-$ final states. According to theoretical predictions, Δ^{++} production should dominate at low W, an expectation confirmed by the experimental results[13] from DESY. At $Q^2 = 0.5$ (GeV/c)^2 we assume production in the ratio of

$$p\pi^+\pi^- \; : \; \Delta^{++}\pi^- \; : \; \Delta^0\pi^+ \;\; = \;\; 1.8 \; : \; 9 \; : 1 \,,$$

values that are consistent with DESY results.

The quality of the CLAS results will far surpass that of the only previous measurement of Δ^{++} electroproduction.

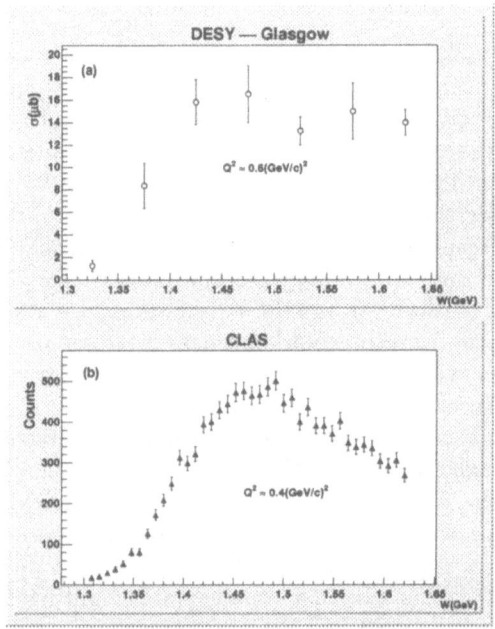

Figure 1. (a) the total cross section for $ep \to e'\Delta^{++}\pi^-$ as a function of W averaged over Q^2 interval 0.3 -1.4 $(GeV/c)^2$. (b) the W distribution of $ep \to e'\Delta^{++}\pi^-$ from CLAS for approximately 5% of the data to be accumulated in this region.

4   CONCLUSION

The measurement near threshold of the cross section for the $ep \to e'\pi^- \Delta^{++}$ reaction with CLAS will provide information of unprecedented quality on the kinematic dependence of the axial vector form factor for the $N \to \Delta$ transition. We will also have the opportunity to test the proposed relation between the axial vector form factors for the nucleon and the $N \to \Delta$ transition.

References

1. N. M. Kroll and M. A. Ruderman: Phys. Rev. **93**, 233 (1954)

2. G.F. Chew, M.L. Goldberger, F.E. Low, Y. Nambu: Phys. Rev. **106** 1337 (1957); *ibid.* **106**, 1345 (1957)

3. S. Fubini, G. Furlan, C. Rossetti: Nuovo Cim. **40**, 1171 (1967)

4. E. Amaldi, S. Fubini, G. Furlan, *"Pion Electroproduction"* , Springer Tracts in Modern Physics, **83**, (1979)

5. S. Choi et al.: Phys. Rev. Lett. **71** 3927 (1993)

6. V. Bernard, N. Kaiser, Ulf-G. Meißner Int. J. Mod. Phys. **E4**, 193 (1995); V. Bernard, J. Gasser, N. Kaiser, Ulf-G. Meißner: Phys. Lett. **B268**, 291 (1991); V. Bernard, N. Kaiser, Ulf-G. Meißner: Z. Phys. **C70**, 483 (1996); Phys: Rev. Lett. **69**, 1877 (1992)

7. P. Carruthers and K.W. Huang: Phys. Lett. **24B**, 464 (1967)

8. T. Ebata: Phys. Rev. **154**, 1341 (1967)

9. P. Narayanaswamy and B. Renner, Nuovo Cim. **53A**, 107 (1968)

10. A. Bartl, N. Paver, C. Verzegnassi, S. Petrarca: Lett. Nuovo Cim. **18**, 588 (1977); A. Bartl, K. Wittmann, N. Paver, C. Verzegnassi: Nuovo Cim. **45A**, 457 (1978)

11. P. Joos et al.: Phys. Lett. **52B**, 481 (1974)

12. P. Joos et al.: Phys. Lett. **62B**, 230 (1976)

13. K. Wacker et al.: Nucl. Phys. **B144**, 269 (1978)

14. Aachen–Berlin–Bonn–Hamburg–Heidelberg–München collaboration: Phys. Rev. **175**, 1669 (1968)

15. V.D. Burkert: In these Proceedings

Few-Body Systems Suppl. 11, 134–139 (1999)

Few-
Body
Systems
© by Springer-Verlag 1999

Exchange Currents and Nucleon Deformation

A. J. Buchmann*

Institute for Theoretical Physics, University of Tübingen, Auf der
Morgenstelle 14, D-72076 Tübingen, Germany

Abstract. Information concerning the intrinsic deformation of the proton can
be obtained by studying the electromagnetic $p \to \Delta^+$ quadrupole transition.
Recent experiments have shown that the electric quadrupole ($E2$) strength in
$\gamma p \to \Delta^+$ is about 10 times larger than predicted by the simple quark model
using only one-body currents. Our analysis provides compelling evidence for
the dominance of exchange currents in the $N \to \Delta$ quadrupole transition, and
identifies the physical mechanism leading to the observed $E2$ strength.

1 Introduction

The quadrupole moment of a particle measures the deviation of its internal
charge distribution from spherical symmetry. However, as a particle with total angular momentum $J = 1/2$, the nucleon cannot have a spectroscopic
quadrupole moment. In order to learn something about the shape of the nucleon one has to electromagnetically excite it, e.g. to the $\Delta(1232)$ resonance,
with total angular momentum $J = 3/2$.

Aside from the dominant $N \to \Delta$ magnetic dipole ($M1$) excitation mode,
in which the spin of a single quark is flipped, there are small transverse electric quadrupole ($E2$), and longitudinal charge quadrupole ($C2$) transitions.
The strengths of these electromagnetic multipoles at photon three-momentum
transfer $\mathbf{q} = 0$, called $G_{M1}(0)$, $G_{E2}(0)$, and $G_{C2}(0)$, have recently been extracted from high precision photo-pion- and electro-pionproduction experi-
ments off the proton [1]. The quadrupole transition strengths $G_{E2}(0)$ and
$G_{C2}(0)$ are a measure of the intrinsic deformation of the nucleon.

The experimental $E2$ strength is, with $G_{E2}^{exp} = 0.133(20)$ [1], quite large.
The corresponding Bohr-Mottelson deformation parameter $\delta = 0.17$ indicates
a prolate deformation of the nucleon. The new experimental results are about
10 times larger than predicted by various quark model calculations using spatial
single-quark currents [2, 3, 4]. In order to explain the measured $E2$ strengths as

---

*E-mail address: alfons.buchmann@uni-tuebingen.de

a single quark transition, one would need D-state probabilities in the nucleon and Δ of the order of 30%. However, there is no known quark-quark interaction that would give rise to such strongly deformed single particle orbits.

2 Single Quark $N \to \Delta$ Quadrupole Transition

Tensor forces coming from effective one-gluon, and one-pion exchange induce D-wave admixtures in the three-quark N and Δ wave functions, which can be schematically written as

$$
\begin{aligned}
|N\rangle &= a_S\,|(S=1/2\,,L=0)J=1/2\rangle + a_D\,|(S=3/2\,,L=2)J=1/2\rangle \\
|\Delta\rangle &= b_S\,|(S=3/2\,,L=0)J=3/2\rangle + b_D\,|(S=1/2\,,L=2)J=3/2\rangle\,. \quad (1)
\end{aligned}
$$

The D-wave probabilities a_D^2 and b_D^2 as calculated by different authors [2, 3] are 0.2% and 0.4%, much smaller than the 5% D-wave admixture in the deuteron. Where does this large difference come from? In the harmonic oscillator model one needs about $2\hbar\omega \sim 600$ MeV to lift a quark into an excited D state, whereas one needs only $2\hbar\omega \sim 4$ MeV to lift a nucleon into a D state. This huge difference is mainly due to the very different sizes of the systems, and partly explains why D-state admixtures in the nucleon wave function are relatively small.

In the single quark transition model, based on the single quark current $\mathbf{J}_{[1]}$, a quark in an excited D state (L=2) of the nucleon can absorb electromagnetic quadrupole radiation (L_γ=2), and thus fall to the ground state (L'=0) of the Δ. The amplitude for this process is proportional to $a_D b_S$. In addition, a quark in the ground state (L=0) of the nucleon can absorb quadrupole radiation (L_γ=2), and jump to an excited D-state (L'=2) in the Δ. This happens with an amplitude $a_S b_D$. Both transitions involve only orbital angular momentum changes, but leave the inner spin states of the particles unchanged. The resulting single-quark $E2$ transition strength is:

$$
G_{E2}(\mathbf{J}_{[1]}) = \frac{1}{\sqrt{5}}\left(a_D b_S + a_S b_D\right). \quad (2)
$$

With the calculated D state admixtures of various quark-quark potential models, one obtains $G_{E2} = 0.003 - 0.01$ [2, 3, 4], i.e. theoretical results that differ by an order of magnitude from the recent experimental results. This suggests that some important dynamical feature has been missing in previous quark model calculations.

3 Exchange Currents and Siegert's Theorem

From the theoretical point of view the conventional explanation outlined above is incomplete because it violates the continuity equation:

$$
\mathbf{q} \cdot \mathbf{J}(\mathbf{q}) = [H, \rho(\mathbf{q})]\,. \quad (3)
$$

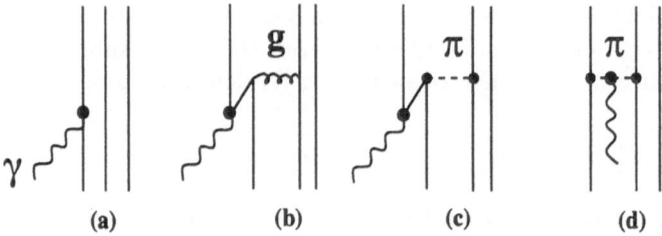

Figure 1. Feynman diagrams of the four vector current $J^\mu = (\rho, \mathbf{J})$: (a) one-body current $J^\mu_{[1]}$, and (b-d) two-body gluon and pion exchange currents $J^\mu_{[2]}$. If the quarks interact via pion and gluon exchange, the two-body exchange currents depicted in diagrams (b-d) must be taken into account. If only diagram (a) is considered, the continuity equation (3) is violated.

After a decomposition of the charge and current operators into one- and two-body terms, i.e., $\rho = \rho_{[1]} + \rho_{[2]}$, and $\mathbf{J} = \mathbf{J}_{[1]} + \mathbf{J}_{[2]}$, one can show that the spatial two-body current $\mathbf{J}_{[2]}$ satisfies the following consistency relation

$$\mathbf{q} \cdot \mathbf{J}_{[2]}(\mathbf{q}) \quad = \quad \left[V_{[2]}, \rho_{[1]}(\mathbf{q}) \right]. \tag{4}$$

Equation (4) connects the quark-quark interactions $V_{[2]}$, which determine the coefficients a_S, a_D, b_S, b_D in Eq. (1) with the two-body currents $\mathbf{J}_{[2]}$ of Fig. (1), which determine the electromagnetic properties of the $N - \Delta$ system. Eq. (4) is violated if the potential contains momentum and/or isospin dependent terms (our case), but the current contains only one-body terms.

An important theorem based on Eq.(3) is Siegert's theorem. In the limit of small momentum transfers it relates the transverse electric $T^{E\,J}$ and longitudinal Coulomb multipoles $T^{C\,J}$:

$$\langle f | T^{E\,J}(|\mathbf{q}| \to 0) | i \rangle = -\frac{\omega}{|\mathbf{q}|} \sqrt{\frac{J+1}{J}} \, \langle f | T^{C\,J}(|\mathbf{q}| \to 0) | i \rangle. \tag{5}$$

Thus, G_{E2} can be calculated via the charge operator (right hand side). The result will be the same as the one based on the spatial current operator (left-hand side).

However, until recently, calculations of the $E2$ transition strength G_{E2} using the spatial one-body current operator have differed considerably (in some calculations by an order of magnitude) from those using the one-body charge operator and Siegert's theorem. This observation was made in Ref.[4], and has been confirmed by other authors [5]. But the physical reason behind this discrepancy has not been recognized. We have recently shown [6] that this difference is almost entirely explained by spatial two-body exchange currents required by Eq.(4) (see Table 3.1). Exchange currents were not explicitly included in previous analyses of this problem.

Table 3.1. The transverse electric quadrupole form factor $G_{E2}(q^2 = 0)$ for the $\gamma + p \rightarrow \Delta^+$ transition calculated with (i) the one-body charge density $\rho_{[1]}$ using Siegert's theorem (first row), (ii) with the spatial current density $\mathbf{J} = \mathbf{J}_{[1]} + \mathbf{J}_{[2]}$ (last row) for various quark models. A comparison of the results in the first and last rows shows that the continuity equation is approximately satisfied, provided that the spatial two-body exchange currents $\mathbf{J}_{[2]}$ required by Eq.(4) are included [6]. The remaining discrepancy between theory and recent experiment is removed by including $\rho_{[2]}$ (see Eq.(10)).

	Ref.[2]	Ref.[3]	Ref.[4]	Ref.[7]	Ref.[8](π)
$G_{E2}(\rho_{[1]})$	0.0192	0.0203	0.0796	0.0177	0.0165
$G_{E2}(\mathbf{J}_{[1]})$	0.0118	0.0092	0.0076	0.0027	0.0058
$G_{E2}(\mathbf{J}_{[2]})$	0.0084	0.0114	0.0561	0.0127	0.0105
$G_{E2}(\mathbf{J})$	0.0202	0.0206	0.0637	0.0154	0.0163

4 Double Spin Flip $N \rightarrow \Delta$ Quadrupole Transition

As we have seen, a calculation using the one-body charge operator $\rho_{[1]}$ and Siegert's theorem yields substantially larger values for G_{E2} as pointed out in Ref.[4]. We understand now why this is so. Most of the $\rho_{[1]}$ contribution to G_{E2} comes from the two-body currents $\mathbf{J}_{[2]}$ (see Table 3.1 and Ref.[6] for further explanation). This finding suggests a different interpretation of the deformation of the nucleon in the quark model. However, there is still a large discrepancy with experiment when only $\rho_{[1]}$ is taken into account. We have shown that a double spin flip transition based on the two-body charge operator $\rho_{[2]}$ gives values for G_{E2} in agreement with experiment [6]. This is explained in the following.

The Coulomb quadrupole operator entering Eq.(5) is:

$$T^{C2}(|\mathbf{q}|) = \frac{1}{4\pi} \int d\Omega_q \, \rho(\mathbf{q}) \, Y_0^2(\hat{\mathbf{q}}), \qquad (6)$$

where $\rho = \rho_{[1]} + \rho_{[2]}$ is the total charge operator. The quadrupole operator projects onto the $Y^2(\hat{\mathbf{q}})$, i.e., the quadrupole component in $\rho(\mathbf{q})$. We discuss the quadrupole components in $\rho_{[1]}$ and $\rho_{[2]}$ separately. After a multipole expansion of the spin-independent one-body quark charge $\rho_{[1]}$ we see that

$$T^{C2}(\rho_{[1]}) \propto Y^2(\hat{\mathbf{r}}_i), \qquad (7)$$

where $\mathbf{r}_i$ is a single-quark position coordinate. The operator in Eq.(7) has non-vanishing matrix elements between the wave functions of Eq.(1) only for the off-diagonal $S \rightarrow D$, $D \rightarrow S$, and the diagonal $D \rightarrow D$ transitions.

On the other hand, the two-body gluon and pion exchange charge densities $\rho_{[2]}$ contain, just like the corresponding potentials, a tensor in spin space, and the quadrupole operator is

$$T^{C2}(\rho_{[2]}) \propto [\boldsymbol{\sigma}_i^1 \times \boldsymbol{\sigma}_j^1]^2. \qquad (8)$$

138

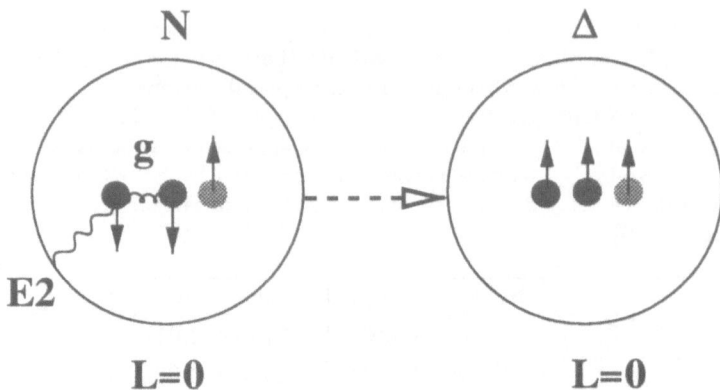

Figure 2. Double spin flip due to, e.g., the *two-body* gluon exchange charge operator $\rho_{[2]}^{g q \bar{q}}$. The probability amplitude for this $S \rightarrow S$ transition is proportional to $a_S b_S \sim 1$, and thus considerably larger than the single-quark $E2$ transition amplitude in Eq. (2), which is proportional to the small D wave admixtures a_D and b_D. The quadrupole strength G_{E2} based on the double spin flip mechanism is in quantitative agreement with experiment [1].

Consequently, the operator in Eq.(8) has a nonvanishing matrix element also for an $S \rightarrow S$ transition. We stress that unlike the single-quark operator in Eq.(7), the two-body quadrupole operator in Eq.(8) does not change the angular momentum of the wave function. However, as a tensor of rank 2 in spin space it simultaneneously flips the spin of two quarks. The probability amplitude for this double spin flip transition is proportional to $a_S b_S \sim 1$, i.e., two orders of magnitude larger than the orbital angular momentum changing one-body transition of Eq.(2). Morpurgo [9] has anticipated the important role of the operator in Eq.(8) for the $\gamma p \rightarrow \Delta^+$ quadrupole transition.

The double spin flip transition due to $\rho_{[2]}$ also affects the quadrupole moment of the deuteron. However, there it contributes only a few percent, while the major contribution comes from the D-wave motion of the nucleons. The interchanged role of the double spin flip mechanism and D-waves in the deuteron as compared to the nucleon is a consequence of the very different sizes of these systems.

The inclusion of two-body exchange currents leads to a heretofore unknown relation between the mean square charge radii of the N and Δ, and the $N \rightarrow \Delta$ transition quadrupole moment [6]:

$$Q_{p \rightarrow \Delta^+} = \frac{1}{\sqrt{2}}(r_p^2 - r_{\Delta^+}^2) = \frac{1}{\sqrt{2}}r_n^2. \tag{9}$$

One should note that no model parameter appears in Eq.(9). Using the empirical charge radius of the neutron, and Siegert's theorem, we obtain using both D waves *and* exchange currents [6]:

$$G_{E2}(0) = G_{E2}(\rho_{[1]}) + G_{E2}(\rho_{[2]})$$

$$= \quad 0.017 + 0.107 = 0.124. \tag{10}$$

This has to be compared to the experimental results which lie between $G_{E2}^{exp} = 0.133(20)$ and $G_{E2}^{exp} = 0.107(17)$ [1]. We also mention that the Z-diagrams in Fig.1 not only explain G_{E2} but also the charge radius of the neutron [8].

5 Summary

We have shown that the most important part of the $E2$ transition strength in $\gamma N \to \Delta$ comes from a two-body spin flip due to exchange currents. The effect of one-body currents on G_{E2} is small. Exchange currents also dominate the quadrupole moment of the $\Delta(1232)$ [6], and are very important for the radiative $E2$ decays of all decuplet baryons [10]. Thus, one has to conclude that the intrinsic deformation of baryons lies mainly in the nonvalence quark degrees of freedom ($q\bar{q}$ pairs, pions, gluons), and not in deformed single-quark orbits (D-waves) as was heretofore assumed. Our prediction of the $E2$ transition strength, which is based on the parameter-independent relation of Eq. (9) and Siegert's theorem, is in very good agreement with experiment.

References

1. R. Beck et al.: Phys. Rev. Lett. **78**, 606 (1997); G. Blanpied et al.: Phys. Rev. Lett. **79**, 4337 (1997); F. Kalleicher et al., Z. Phys. **A359**, 201 (1997); A. M. Bernstein, S. Nozawa, and M. A. Moinester, Phys. Rev. **C47**, 1274 (1993)

2. N. Isgur, G. Karl, and R. Koniuk: Phys. Rev. **D25**, 2394 (1982)

3. S.S. Gershtein and G.V. Dzhikiya: Sov. J. Nuc. Phys. **34**, 870 (1982); J. Dey and M. Dey: Phys. Lett. **B138**, 200 (1984)

4. D. Drechsel and M.M. Giannini: Phys. Lett. **B143**, 329 (1984)

5. M. Bourdeau and N.C. Mukhopadhyay: Phys. Rev. Lett. **58**, 976 (1987); S. Capstick and G. Karl: Phys. Rev. **D41**, 2767 (1990)

6. A.J. Buchmann, E. Hernández, and Amand Faessler: Phys. Rev. **C55**, 448 (1997); A.J. Buchmann, E. Hernández, U. Meyer and Amand Faessler: Phys. Rev. **C58**, 2478 (1998)

7. M. Weyrauch and H. J. Weber: Phys. Lett. **B171**, 13 (1986)

8. A. Buchmann, E. Hernández, and K. Yazaki: Phys. Lett. **B269**, 35 (1991); Nucl. Phys. **A569**, 661 (1994)

9. G. Morpurgo: Phys. Rev. **D40**, 2997 (1989)

10. Georg Wagner, A.J. Buchmann, Amand Faessler: Phys. Rev. **C58**, 1745 (1998)

Few-Body Systems Suppl. 11, 140–143 (1999)

Few-
Body
Systems
© by Springer-Verlag 1999

The N-Δ(1232) Electromagnetic Transition

R.M. Davidson

Department of Physics, Applied Physics and Astronomy
Rensselaer Polytechnic Institute, Troy, NY, 12180, U.S.A.

Abstract. An effective Lagrangian approach is used to analyze the available exclusive pion electroproduction data in the Δ(1232) resonance region. As expected from studies of inclusive measurements, G_M^* for the N-Δ electromagnetic transition is falling faster than a dipole up to (at least) a Q^2 of 4 GeV^2. At the highest available Q^2, the ratio E2/M1 is found to be small and negative, indicating that the pQCD domain has not yet been reached for this reaction.

Although the N-Δ electromagnetic transition has been studied for decades, it remains a topical subject as was evidence by this workshop. Recent pertinent experiments have been done at Mainz [1] and BNL [2] at the real photon point, and at JLab [3] for nonzero Q^2. At the same time, theorists [4] have made progress in understanding the degree of model dependence inherent in the extraction of the transition amplitudes from the data, and serious efforts are underway [5] to solve the long-standing problem of why the magnetic dipole transition amplitude, M1, predicted in quark models is about 30% lower than the extracted value.

The N-Δ transition form factors are expected to have a dramatic Q^2 dependence. At the real photon point, the ratio of E2/M1 is about -3% [4, 6], while at some unknown large Q^2, this ratio should approach the pQCD value of +100%, that is, E2 and M1 become equal. Apart from the interest in the value of Q^2 where pQCD is applicable for this reaction, one is also interested in the asymptotic values of M1 and E2. In addition, one would like to know if there is any interesting structure between a Q^2 of zero and infinity. For example, analyses of the older data based on M1 dominance suggest that the E2/M1 ratio becomes positive at about 0.2 GeV^2 and negative again at about 0.7 GeV^2. If this behavior is confirmed, for example by experiments at Mainz and Bates, then it will be very difficult to explain in quark models.

Before discussing the model used in this analysis, it is first useful to give a precise definition of the transition amplitudes and to point out why, at present, a model is needed to extract these amplitudes from the data. Recall that the amplitude for πN elastic scattering has a pole at s = M^2, M the nucleon mass, with a residue proportional to g^2, where g is the *physical* πNN coupling constant. Therefore, it is reasonable to associate residues of poles with physical couplings. In the case of Δ excitation, there are two poles of interest; the K- and T-matrix poles. In this case, the K-matrix has the slight advantage that the pole occurs at a physical energy, and therefore no extrapolation into the complex W plane is needed. Thus, in terms of the K-matrix residues, one obtains [7]

$$\text{M1} = \text{Im} M_{1+}^3 \sqrt{\frac{8\pi M_\Delta q \Gamma_\Delta}{3 M K_C}} \,, \tag{1}$$

with similar expressions holding for E2 and C2. In (1), M_Δ is the Δ mass, Γ_Δ is the Δ width, $K_C = (M_\Delta^2 - M^2)/(2M_\Delta)$ and q is the pion three-momentum in the cm frame evaluated at $W = M_\Delta$.

One sees in (1) that if the isospin 3/2 multipole M_{1+}^3 can be determined model-independently, then M1 can also be determined model-independently. In fact, it is in principle possible to determine M_{1+}^3 model-independently, but to do that one would need a complete set of experiments [8] for both $p\pi^0$ and $n\pi^+$ with sufficient angular coverage to accurately project out this multipole from the Walker helicity amplitudes, H_i. As a complete set of experiments does not yet exist, one is forced to model the energy and angular dependence of the production amplitude. Fortunately, as recent work shows [4], at the real photon point this model dependence seems to be well under control. For electroproduction, the situation is worse since there are practically no polarization data and at most Q^2 there are only $p\pi^0$ data so that one must also model the isospin structure of the production amplitude. Additional model dependence arises if one tries to "undress" the M1 amplitude.

The model used in this analysis is the effective Lagrangian model of Davidson, Mukhopadhyay, and Wittman (DMW) [9] which has been extended to nonzero Q^2. The model consists of the pseudovector nucleon Born terms and $s-$ and $u-$channel Δ exchanges. The role of ρ and ω exchanges were also investigated, but found to be small. For neutral pion production, the nucleon Born terms pick up the Dirac and Pauli form factors, F_1 and F_2, and the amplitude is gauge invariant. For charged pion production, one also needs the pion electromagnetic form factor, F_π, and the form factor $F_A(Q^2) = g_A(Q^2)/g_A(0)$, g_A being the axial-vector form factor. Based on current-algebra and PCAC, this latter form factor multiplies the so-called "seagull" diagram. In this case, there are problems with gauge invariance, which we overcome with the simple prescription of adding terms proportional to $K \cdot \epsilon$, K being the four-momentum of the virtual photon and ϵ its polarization. There is no deep theoretical justification for this procedure, but the resulting amplitude is gauge invariant, has the correct pole structure, and obeys the low-energy theorems to order μ/M, μ being the pion mass. Any other prescription used to restore gauge invariance must have the same pole structure as our amplitude, in other words, the various prescriptions agree up to a priori unknown gauge invariant "contact" terms. In fact, due to our off-shell parameters, our amplitude does contain such terms of a priori unknown strength.

Although no free parameters appear in the nucleon Born sector, there are unknown parameters describing the Δ contribution. Obviously, there are three parameters describing the transition amplitudes M1, E2 and C2. Since we treat the Δ as a relativistic spin 3/2 particle, there is an "off-shell" parameter associated with each of the electromagnetic three-point vertices [10]. In addition, there is an off-shell parameter coming from the $\pi N\Delta$ vertex. As the mass and width of the Δ are taken from analyses of elastic πN scattering, the model contains a total of five parameters for photoproduction and seven parameters for electroproduction, which are determined by fits to the data. Note that no *a priori* assumptions are made about the values of these parameters, e.g., M1 dominance is not assumed. The s-channel Δ contribution is pure isospin 3/2 while the u-channel contribution is both isospin 3/2 and 1/2. However, the parameters that describe the isospin 3/2 sector also describe the isospin 1/2 sector, and thus the isospin structure of the amplitude is constrained.

To satisfy unitarity, the s and p wave multipoles are projected out and unitarized via a K-matrix approach. After the s and p waves are unitarized (the d and higher l waves have very small phases in this energy region), the CGLN $\mathcal{F}$'s are reconstructed by adding in the unitarized multipoles and subtracting off the tree level s and p wave multipoles (to avoid double counting). In this manner, all multipoles are kept in the model and the pole structure of the amplitude is maintained.

The model described above has been used to analyze the available exclusive electroproduction data in the $\Delta(1232)$ region up to $Q^2 = 4.0$ GeV^2. The chi-squared per degree of freedom are between 1.5 and 2.5. The results for G_M^* tend to be slightly smaller than those obtained from analyses of inclusive data. This is not too surprising since from inclusive data it is not possible to separate out G_M^*, G_E^* and G_C^*. The earlier analyses assumed that it was pure G_M^*. Starting at about 0.2 GeV^2, $G_M^*(Q^2)$ is well described by the function

$$G_M^*(Q^2) = \frac{2.53 \left(1 + \frac{Q^2}{(3M_\Delta+M)(M_\Delta+M)}\right)}{\left(1 + \frac{Q^2}{\Lambda^2}\right)^3} ,\tag{2}$$

with $\Lambda = 1.08$ GeV. For the E2/M1 ratio, this analysis of the older data also indicates sign changes around 0.2 and 0.7 GeV^2. However, the systematics have not been taken into account, so no firm conclusion about these sign changes can be made. The magnitude of E2/M1 remains small, roughly less than 5%, up to 4 GeV^2. The C2/M1 ratio is fairly constant up to 4 GeV^2 with a value of about -10%.

Of particular interest are the results from the recent JLab experiment [3] at 2.8 and 4.0 GeV^2. For many exclusive reactions, for example, the proton elastic form factor, the pQCD counting rules seem to hold at a Q^2 as small as 4 GeV^2. For the $\gamma N\Delta$ transition, this turns out not to be the case. At 2.8 GeV^2, the results of this analysis are M1 = 46.5$\pm$ 0.6 $\pm$ 0.6 $\times$ 10^{-3} GeV$^{-1/2}$, E2/M1 = -(2.1 $\pm$ 1.0 $\pm$ 0.3)% and C2/M1 = -(10.2 $\pm$ 1.3 $\pm$ 0.3)%, where the first error is statistical and the second is systematic. The χ^2 per degree of freedom is 1.6. At 4 GeV^2 the results are M1 = 23.3 $\pm$ 0.4 $\pm$ 0.3 $\times$ 10^{-3} GeV$^{-1/2}$, E2/M1 = -(3.3 $\pm$ 1.2 $\pm$ 0.1)%, and C2/M1 = -(14.9 $\pm$ 1.9 $\pm$ 0.2)% , with a χ^2 per degree of freedom of 1.45. Results based on M1 dominance give similar results [3]. Thus, for this reaction, pQCD is clearly not directly applicable at 4 GeV^2.

I would like to thank the organizers of this workshop for their invitation and for their efforts in making it successful. I am indebted to V. Frolov and P. Stoler for making the E94-014 data available to me, and to N.C. Mukhopadhyay for many stimulating discussions. This work was supported by the US DOE grant number DE-FG02-88ER40448.

References

1. R. Beck et al.: Phys. Rev. Lett. **78**, 606 (1997); R. Beck and H. P. Krahn, *ibid*, **79**, 4510 (1997)

2. G. Blanpied et al.: Phys. Rev. Lett. **69**, 1880 (1992); *ibid* **79**, 4337 (1997)

3. V. Frolov, Ph.D. thesis, RPI (1998); V. Frolov et al.: in preparation.

4. R.M. Davidson et al.: submitted to PRL (1998).

5. See Harry Lee's contribution to this workshop.

6. R.M. Davidson and N.C. Mukhopadhyay: Phys. Rev. Lett. **79**, 4509 (1997)

7. R.M. Davidson and N.C. Mukhopadhyay: Phys. Rev. **D42**, 1546 (1990)

8. For a general discussion of amplitude analysis, see Ron Workman's contribution to this workshop.

9. R.M. Davidson, N.C. Mukhopadhyay and R.S. Wittman, Phys. Rev. **D43**, 71 (1991).

10. M. Benmerrouche, R.M. Davidson and N.C. Mukhopadhyay: Phys. Rev. **C39**, 2339 (1989)

Few-Body Systems Suppl. 11, 144–156 (1999)

Few-
Body
Systems
© by Springer-Verlag 1999

Multipole Analyses for p(γ, π) and p(γ, γ) in the Region of the P_{33} Δ Resonance

A.M. Sandorfi[1], S. Hoblit[1] and J. Tonnison[2,1]

[1] Physics Dept., Brookhaven National Lab, Upton, NY 11973
[2] Physics Dept., VPI & SU, Blacksburg, VA 24061

Abstract. Multipole analyses of the p(γ, π°), p(γ, π^+) and p(γ, γ) reactions are carried out using different data sets. With sufficient constraints from polarization observables, the ratio of E2/M1 transition amplitudes for N $\to$ Δ (EMR) appears to be largely insensitive to differences between recent p(γ, π°) cross section measurements. We deduce a current best estimate of $EMR = -(2.85 \pm 0.34 \pm 0.21)\%$. Back angle Compton cross sections require a value for the *backward spin polarizability* δ_π that is significantly lower than previous expectations, with a magnitude that is coupled to the (γ, π) cross sections.

Elastic photon (Compton) scattering and pion photo-production in the energy region of the P_{33} $\Delta(1232)$ resonance are both rich sources of nucleon structure information. The proton's first order scattering response is fixed by its static properties of mass, charge, magnetic moment and spin. The leading corrections to this *point* scattering come from the dynamic rearrangement of constituent charges and spins within the proton, and are expressed in terms of six *polarizability* parameters [1, 2]. These fundamental properties of the proton can be compared to QCD through, for example, the calculational techniques of Chiral perturbation theory (χPT) [3, 4, 5].

Although the lifetime of the $\Delta(1232)$ precludes scattering measurements, the N $\to$ Δ transition amplitudes carry structure information. While this transition is dominantly M1 quark spin-flip, a small E2 component is expected from interactions with pions (either in a cloud surrounding the proton [6, 7, 8], or as $q\bar{q}$ exchange currents between constituent quarks [9]). Since nucleon models differ greatly on the mechanisms used to generate these components, the E2 and M1 transition amplitudes provide another sensitive testing ground.

Compton scattering, pion photo-production, and pion-nucleon scattering are related by unitarity through a common *S-matrix*. Below 2π threshold, E_γ = 309 MeV lab, Watson's theorem requires the (γ, π) and (π, π) channels to have a common phase [10], and *K-matrix* theory can be used to provide a consistent, albeit model dependent, extension of this unitarity relation to higher energies [11]. Once the (γ, π) multipoles are specified, the imaginary parts of the Compton amplitudes are completely determined by unitarity and a dispersion calculation involving integrals of the pion multipoles can be used to generate their real parts with the only unknowns being the nucleon polarizabilities [12, 13].

At any given energy, a minimum of 8 independent observables (for each pion charge state) are necessary to specify the 4 photo-pion helicity amplitudes [14]. Such complete information has never been available and most analyses have relied almost exclusively on only four, the cross section and the three single polarization asymmetries, Σ (linearly polarized beam), T (target) and P (recoil nucleon). Very recently, we have used Compton scattering to provide both two new constraints on the photo-pion multipoles as well as information on the proton polarizabilities [17, 13]. Specific multipoles such as the very interesting isospin $\tau = 3/2$ M1 and E2 components can be extracted from fits to a multipole expansion of the amplitude. But since such expansions must necessarily be truncated at some point, constraints from many observables are needed to avoid Donnachie's ambiguity of higher partial wave strength appearing in lower partial waves, and vice versa [15].

A new experiment at LEGS has reported cross sections and linear beam polarization asymmetries for the $p(\gamma, \gamma)$, $p(\gamma, \pi^{\circ})$ and $p(\gamma, \pi^+)$ reactions [16, 17]. Recent experiments at Mainz and at Bonn have also reported results on Compton scattering and π-production [18, 19, 20, 21]. At energies below the Δ (for E_{γ} less than about 270 MeV) the results from the three labs are in substantial agreement for all three channels. However, while the Mainz Compton cross sections are in quite good agreement with LEGS results at all overlapping energies, the LEGS π° cross sections rise above those from Mainz in the vicinity of the Δ and are about 10% higher at the resonance peak. (The LEGS π^+ cross sections also tend to be slightly higher than those from Mainz and Bonn, but the differences in this channel are not as pronounced.)

In this paper, we examine how the cross section differences among recent π-production data sets influence the multipole decomposition of the pion amplitude, as well as the extraction of the polarizabilities that rely on these multipoles for the computation of dispersion integrals.

The problem with the pion cross sections is illustrated in Fig. 1 where we plot the $p(\gamma, \gamma)$ and $p(\gamma, \pi^{\circ})$ results from LEGS and Mainz at 90°center of mass (c.m.). The error bars on the Mainz points are purely statistical. Most of the systematic effects are angle and energy dependent, and for the LEGS results these have been evaluated point by point and have been combined with the statistical error to produce the net uncertainty bars. The residual systematic scale uncertainties associated with the two measurements ($\pm\sigma_{sys}$) are indicated by bands (solid for LEGS and dotted for Mainz). Considering these, the net accuracy of the two experiments is comparable. Because the LEGS $p(\gamma, \gamma)$ and $p(\gamma, \pi^{\circ})$ measurements were made simultaneously, in fact in the same detector, there is no possibility of independent normalizations which could improve the agreement in the π° channel without destroying the agreement in the Compton channel.

146

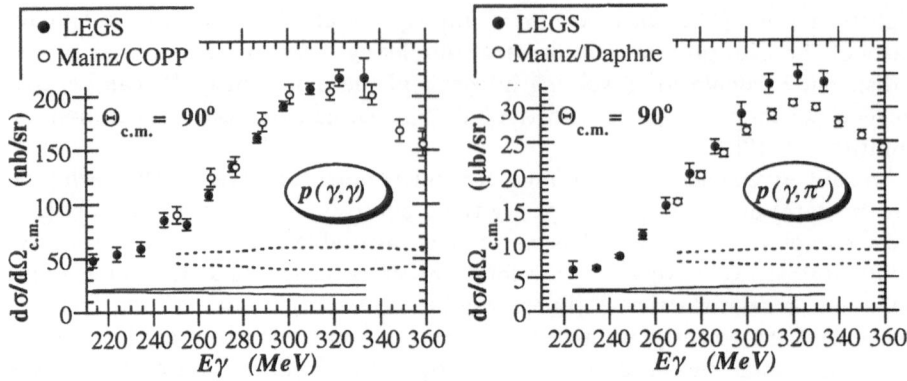

Figure 1. Compton cross sections from LEGS [17] and Mainz [18] are compared in the left panel, while π° production cross sections from [17] and [20] are shown in the right panel. The width of the solid and dotted bands at the bottom of the figures indicate the systematic scale uncertainties for LEGS (solid) and Mainz (dotted).

The experimental agreement is better for the beam asymmetries, as shown in Fig. 2. The plotted errors include all statistical and *polarization-dependent* systematic uncertainties, with polarization-independent systematic errors canceling out of these ratios.

To understand the extent to which these data constrain the photo-pion multipoles, we have performed a series of energy-dependent analyses, expanding the π-production amplitude into electric and magnetic partial waves, $E^\tau_{\ell\pm}$ and $M^\tau_{\ell\pm}$ with relative πN angular momentum ℓ, and intermediate-state spin $j = \ell\pm 1/2$ and isospin $\tau = 1/2$ or $3/2$. The (γ, π) multipoles have been parameterized with a K-matrix-like unitarization of the form,

$$A^\tau_{\ell\pm} = \{\text{Born}_{s,u} + \text{Born}(\rho/\omega)_t + \mathcal{P}(\alpha \cdot \varepsilon_\pi\} \left(1 + iT^\ell_{\pi N}\right) + \beta \cdot T^\ell_{\pi N}. \qquad (1)$$

In addition to the $s-$ and $u-$channel Born terms, and $t-$channel ρ and ω exchange, a low-order polynomial $\mathcal{P}$ in the pion energy ε_π has been included to allow for other possible terms that are expected from contributions such as $u-$channel resonance graphs and pion rescattering [17]. The VPI[SM95] values have been used for the πN scattering T-matrix elements [23]. Below 2π threshold (309 MeV) these reduce to $\sin(\delta_\ell)e^{i\delta_\ell}$, $\delta_\ell(E_\gamma)$ being the elastic πN phase shift. When a single $s-$channel resonance dominates a partial wave having only one open decay channel the last term in Eq. (1) exactly reduces to a Breit-Wigner energy dependence.

Once the (γ, π) multipoles are fixed by the choice of the α and β parameters in Eq. (1) the imaginary parts of the six Compton helicity amplitudes are completely determined by unitarity, and dispersion integrals can be used to calculate their real parts. For the latter, we have followed the theory of L'vov

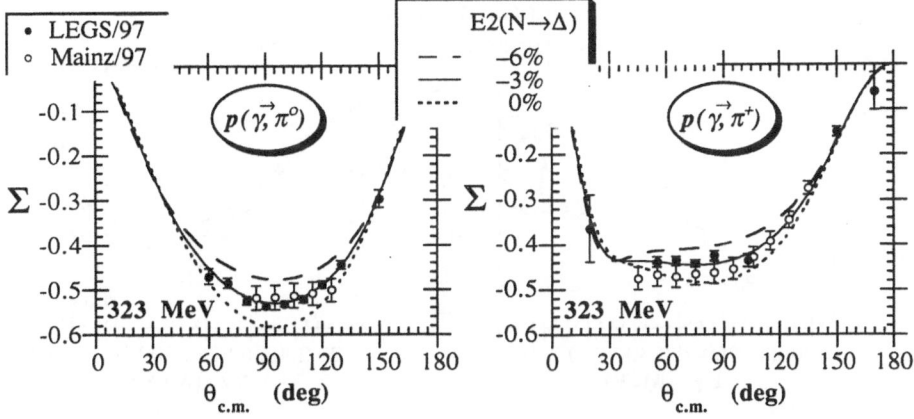

Figure 2. Linear polarization beam asymmetries for p(γ, π) as measured at LEGS [17] and Mainz [20, 22]. Error bars reflect combined statistical and polarization-dependent systematic uncertainties. Curves are predictions with LEGS multipoles for different values of $\beta(E_{1+}^{3/2})$ in Eq. (1).

[12], writing the real part of the scattering amplitude as

$$\mathrm{Re}A_i(v,t) = A_i^B(v,t) + \frac{2}{\pi}P\int_{v_0}^{v_{max}} \frac{v'\mathrm{Im}A_i(v',t)}{v'^2 - v^2}dv' + A_i^{as}(t), \qquad (2)$$

where $v = \frac{1}{4M}(s-u)$, M is the nucleon mass, and A_i^B denotes the Born contribution. The Principal value integral in (2) is calculated from v_0 (corresponding to photopion threshold) up to a moderately high energy ($v_{max} = 1.5GeV$), and the A_i^{as} are the residual asymptotic components above v_{max}.

At energies below 2π threshold, the unitarity connection between the imaginary parts of the Compton amplitudes appearing in Eq. (2) and the photo-pion multipoles of Eq. (1) is unambiguous. As E_γ approaches 309 MeV, these single π-production contributions to ImA_i become very large, while 2π contributions are quite small below 400 MeV and at higher energies are suppressed by the energy denominator in the principle value integral of Eq. (2). As a result, there is in fact very little freedom in the scattering amplitude up to the Δ peak. This allows the Compton observables to be used as an effective constraint on the pion multipoles without incurring significant model dependent uncertainties, provided that we restrict their use to energies below the onset of appreciable $(\gamma, 2\pi)$ strength. A reasonable set of multipoles is needed to extend the computation of the integrals in Eq. (2) up to 1.5 GeV, and for this we have used VPI[SM95] [23], but the particular choice of the multipole solution used for this extension has little effect on the evaluation of the amplitudes at energies below 350 MeV. The only remaining degree of freedom in Eq. (2) lies in the A_i^{as} asymptotic components. These fix the proton polarizabilities which are determined by the $s - u = t = 0$ limits of the non-Born parts of Eq. (2) [12, 13].

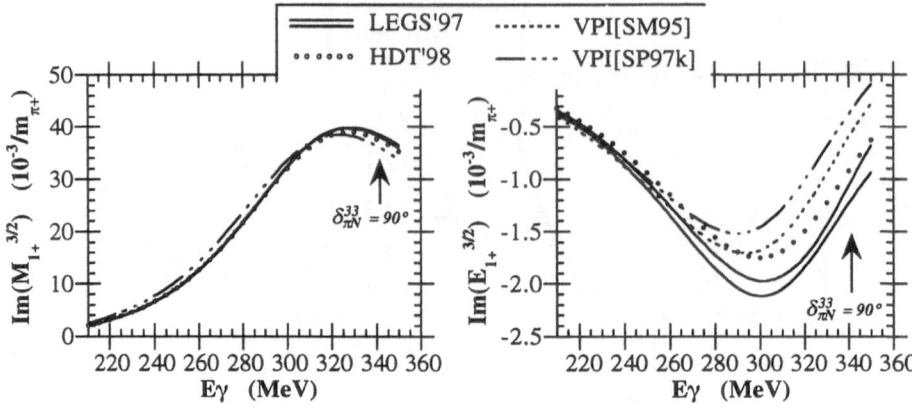

Figure 3. The multipole solution obtained using pion σ and Σ data from LEGS [17] is shown as solid curves which indicate the uncertainty band. For comparison, solutions from VPI [23] and from Mainz [35] are also shown.

With these considerations in mind, we have performed fits to the proton polarizabilities and to the pion-multipole parameters of Eq. (1), allowing non-Born contributions up to F-waves. We have minimized χ^2 for both the $p(\gamma, \gamma)$ and the (γ, π) observables using data in the energy region from 200 MeV to 350 MeV. When combining data from different experiments, relative cross section normalizations must be fitted. (To neglect this would ignore the systematic uncertainties that are present in every experiment and thus assume an unphysical level of accuracy.) We have followed the procedure of [24], multiplying all data from a set with a systematic scale error (σ_{sys}) by a common factor (f) while adding $(f-1)^2/\sigma_{sys}^2$ to the χ^2. The latter term weights the penalty for choosing a normalization scale different from unity by the systematic uncertainty of the measurement.

There is good agreement among all modern Compton data and we have included in the multipole fits all data below 350 MeV from [17, 18, 19, 25, 26, 27, 28, 29]. For π-production, in addition to $p(\gamma, \pi^\circ)$ and (γ, π^+) cross sections and beam asymmetries, we have included in these analyses T data from [30, 31, 32], P data from [31, 32], and the small amount of G and H beam-target double-polarization data available from [33].

Since the (γ, π) cross section differences evident in Fig. 1 are energy dependent, they cannot be reconciled with a simple shift in normalization scales. Combining them all in one multipole analysis would produce an average result that would correspond to neither data set. Instead, we have performed successive analyses using π-production cross sections from either LEGS, or Mainz and Bonn, but not both simultaneously.

The two most interesting results from these analysis are the M1 and E2 $N \to \Delta$ transition amplitudes and the proton polarizabilities. These quantities turn out to be fairly decoupled, and we discuss each in turn.

Table 1. Evolution of the N $\rightarrow$ Δ EMR, for the interval ($200 \leq E_\gamma \leq 350$), starting with fits to the (γ, π) cross sections and beam asymmetries from LEGS, and expanding the data base in subsequent rows by adding data on other observables as indicated. The number of partial waves with fitted non-Born components is increased in successive columns to the right.

Data included successively	EMR (%) $\ell_\pi = S - P$	EMR (%) $\ell_\pi = S - D$	EMR (%) $\ell_\pi = S - F$
(γ, π): $\{\sigma, \Sigma\}$[17]	$-(2.16 \pm 0.43)$	$-(4.22 \pm 1.08)$	$-(4.03 \pm 1.34)$
(γ, π): $+ \{T\}$[30] $+ \{T, P, G, H\}$[31-33]	$-(2.61 \pm 0.29)$	$-(2.74 \pm 0.28)$	$-(2.82 \pm 0.29)$
(γ, γ): $+ \{\sigma, \Sigma\}$[17] $+ \{\sigma\}$ *world*	$-(2.77 \pm 0.29)$	$-(2.90 \pm 0.28)$	$-(3.00 \pm 0.27)$

The multipole solution obtained by taking (γ, π) cross sections and beam asymmetries from the LEGS data [34, 17] yields a reduced χ^2 of $1093/(734 - 36) = 1.57$. (This is the solution corresponding to row 3 of Table 1 in [13].) Fitted normalization scales are all within about one standard deviation of the systematic scale uncertainties associated with the various data sets included. The imaginary parts of the resulting $\tau = 3/2$ M_{1+} and E_{1+} photo-pion multipoles are shown in Fig. 3 as pairs of solid lines denoting the uncertainty band. This is compared with two recent solutions from VPI and one from Mainz, labeled HDT [35]. The LEGS, VPI[SM95] and HDT solutions are all in agreement for the M_{1+}, but vary significantly in the small E_{1+} multipole. The energy at which the P_{33} phase passes through 90°is indicated. There, the LEGS and HDT solutions are fairly close. The EMR for N $\rightarrow$ Δ is just the ratio of the fitted β coefficients in Eq. (1) for the $E_{1+}^{3/2}$ and $M_{1+}^{3/2}$ multipoles, $-(3.00 \pm 0.27)\%$.

In a remarkably thorough but rarely quoted review article, Donnachie has pointed out potential ambiguities that can occur when truncating a multipole expansion [15]. These can only be mitigated by the use of many independent observables as constraints. This is illustrated in Table 1 which shows the evolution of the fitted EMR to its final value. The number of partial waves with fitted non-Born contributions increases to the right in the columns while the number of observables is increased in successive rows. If only LEGS (γ, π) cross sections and beam asymmetries [17] are used as constraints (row 1), the result is unstable and strongly depends on the number of partial waves included in the fit. But as soon as additional (γ, π) polarization asymmetries are added (row 2), the extracted EMR value stabilizes. Further addition of the Compton observables has only small effects (row 3). It is by now well known that the asymmetry in ($\gamma, \pi°$) is particularly sensitive to the N $\rightarrow$ Δ EMR [34, 20]. The predictions corresponding to the -3.00% solution from the final analysis of Table 1 (the lower right-hand entry) are shown in Fig. 2 as solid curves. Setting

$\beta(E_{1+}^{3/2})$ to 0 or -6% in Eq. (1) gives the dotted and dashed curves, respectively. Despite the sizeable separation between these curves, it should be clear from the exercises in Table 1 that this observable alone is insufficient to guarantee an EMR that is free from ambiguities.

We have also tracked the evolution of a multipole solution starting with the Mainz (γ, π) cross sections and beam asymmetries [20, 22]. This is shown in Table 2. When only σ and Σ observables are fit (row 1), the resulting EMR again varies with the number of fitted partial waves. But when additional polarization asymmetries are included in the fit (row 2), the EMR value stabilizes. This is the same phenomenon encountered in Table 1.

It should be noted that all of these analyses include both (γ, π°) and (γ, π^+) data, and so are different from the treatment of [20] which relies on only the (γ, π°) channel. The EMR values in row 2 of Table 2 are smaller than the result of [20], and a contributing factor to this difference is our inclusion of (γ, π^+) beam asymmetry data from [22]. The centroid values of the Mainz (γ, π^+) beam asymmetries tend to be more negative than the corresponding data from LEGS. As illustrated with the calculations in Fig. 2, this favors a smaller EMR. The LEGS and Mainz beam asymmetry data are in experimental agreement (error bars from the two measurements always at least touch), so it is appropriate to include the LEGS beam asymmetry data into this analysis. When this is done, row 3 of Table 2, the resulting EMR value doubles. This is simply because the errors on the LEGS asymmetry data are considerably smaller than those from the Mainz measurements and thus dominate the χ^2 fit. The further addition of Compton data, row 4, produces only small alterations, although this is achieved in the fit with polarizabilities that are different from the solution of Table 1, row 3. (This is discussed further below.)

The third row of Table 1 and the forth row of Table 2 essentially agree, so that at this point it would appear that the final $N \to \Delta$ EMR is sufficiently constrained by the polarization asymmetry observables so as to be independent of the (γ, π) cross section problems of Fig. 1. Although this would be a highly desirable conclusion, there is still one complication. The Mainz data of [20, 22] were restricted to the $(45° \le \theta \le 135°)$ central angular range. In this range, these data agree with earlier measurements from Bonn [36, 37] that covered a much wider angular range $(10° \le \theta \le 180°)$. If all of these Bonn data are also included in the fit (Table 2, row 5), the resulting EMR drops by a factor of two.

The angular dependence of the differential cross section is influenced by all multipoles. The cross sections at extreme angles are particularly sensitive to interfering multipoles of opposite parity and can cause a significant rearrangement of multipole strength. This, and the large number of data points from Bonn which can overwhelm a χ^2 fit, result in the substantial EMR shifts appearing in the last row of Table 2. On the other hand, while acceptances and efficiencies are always angle dependent, few experiments report the angular dependence of the systematic uncertainty and none have provided the correlation of this error with angle. Angle-independent systematic errors allow adjustments of the over all scale, but not the shape of angular distributions, and it is the latter that affects the multipole decomposition. For that reason, we prefer to assign the Bonn data a much lower weight in the analyses. With that philosophy, we take the current best estimate of the N $\rightarrow \Delta$ EMR as the mean of the $S - F$ results of Table 1, row 3, and Table 2, row 4. The uncertainties reported in the tables are unbiased estimates [38] of the fitting errors that combine *statistical* and *systematic* scale uncertainties. Additional *model*-dependent errors associated with the multipole analyses have been calculated at $\pm 0.21\%$ [17]. Thus, we take the current best estimate for the EMR,

$$EMR = -(2.85 \pm 0.34 \pm 0.21)\%.\qquad(3)$$

Unfortunately, no such mean result can be derived for the individual M1 and E2 transition amplitudes. These depend on the (γ, π) cross sections and must await a resolution to the problems evident in Fig. 1.

Table 2. Evolution of the N $\rightarrow \Delta$ EMR, for the interval $(200 \leq E_\gamma \leq 350)$, starting with fits to the (γ, π) cross sections and beam asymmetries from Mainz, and expanding the data base in subsequent rows by adding data on other observables as indicated. The number of partial waves with fitted non-Born components is increased in successive columns to the right.

Data included successively	EMR (%) $\ell_\pi = S - P$	EMR (%) $\ell_\pi = S - D$	EMR (%) $\ell_\pi = S - F$
(γ, π): $\{\sigma, \Sigma\}$[20]	$-(1.8 \pm 0.5)$	$-(3.4 \pm 0.8)$	$-(2.2 \pm 0.8)$
(γ, π): $+ \{T\}$[30] $+ \{T, P, G, H\}$[31, 32, 33]	$-(1.7 \pm 0.4)$	$-(1.5 \pm 0.4)$	$-(1.4 \pm 0.5)$
(γ, π): $+ \{\Sigma\}$[17]	$-(3.0 \pm 0.2)$	$-(2.9 \pm 0.2)$	$-(2.8 \pm 0.2)$
(γ, γ): $+ \{\sigma, \Sigma\}$[17]	$-(2.8 \pm 0.2)$	$-(2.7 \pm 0.2)$	$-(2.7 \pm 0.2)$
(γ, π): $+ \{\sigma\}$[37, 36]	$-(1.4 \pm 0.1)$	$-(1.5 \pm 0.1)$	$-(1.3 \pm 0.1)$

We return now to the discussion of the proton polarizabilities that are extracted from our analyses when Compton data are included in the χ^2 minimization, as in the solutions of Table 1, row 3, and Table 2, rows 4 and 5. While four of the six Compton amplitudes of Eq. (2) converge rapidly with energy, the two associated with 180° photon helicity-flip (A_1 and A_2) can have appreciable asymptotic parts. In earlier analyses of data below single π-production threshold [12, 25, 26, 27], t-channel π^0-exchange was assumed to completely dominate A_2^{as}, which is then evaluated in terms of the $F_{\pi^0 \gamma\gamma}$ coupling. This *ansatz* left only A_1^{as} to be varied in a fit to data. This determined the difference of the electric and magnetic dipole polarizabilities, $\bar{\alpha} - \bar{\beta}$, since the $s - u = t = 0$ limit of the A_1 amplitude is just,

$$\bar{\alpha} - \bar{\beta} = -\frac{1}{2\pi} A_1^{nB}(0, 0).\qquad(4)$$

Here, the nB superscript denotes the non-Born contributions from the *integral* and *asymptotic* parts of Eq. (2). This has led to a consistent description of Compton scattering up to π-production threshold ($E_\gamma \sim 150$ MeV lab), with a global average from all data [26] of $\bar{\alpha} - \bar{\beta} = 10.0 \pm 1.5(\text{stat} + \text{sys}) \pm 0.9(\text{model})$, in units of $10^{-4} fm^3$.

Although this had been accepted as a standard treatment of Compton scattering, we have observed that as higher energy data were added to the fit, the deduced value of $\bar{\alpha} - \bar{\beta}$ dropped [13], becoming even negative when Compton data up to 2π threshold were included from LEGS and Mainz. We have recently proposed that the weak link here is the *ansatz* of no additional contributions to the asymptotic part of the A_2 amplitude beyond those from π° t-channel exchange. We have model corrections to A_2^{as} with an additional exponential t-dependent term having one free parameter, the derivative at $t = 0$. Fitting all modern Compton data, we have found that this addition restores consistency in $\bar{\alpha} - \bar{\beta}$ values deduced from all data up to 2π threshold [13].

Another consequence of adding a term to A_2^{as} is to alter the expected value for a linear combination of the proton spin polarizabilities that characterizes backward scattering. This *backward spin polarizability*, δ_π, is determined by the $s - u = t = 0$ limits of A_2 and A_5,

$$\delta_\pi = \frac{1}{2\pi M} \left[A_2^{nB}(0,0) + A_5^{nB}(0,0) \right]. \tag{5}$$

Without the new variable term in A_2^{as}, the expected value for δ_π is 36.6 (in units of $10^{-4} fm^4$). If δ_π is held to 36.6, the calculated Compton cross section always falls below the back angle data. This is shown for two beam energies as curves denoted by plus signs in Fig. 4. Allowing A_2^{as} to vary reduces δ_π and brings the back angle predictions up in agreement with data (solid curves in Fig. 4).

The fitted value of δ_π deduced from data up to 2π threshold (309 MeV) is 27.1±2.2 (stat+sys), with an additional model-dependent uncertainty of +2.8/-2.4 [13]. (The value of $\bar{\alpha} - \bar{\beta}$ from this fit is 10.11 ± 1.74, in excellent agreement with the low energy experiments.) If data up to 350 MeV are included, as in row 3 of Table 1, the deduced δ_π is 25.1 ± 2.1 [13]. These values for δ_π are appreciably different from the π°-dominated expectation.

To examine the sensitivity of the deduced value of δ_π upon the (γ, π) multipole solution we have refit the Compton data from LEGS, Mainz and SAL [17, 18, 19, 25] using the HDT multipoles from [35] and the SP97k solution from VPI [23]. The results are listed in the first row of Table 3. These two solutions were fitted to the Mainz and Bonn (γ, π) data. If δ_π is fixed to 37, the Compton predictions using either of these are lower than the plus-sign curves of Fig. 4. So a value for δ_π even lower than 27 is needed to raise the predictions up to the scattering data. This is a general feature of multipoles that are fit to the lower (γ, π) cross sections of Fig. 1. Our fits in Table 2 give 21 for δ_π with the multipoles of row 4, and 19 when the Bonn (γ, π) data are included in row 5. (The result for δ_π is almost independent of the number of partial waves, varying by at most 2 across the columns of Table 2.)

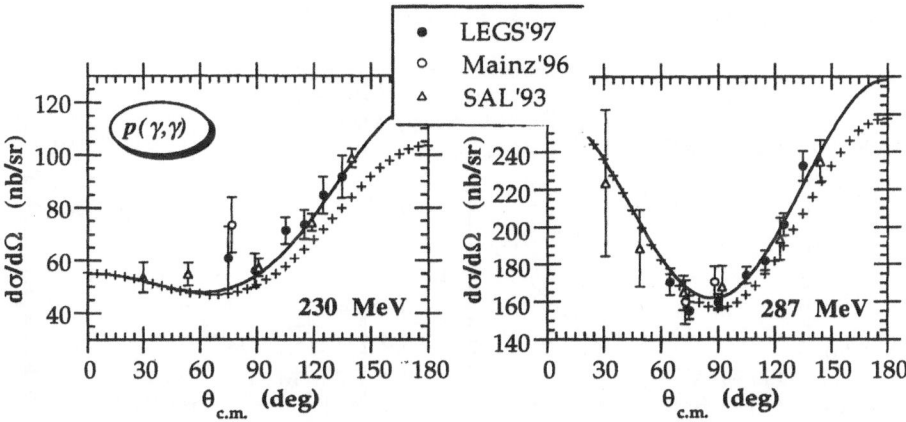

Figure 4. Predictions with LEGS multipoles from the simultaneous fits to $p(\gamma, \gamma)$ and $p(\gamma, \pi)$ are shown as solid curves and compared with recent data at two energies. For these, the extracted value for δ_π is 27. Curves denoted by plus signs used the same multipoles but held δ_π fixed at 37.

Table 3. Results of fits to Compton data up to 2π threshold [17, 18, 19, 25] using different (γ, π) multipoles from [13, 35, 23]. In all cases, $\bar{\alpha} - \bar{\beta}$ is fixed at 10 and $\bar{\alpha} + \bar{\beta}$ to 13.7 $(10^{-4} fm^3)$. For fits in the first row, the σ mass was fixed at 600 MeV and δ_π was varied. For the second row, δ_π was fixed at 37 $(10^{-4} fm^3)$, and the σ mass was varied. The χ^2/point for all fits is less than 1.4.

| | (γ, π) multipoles | | |
	LEGS '98	HDT '98	SP97K
$m_\sigma = 600$	$\delta = 27.1 \pm 2.2$	$\delta = 21.4 \pm 0.9$	$\delta = 20.9 \pm 0.8$
$\delta = 37$	$m_\sigma = 217 \pm 6$	$m_\sigma = 82 \pm 20$	$m_\sigma = 58 \pm 23$

There has been a recent suggestion [39] of a possible way to fit the Compton data while leaving the value of δ_π at its π°-dominated expectation of 37. The asymptotic part of the A1 amplitude is assumed to be dominated by t-channel σ-exchange, with σ being the correlated s-wave 2π object required in analyses of N-N scattering [12]. Since its couplings are poorly known they are simple treated as a free parameter in fitting A_1^{as}. In this procedure we have set the σ mass to 600 MeV, an average of several N-N analyses. The authors of [39] have pointed out that reducing m_σ changes the t-dependence in such a way as to raise the back angle cross section so that one might be able to reconcile predictions with data in this way while leaving δ_π fixed at 37. We have investigated this suggestion, and the results of refitting the Compton data, varying m_σ while fixing $\delta_\pi = 37$, are shown in row 2 of Table 3. Good fits can indeed be obtained in this way, but only with a value for m_σ that is substantially less than the

154

mass of two pions. This does not seem a realistic alternative.

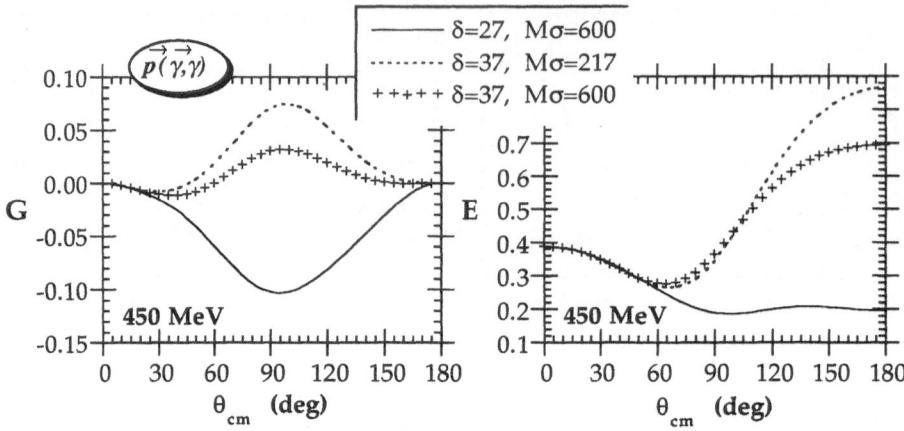

Figure 5. Predictions from the LEGS multipoles for Compton double-polarization observables, G linearly-polarized and E circularly-polarized beam, both on longitudinally-polarized proton targets.

A value of δ_π appreciably lower than 37 is difficult to accommodate within existing theories. Although χPT cannot be expected to directly predict Compton observables at the high energies included in these dispersion analyses, it should be able to reproduce the polarizabilities obtained by evaluating the fitted amplitudes at $s - u = t = 0$. Nonetheless, existing $O(\omega^3)$ calculations remain close to the π°-dominated value [5]. Since our result for δ_π would indicate some new contribution from the low-energy spin structure of the proton, it is highly desirable to verify this in some independent way. As pointed out in [39], beam-target double-polarization observables are sensitive to both δ_π and the σ mass. In Fig. 5 we plot angular distributions predicted with the LEGS multipoles for two such observables: the G-asymmetry obtained with linearly polarized beam on longitudinally polarized protons (the Σ_{1Z} observable in [39]), and the E-asymmetry from circularly polarized beam on a longitudinally polarized target (Σ_{2Z} in [39]). Checking either the $\delta_\pi = 37$, or reduced m_σ predictions should be quite straight forward, and measurements of these quantities are expected in the near future. (Unfortunately, the large sensitivities evident in Fig. 5 only occur for energies above the P_{33} resonance. Since this is now closer to the onset of appreciable $(\gamma, 2\pi)$ strength, the model dependence will increase. But one should be able to estimate this effect using data on both E and G.) Since the value of δ_π does depend upon the (γ, π) cross sections, constraining δ_π will in turn increase the effectiveness of Compton scattering as a constraint on π-production.

Acknowledgement. We are indebted to Dr. B. Preedom for his critical comments. This work was supported by the U.S. Department of Energy under contract No. DE-AC02-98CH10886, and by the National Science Foundation.

References

1. V.A. Petrunkin: Sov. J. Part. Nucl. **12**, 278 (1981)

2. S. Ragusa: Phys. Rev. **D47**, 3757 (1993)

3. V. Bernard, N. Kaiser, Ulf-G. Meißner and A. Schmidt: Z. Phys. **A348**, 317 (1994)

4. D. Babusci, G. Giordano and G. Matone: Phys. Rev. **C55**, R1645 (1997)

5. T. Hemmert, B. Holstein, J. Kambor and G. Knöchlein: Phys. Rev. **D57**, 5746 (1998)

6. A. Wirzba and W. Weise: Phys. Lett. **B188**, 6 (1987)

7. R. Bijker, F. Iachello and A. Leviatan: Ann. Phys. **236**, 69 (1994)

8. T. Sato and T.-S.H. Lee: Phys. Rev. **C54**, 2660 (1996)

9. A.J. Buchmann, E. Hernndez and A. Faessler: Phys. Rev. **C55**, 1 (1997)

10. K. Watson: Phys. Rev. **95**, 228 (1954)

11. R. Davidson, N. Mukhopadhyay, R. Wittman: Phys. Rev. **D43**, 71 (1991)

12. A.I. L'vov: Sov. J. Nucl. Phys. **34**, 597 (1981); A.I. Lvov, et al.: Phys. Rev. **C55**, 359 (1997)

13. J. Tonnison, A.M. Sandorfi, S. Hoblit and A.M. Nathan: Phys. Rev. Lett. **80**, 4382 (1998)

14. W. Chiang and F. Tabakin: Phys. Rev. **C55**, 2054 (1997)

15. A. Donnachie: Rep. Prog. Phys. **36**, 695 (1973)

16. LEGS Collaboration, G. Blanpied et al.: Phys. Rev. Lett. **76**, 1023 (1996)

17. LEGS Collaboration, G. Blanpied et al.: Phys. Rev. Lett. **79**, 4337 (1997)

18. C. Molinari et al.: Phys. Lett. **B371**, 181 (1996)

19. J. Peise et al.: Phys. Lett. **B384**, 37 (1996)

20. R. Beck et al.: Phys. Rev. Lett. **78**, 606, (1997)

21. K. Büchler, et al.: Nucl. Phys. **A570**, 580 (1994)

22. H.-P. Krahn: thesis, U. Mainz (1996); R. Beck: private comm.

23. *SAID code;* R. Arndt, I. Strakovsky and R. Workman: Phys. Rev. **C53**, 430 (1996)

156

24. G. D'Agostini: Nucl. Inst. Meth. **A346**, 306 (1994)

25. E.L. Hallin et al.: Phys. Rev. **C48**, 1497 (1993)

26. B.E. MacGibbon et al.: Phys. Rev. **C52**, 2097 (1995)

27. A. Zieger et al.: Phys. Lett. **B278**, 34 (1992)

28. P.S. Baranov et al.: Sov. J. Nucl. Phys. **21**, 355 (1975); only 90°data are included [13].

29. F.J. Federspiel et al.: Phys. Rev. Lett. **67**, 1511 (1991)

30. H. Dutz et al.: Nucl. Phys. **A601**, 319 (1996); Gisela Anton, priv. comm.

31. V.A. Get'man et al.: Nucl. Phys. **B188**, 397 (1981)

32. A. Belyaev et al.: Nucl. Phys. **B213**, 201 (1983)

33. A.A. Belyaev et al.: Sov. J. Nucl. Phys. **40**, 83 (1984); *ibid*, **43**, 947 (1986)

34. LEGS Collaboration, G. Blanpied et al.: Phys. Rev. Lett. **69**, 1880 (1992)

35. O. Hanstein, D. Drechsel and L. Tiator: Nucl. Phys. **A632**, 561 (1998)

36. H. Genzel et al.: Z. Physik **A268**, 43 (1974)

37. G. Fischer et al.: Z. Physik **253**, 38 (1972)

38. J.R. Wolberg, *"Prediction Analysis"*, Van Nostrand Co., NY (1967), p. 54-66

39. D. Babusci, G. Giordano, A. Lvov, G. Matone and A. Nathan: Phys. Rev. **C58**, 1013 (1998)

Few-Body Systems Suppl. 11, 157–164 (1999)

Few-
Body
Systems
© by Springer-Verlag 1999

Theories of Electromagnetic Productions of Pions

T.-S. H. Lee *

Physics Division, Argonne National Laboratory, Argonne, Illinois 60439, USA

Abstract. The current theoretical approaches for investigating the $\gamma N \to \pi N$ and $N(e, e'\pi)$ reactions are reviewed. It is shown that the dynamical approach can be used to test the predictions from hadron structure calculations. Some N^* form factors calculated using various chiral constituent quark models are presented.

1 Introduction

The opportunities for using electromagnetic probes to explore the nonperturbative QCD aspects of hadron and nuclear dynamics will be unprecedented in the next few years, mainly owing to the developments of several new electron beam facilities. All of the existing theoretical studies of πN and γN reactions can be classified into two different formulations. The first one is the dispersion-relations approach. In this approach, the dynamics is defined by imposing the crossing symmetry, analyticity, high-energy behavior, and appropriate subtraction terms on the reaction amplitudes. This approach was developed by Chew, Goldberger, Low and Nambu [1] in the 1950's and has been revived recently by the Mainz group [2]. The second formulation is based on effective Lagrangians. Within this framework, there are three different approaches. The first one is the K-matrix approach [3] which uses only the πN scattering phase shifts to account for the πN final state interactions. The second one is Chiral Perturbation Theory [4] which is applicable mainly at energies near thresholds and for the low momentum-transfer processes. The third approach is the dynamical approach [5] which is aimed at testing the QCD-based hadron structure models. In this talk I will focus on the development of the dynamical approach and present our recent results [6].

---

*This work is supported by the U.S. Department of Energy, Nuclear Physics Division, under Contract No. W-31-109-ENG-38

2 Dynamical Approach

The starting point of a dynamical approach is a hadron structure calculation which predicts the masses of the excited states of the nucleon. In practice, this calculation can be done accurately only in the absence of the couplings with the decay channels. These masses are therefore called the *bare* masses which do not correspond to the resonance positions determined in the empirical amplitude analyses.

The next step is to introduce the couplings of these *bare* states with the meson and photon fields. In Ref. [5], we focus on the $N \to \Delta$ transition. By including the interactions with π, ρ, ω and γ fields and applying a unitary transformation method, we obtain an effective Hamiltonain for describing πN and γN reactions up to the Δ excitation energy region. It takes the following form:

$$H_{eff} = H_0 + \sum_{\alpha = \gamma N, \pi N} \Gamma_{\Delta \leftrightarrow \alpha} + \sum_{\alpha, \beta = \gamma N, \pi N} v_{\alpha, \beta}, \tag{1}$$

where the two-body interactions $v_{\alpha\beta}$ are defined by the meson-exchange mechanisms. It is straightforward to derive from Eq. (1) the scattering amplitudes for $\pi N \to \pi N$ and $\gamma N \to \pi N$ reactions. The resulting scattering amplitudes can be cast into the following form:

$$T_{\alpha, \beta}(E) = t_{\alpha, \beta}(E) + \bar{\Gamma}_{\alpha \to \Delta}(E) \frac{1}{E - m_\Delta - \Sigma_\Delta(E)} \bar{\Gamma}_{\Delta \to \beta}(E). \tag{2}$$

The first term is the nonresonant amplitude which is generated from a relativistic Lippman-Schwinger equation with the two-body meson-exchange interactions $v_{\alpha, \beta}$. The second term of Eq. (2) is the resonant amplitude determined by the dressed Δ mass and the dressed vertex functions. They are defined by

$$\bar{\Gamma}_{\pi N \to \Delta}(E) = \Gamma_{\pi N \to \Delta} + t_{\pi N, \pi N}(E) G_{\pi N}(E) \Gamma_{\pi N \to \Delta}, \tag{3}$$

$$\bar{\Gamma}_{\gamma N \to \Delta}(E) = \Gamma_{\gamma N \to \Delta} + v_{\gamma N, \pi N}(E) G_{\pi N}(E) \bar{\Gamma}_{\pi N \to \Delta}(E), \tag{4}$$

where $G_{\pi N}(E)$ is the free πN propagator and

$$\Sigma_\Delta(E) = \Gamma_{\pi N \to \Delta} G_{\pi N}(E) \bar{\Gamma}_{\Delta \to \pi N}(E). \tag{5}$$

In Ref. [5], we did not rely on a hadron structure calculation. Therefore the bare mass of the Δ and the form factors of the effective Hamiltonian are treated phenomenologically. Our first task was to determine the parameters of the hadronic part of the effective Hamiltonian, Eq. (1), by fitting the πN phase shifts. Our results were presented and discussed in Ref. [5]. For this talk, it is only necessary to emphasize that the resulting bare mass of the Δ is ~ 1300 MeV which is considerably higher than the resonance position 1236 MeV of the πN cross section data.

We now turn to discussing our results of pion photoproduction. Here we also need to include the ω-exchange which does not play a role in πN scattering. Since the coupling constant $g_{\omega NN}$ is not well determined in the literature, it is treated as a free parameter in our study. Thus, our investigation of pion photoproduction has three adjustable parameters: $G_M(0)$ and $G_E(0)$ of the bare $\Delta \leftrightarrow \gamma N$ vertex, and the coupling constant $g_{\omega NN}$ of ω exchange. With $G_M(0) = 1.85, G_E(0) = +0.025$ and $g_{\omega NN} = 10.5$, we find that all of the available $\gamma N \rightarrow \pi N$ data can be described very well, as illustrated in Fig. 1 for the LEGS data [9]. We also find that the data can be fitted equally well with $G_M(0) = 1.95$, $G_E(0) = -0.025$, and $g_{\omega NN} = 7.0$.

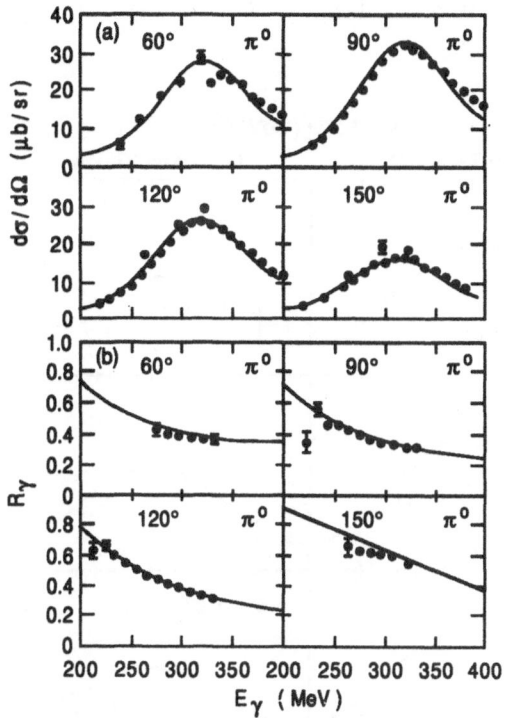

Figure 1. The calculated differential cross sections $d\sigma/d\Omega$ and $R_\gamma = d\sigma_\parallel/d\sigma_\perp$ of $\gamma p \rightarrow \pi^\circ p$ are compared with the LEGS data.

We now focus on the theoretical interpretations of the $\Delta \leftrightarrow \gamma N$ vertex. Eq. (2) indicates that it could be misleading to compare the branching ratios predicted by the constituent quark models [7, 8] with the values extracted from the amplitude analyses. This can be seen by comparing in Table 2.1 the predicted helicity amplitudes with the values listed by the Particle Data Group(PDG) [10]. We notice that our bare values are close to the constituent quark model predictions [7, 8], and the dressed values are close to the values of PDG [10]. This suggests that the large differences between the constituent quark model predictions and the PDG values are due to the nonresonant meson-exchange production mechanisms (the second term of Eq. (2)). Similar considerations must be taken into account in comparing the PDG values with the predictions of other hadron structure calculations and in investigating the higher mass N^* resonance parameters.

Table 2.1. Helicity amplitudes of the $\Delta \to \gamma N$ transition at $W = 1236$ MeV are compared with the values from Particle Data Group (PDG) [10] and the predictions of constituent quark models of Refs. [7,8]. The amplitudes are in unit of 10^{-3} $(\text{GeV})^{-1/2}$.

A	PDG	Dressed	Bare	Ref. [8]	Ref. [7]
$A_{3/2}$	-257 ± 8	-228	-153	-157	-186
$A_{1/2}$	-141 ± 5	-118	-84	-91	-108

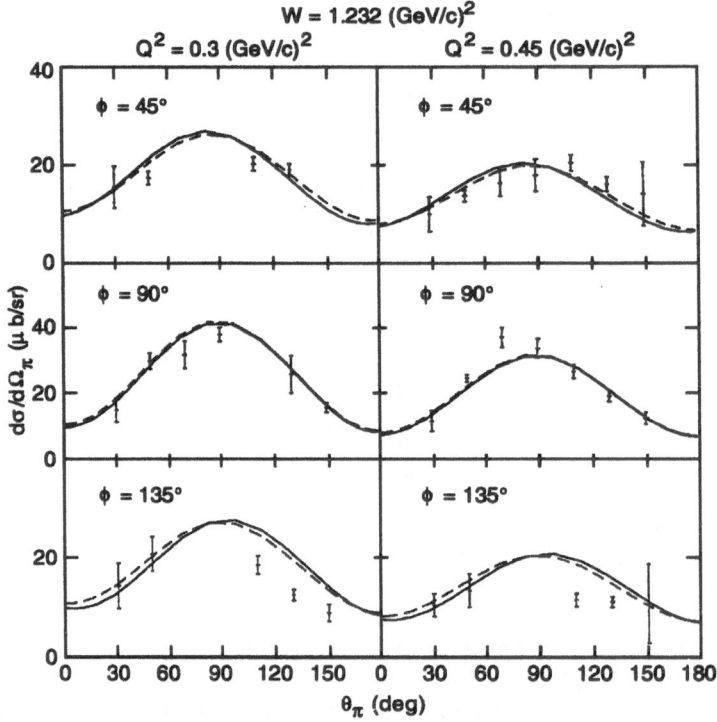

Figure 2. The predicted differential cross sections of $p(e, e'\pi^\circ)$ reactions are compared with the data.

The dynamical model of Ref. [5] has been extended to make predictions for $N(e, e'\pi)$ reactions. As a first step to help the experimental efforts in the past few years, we had made predictions based on the simplest assumption that the electromagnetic form factors for the $\gamma N \to \Delta$ is proportional to the the well measured proton form factor. The charge component $G_C(q^2)$ of the $\gamma N \to \Delta$ form factor is calculated from $G_M(q^2)$ and $G_E(q^2)$ using the long wavelength approximation. In Fig. 2, we see that our predictions are in good agreement with the available data. The comparison of our predictions of $A_{tl} = [d\sigma(\phi =$

$0) - d\sigma(\phi = 180^0)]/[d\sigma(\phi = 0) + d\sigma(\phi = 180^0)]$ with the most recent data from MIT-Bates are shown in Fig. 3. Here we see some significant deviations from the data at $W = 1236$ MeV. This could be mainly due to our use of the very naive form factors for the $\gamma N \rightarrow \Delta$ transition.

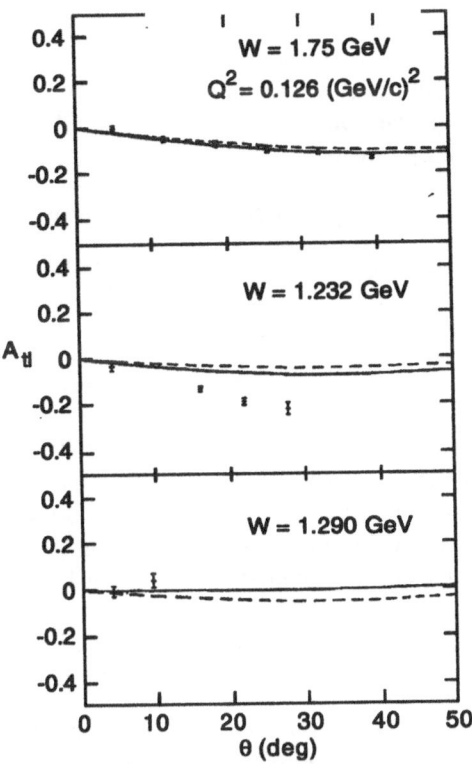

Figure 3. The predicted A_{tl} are compared with the recent data from MIT-Bates.

3 Chiral Constituent Quark Model Calculations

Encouraged by the results of Ref. [5], as briefly described above, we [6] have started to apply our dynamical approach to test various constituent quark models in conjuction with the new experiments being conducted at MIT-Bates and Jefferson Laboratory. We start with the now well accepted assumption that the chiral constituent quark model can emerge from QCD due to the breaking of chiral symmetry. This is supported by a recent Lattice QCD calculation [11]. It is therefore reasonable to assume that a Hamiltonian for relating the πN and γN reactions to the quark-substructure of N and Δ is of the following form:

$$H = H_B + \sum_{B,B'} \left[\left(h_{\pi B, B'} + h_{\gamma B, B'} \right) + (h.c.) \right], \tag{6}$$

where B and B' are the eigenstates of an one-baryon Hamiltonian H_B. In the simplest nonrelativistic model, it is defined by

$$H_B = \sum_i \left(m_q + \frac{p_i^2}{2m_q} \right) + V + V_{conf} , \tag{7}$$

where $V_{conf} = \sum_{i>j} \alpha_c r_{ij}$ is the usual linear confinement potential. The quark-quark residual interaction V is due to the exchange of Goldstone bosons and gluons. The vertex interactions in Eq. (6) are calculated from the baryon wave-functions and quark operators

$$f_{B,\pi B'} = \langle B | \sum_i f_{\pi q,q}(i) | B' \rangle , \tag{8}$$

$$f_{B,\gamma B'} = \langle B | \sum_i f_{\gamma q,q}(i) | B' \rangle . \tag{9}$$

In the simplest nonrelativistic model, the above quark-meson and quark-photon operators are well known and are used in our calculations. Within the dynamical approach of Ref. [5], the vertices defined by Eqs. (8)-(9) for $B = \Delta$ and $B' = N$ are identified with the bare vertices $\Gamma_{\Delta,\gamma N}$ and $\Gamma_{\Delta,\pi N}$ in Eq. (1). Accordingly the Δ mass generated from the model Eq. (7) must be identified with the mass 1300 MeV of the bare Δ. To make contact with the previous works, we assume that the interactions between constituent quarks can be represented by the following form:

$$V = f_G V_{oge} + f_M V_{ome} , \tag{10}$$

where V_{oge} is the usual one-gluon-exchange(oge) potential and V_{ome} is the one-meson-exchange (ome) potential due to the exchange of pion and eta mesons. Both the oge and ome potentials are regularized by a form factor of the form $F_i^2(q^2) = (\Lambda_i^2/(q^2 + \Lambda_i^2))$.

With the above formulation, we can investigate whether the πN and γN reactions can distinguish three possible constituent quark models. For $f_G = 1$ and $f_M = 0$, the above equations define the traditional constituent quark model (oge). For $f_G = 0$ and $f_M = 1$ we have an one-meson-exchange model(ome) that is similar, but not identical, to the form proposed by Glozman and Riska [12]. The model with $f_G = f_M = 1$ contains both the boson-exchange and gluon-exchange interactions between constituent quarks. This model ($ogme$) was briefly investigated in Ref. [14].

To be consistent with the dynamical model of Ref. [5], the parameters of each considered model must be chosen to reproduce (1) $m_\Delta - m_N = (1300 - 938.5)$MeV, (2) $G_M(0) \sim 2.0, G_E(0) \sim \pm 0.025$, (3) the magnetic moment μ_p of the proton, (4) the proton charge radius $< r >_p \sim 0.8$ fm, (5) the form factors $f_{\pi N,\Delta}(k)$ and $f_{\pi N,N}(k)$ with a cutoff of about 650 MeV/c in dipole form. By using the variational method developed in Ref. [15], we solve the three-body bound state problem by diagonalizing the model Hamiltonian, defined by

Eq. (7), in the space spanned by the harmonic oscillator wavefunctions. The convergent solutions are obtained when up to about N=9 orbitals are included. As a check of our calculation, we are able to reproduce the results of Glozman, Rapp and Plessas [13] if their short-range form of Goldstone boson exchange potential is used in our calculation.

We first adjust the parameters to fit the data (1)-(3) listed above. It turns out that all of the possible consitutent quark models fitted to these data yield a proton radius of only about 0.4 fm. Consequently, the predicted πNN and $\pi N\Delta$ form factors are too hard in comparison with those of Ref. [5]. It is not clear how to remove this difficulty. One obvious possibility is to carry out a relativistic calculation and to consider interaction currents. Nevertheless, our results so far at least demonstrate that the πN and γN reactions can be useful in distinguishing the three considered constituent quark models, since their predicted form factors have significant differences. This is illustrated in Fig. 4 for the predicetd $N^*(S_{11}) \to \pi N, \eta N, \pi\Delta$ form factors.

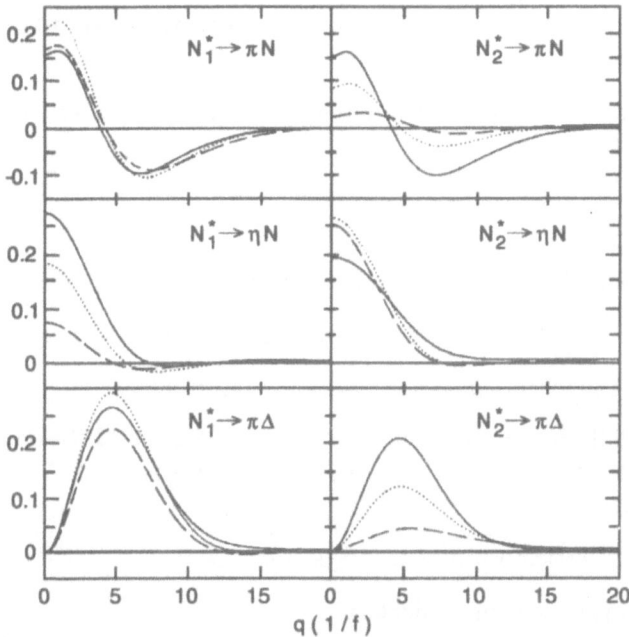

Figure 4. The $N^* \to \pi N, \eta N, \pi\Delta$ form factors. The solid, dotted and dashed curves are respectively from the *oge*, *ome*, and *ogme* models.

Clearly, much has to be done before we can use the $N(e, e'\pi)$ data, like the data of A_{tl} shown in Fig. 3, to distinguish different constituent quark models. This is the focus of our current effort in order to confront the forthcoming data with high precision. This theory-experiment joint effort is needed to give

164

a crucial test of the chiral constituent quark model and its dynamical content.

4 summary

We have reviewed the development of a dynamical approach to study the photo-production and electroproduction of pions on the nucleon. Our current effort is to test the chiral constituent quark model which can be related to QCD. Some preliminary results for the $N^*(S_{11})$ have been presented to indicate that the predicted mesonic and electromagnetic form factors can be tested in the studies of πN and γN reactions. Our complete results will be published elsewhere [6].

References

1. G.F. Chew, M.L. Goldberger, F.E. Low and Y. Nambu: Phys. Rev. **106**, 1345 (1957)

2. O. Hanstein, D. Drechsel and L. Tiator: Nucl. Phys. **A632**, 561 (1998)

3. R.M. Davidson, N.C. Mukhopadhyay and R.S. Wittman: Phys. Rev. **D43**, 71 (1991), and references therein

4. V. Bernard, N. Keiser and U. Meissner: Phys. Lett. **B382** , 19 (1996), and references therein

5. T. Sato and T.-S. H. Lee: Phys. Rev. **C54**, 2660 (1996), and references therein

6. T. Yoshimoto, T.Sato, M. Arima and T.-S. H. Lee: *in preparartion*

7. S. Capstick: Phys. Rev. **D46**, 2864 (1992)

8. R. Bijker, F. Iachello, and A. Leviatan: Ann. of Phys. **236**, 69 (1994)

9. LEGS Collaboration, A.M. Sandorfi et al.: Few Body Systems, Suppl. **7**, 317 (1994); LEGS Data Release L7a8.0 (April '94); A.M. Sandorfi: private communication.

10. Review of Particle Properties: Phys. Rev. **D50**, 1710 (1994)

11. K.F. Liu: private communication

12. L. Ya. Glozman and D.O. Riska: Phys. Rep. **268**, 263 (1996)

13. L. Ya. Glozman, Z. Rapp, and W. Plessas: Phys. Lett. **B381**, 311 (1996)

14. P.N. Shen, Y.-B. Dong, Z.-Y. Zhang, Y.-W. Yu and T.-S. H. Lee: Phys. Rev. **C55**, 2024 (1997)

15. T. Ogaito, M. Arima and and T. Sato: to be published

Few-Body Systems Suppl. 11, 165–168 (1999)

Few-
Body
Systems
© by Springer-Verlag 1998

Single Pion Electroproduction in the $\Delta(1232)$ Resonance from CLAS Data at Jefferson Lab

Kyungseon Joo *

Department of Physics, University of Virginia, Charlottesville VA 22901, USA

and the

CLAS collaboration

Abstract. In this contribution, several physics topics relevant to the $\Delta(1232)$ resonance are discussed and some preliminary experimental results with CLAS for single pion electroproduction in the $\Delta(1232)$ resonance region are presented.

1 Introduction

Baryon electroproduction amplitudes serve as important tests for models of the interaction of quarks and gluons in confined systems. Among the most interesting of resonance studies, is the $\Delta(1232)$ resonance which is one of the most prominent states over all Q^2. Single pion electroproduction in the $\Delta(1232)$ resonance provides information which is sensitive to models, which make specific statements about the internal motion of baryon constituents. Three multipoles are related to the $\gamma N \rightarrow \Delta$ transition: the electric, magnetic and scalar/longitudinal multipoles E_{1+}, M_{1+} and S_{1+}. At low values of Q^2, the focus has been on incorporating relativistic effects into the constituent quark model (CQM). This model, which is based on $SU(6) \otimes O(3)$ symmetry, explains the Δ excitation by a simple spin-flip of one of the constituent quarks in the $L_{3q} = 0$ state of the 3-quark wave function[1, 2] corresponding to a pure magnetic dipole transition, M_{1+}. Thus the ratio $R = E_{1+}/M_{1+} = 0$. A tensor force term can give rise to a non-zero electric dipole[3]. Models incorporating one gluon exchange[4, 5] predict R to be small for low and moderate Q^2. At higher values of Q^2 where the small distance behavior of the quark-gluon confining interaction is probed, QCD sum rule calculations, valence QCD and perturbative QCD (pQCD) have been employed. The latter one predicts $R \rightarrow 1$. The applicable range in Q^2 for this approach is not known.

The $\Delta(1232)$ resonance is an intermediate state of reactions such as single pion electroproduction, $ep \rightarrow e'p\pi^0$ or $ep \rightarrow e'\pi^+ n$. The differential cross

---

* *E-mail address:* kjoo@cebaf.gov

section for single pion electroproduction can be written as:

$$\sigma = \sigma_T + \sigma_L + \sigma_{TT} \cos 2\phi + \sqrt{\frac{\epsilon(1+\epsilon)}{2}} \sigma_{TL} \cos \phi \tag{1}$$

With a polarized beam or/and polarized target, additional observables can be measured. The imaginary part of the response tensor terms can be measured only through these polarization observables.

The CLAS experiments at Jefferson Lab are to measure $\gamma N \to \Delta$ transition amplitudes over a large Q^2 range up to 4 GeV^2 using both unpolarized/polarized electron beam and target.

2 Analysis status of the first data from the CLAS

Data taking began with CLAS with unpolarized electron beam and target in December, 1997. A 4-cm liquid hydrogen target was used and data were taken at a luminosity of 10^{33} cm^{2}s^{-1}.

Electrons are identified by matching negative tracks with the hits in the Cerenkov counter and a shower in the electromagnetic calorimeter. In addition the energy measured by the calorimeter must match the track momentum. Once the electron is identified, its path length is used with its arrival time in the time-of-flight scintillator to calculate the start time for the event. This start time is used to calculate the time-of-flight for each of the remaining particles. Once tracks, momenta and time-of-flight have been determined, the mass of the charged particle can be calculated using its velocity and momentum. This provides excellent particle identification[6] between π^+ and proton as shown in Figure 1 where β is plotted as a function of momentum for positive particles. Also, the plot shows a band due to deuterons produced in the end caps of the target. Once events with a scattered electron and a proton are selected, the missing mass technique can be used to select the exclusive reaction, $ep \to e'p\pi^0$ from $ep \to e'pX$ data. Figure 2 shows the square of missing mass for $ep \to e'pX$ for data taken at 1.645 GeV. The resolution is not yet sufficient to separate the elastic radiative tail ($ep \to e'p\gamma$) from single π^0 production just using missing mass separation. However, it is well known that radiative effects from elastic scattering are dominated by photons emitted along the directions of either the incident or scattered electron. By cutting on the angular distributions of these two directions, it is possible to eliminate a large fraction of the radiative elastic scattering. Figure 2 shows the square of missing mass spectrum before and after the cuts. The same missing mass technique can be used to identify the reaction $ep \to e'\pi^+n$ once the events with a scattered electron and a π^+ are identified. This reaction has no background from the radiative elastic tail, but the reaction phase space included in the CLAS acceptance is quite different from that for π^0 production because the laboratory direction of the π^+ is unconstrained. Moreover, the consequences of pion decay must be considered.

The spectra of W for the reactions $ep \to e'p\pi^0$ and $ep \to e'\pi^+n$ are shown in Figure 3. Because these spectra are not corrected for the difference in acceptance, a quantitative comparison is not possible. However, the $\Delta(1232)$ peak

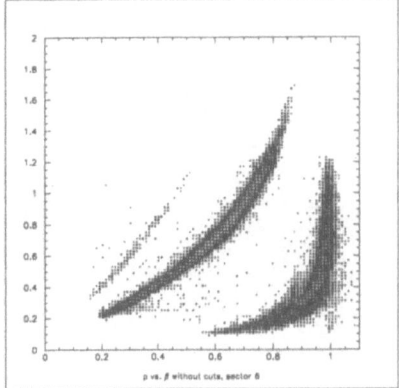

Figure 1. β vs. momentum for positive tracks. The size of the box for each bin is proportional to the logarithm of the number of counts.

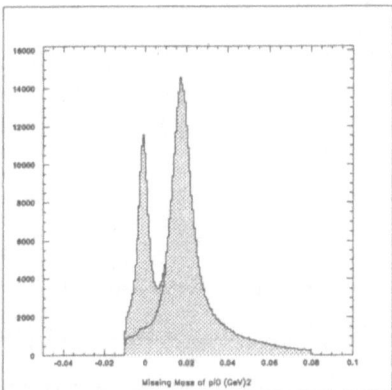

Figure 2. The square of the missing mass spectrum. Radiative processes in elastic scattering have been subtracted from $ep \rightarrow e'pX$ to extract $ep \rightarrow e'p\pi^0$.

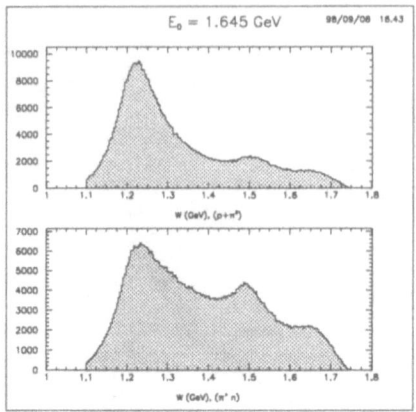

Figure 3. Invariant mass W spectra for measured exclusive reactions. The top figure is for $ep \to e'p\pi^0$ and the bottom is for $ep \to e'\pi^+ n$.

is less dominant for π^+ than for π^0 production, as is expected from isospin arguments. Isospin requires the cross section for $ep \to e'\pi^+ n$ in the $\Delta(1232)$ excitation to be lower by a factor of 2. Background from Born contributions to π^+ production also add to the differences in the two reactions.

3 Conclusion

The preliminary results from the first data from CLAS show that the CLAS is already performing at a level close to its design specifications. It is quite clear that CLAS with high luminosity and large acceptance will provide precise and high statistics measurements on the $\Delta(1232)$ resonance and higher resonances for a large Q^2 range.

References

1. C. Becchi and G. Mopurgo, Phys. Rev. **140B** 687 (1965).

2. H. Harari and H. Lipkin, Phys. Rev. **140B** 1617 (1965).

3. R. Davidson, N. C. Mukhopadyay and R. Wittman, Phys. Rev. Lett. **56** 804 (1986).

4. N. Isgur and G. Karl, Phys. Rev. **D4** 1466 (1971).

5. N. Isgur and G. Karl, Phys. Lett. **72B** 109 (1977).

6. V. Burkert, H. Egiyan, L.Elouadrhiri and S. Stepanyan, CLAS-NOTE-98-004 (1998).

Few-Body Systems Suppl. 11, 169–172 (1999)

Few-
Body
Systems
© by Springer-Verlag 1998

Pion Photo- and Electroproduction on the Proton

S. S. Kamalov*, D. Drechsel, O. Hanstein, and L. Tiator

Institut für Kernphysik, Universität Mainz, 55099 Mainz, Germany

Abstract. Using a unitary isobar model a new operator for pion photo- and electroproduction has been developed for nuclear applications at photon equivalent energies up to 1 GeV. The model contains Born terms, vector mesons and nucleon resonances ($P_{33}(1232)$, $P_{11}(1440)$, $D_{13}(1520)$, $S_{11}(1535)$, $F_{15}(1680)$, and $D_{33}(1700)$). Within this model we have obtained good agreement with the experimental data for pion photo- and electroproduction on the nucleon for both differential cross sections and polarization observables.

1 Introduction

The main goal of our present work is to develop a simple model for nuclear applications, which should be consistent with recent data for pion photo- and electroproduction on nucleons at photon equivalent energies up to 1 GeV. As a starting point we will use the prescriptions of the isobar model [1, 2], assuming that resonance contributions in the relevant multipoles have Breit-Wigner form. The nonresonant contributions will be described using standard Born terms with a mixed pesudovector-pseudosclar πNN coupling and vector meson exchange. The final amplitude will be unitarized by extending the procedure developed by Olsson [3] to the case of virtual photons and higher resonances.

2 Background

The base line for a correct description of the resonance contributions is, of course, a reliable description of the nonresonant part of the amplitude (nonresonant background). This part of our model we describe by standard Born and vector meson exchange terms.

In general, the following ingredients for the construction of the background are not well defined: 1) the type of the πNN coupling and 2) the coupling constants of the vector meson exchange contributions. The best way to fix them is

---

*Permanent address: Laboratory of Theoretical Physics, JINR Dubna, Head Post Office Box 79, SU-101000 Moscow, Russia.

to analyse nonresonant s- and p-wave multipoles. To fix the coupling constants of the vector mesons we consider those multipoles which are independent of the type of the πNN coupling. These are $_{p,n}M_{1+}^{(1/2)}$ and $_{p,n}E_{1+}^{(1/2)}$, of which the first one is especially sensitive to ω exchange contributions. The real parts of the nonresonant multipoles $E_{0+}^{(3/2)}$ and $M_{1-}^{(3/2)}$ are more appropriate to find the optimum parametrization for the πNN coupling. In our model we use an effective Lagrangian for πN interaction with an energy dependent superposition of both PV and PS couplings in the form

$$\mathcal{L}_{\pi NN} = \frac{\Lambda_m^2}{\Lambda_m^2 + \mathbf{q}_0^2} \mathcal{L}_{\pi NN}^{PV} + \frac{\mathbf{q}_0^2}{\Lambda_m^2 + \mathbf{q}_0^2} \mathcal{L}_{\pi NN}^{PS}, \tag{1}$$

where $\mathbf{q}_0$ is the asymptotic pion momentum in the πN cm frame which depends only on W and is not an operator acting on the pion field. The mixing parameter $\Lambda_m = 450$ MeV. We have found that this is a very economical way to describe the energy dependence of the nonresonant multipoles at photon lab energies up to 1 GeV (see, Fig. 1). The Lagrangian (1) combined with the requirement of unitarity for s-wave multipoles in the form $E_{0+} = E_{0+}(Born)\, e^{i\delta_{\pi N}} \cos\delta_{\pi N}$ provides a very good agreement with recent results of Mainz [4] and VPI [5] multipole analyses.

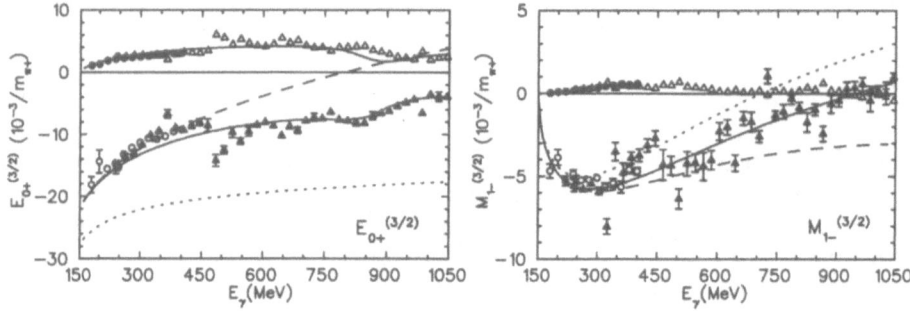

Figure 1. Nonresonant $E_{0+}^{(3/2)}$ and $M_{1-}^{(3/2)}$ multipoles calculated with pure pseudovector (dashed curves) and pure pseudoscalar (dotted curves) πNN couplings. The solid curves are the results for the real and imaginary parts obtained using the Lagrangian (1). The full and open triangles are real and imaginary parts from the VPI analysis[5].

3   Resonance contributions

The resonance contributions in the direct channel we parametrize using standard Breit-Wigner forms for the relevant multipoles

$$A_{l\pm}(W) = F_{\gamma N^*}^{l\pm}(W, Q^2) \frac{\Gamma_{tot} W_R\, e^{i\phi}}{W_R^2 - W^2 - iW_R\Gamma_{tot}} F_{\pi N^*}^l(W), \tag{2}$$

where $F_{\pi N^*}^l$ and $F_{\gamma N^*}^{l\pm}$ are the form factors describing the πNN^* and γNN^* vertices respectively.

Real photons. In the case of the Δ resonance the main role of the unitary phase ϕ is to adjust the phase of the total multipole (background+resonance) to the corresponding pion-nucleon scattering phase. For the resonances in the second and third resonance region we consider ϕ as a free parameter that we take as a constant extracted from the observed ratio between the imaginary and real parts at the resonance peak.

The obtained results in the first resonance region are the same as the results of the recent dispersion relation analysis of Mainz [4]. At higher energies, where contributions from the resonances in the second and third resonance regions become increasingly important, our model agrees well with the experimental data for the differential cross sections and single polarization observables up to 1 GeV.

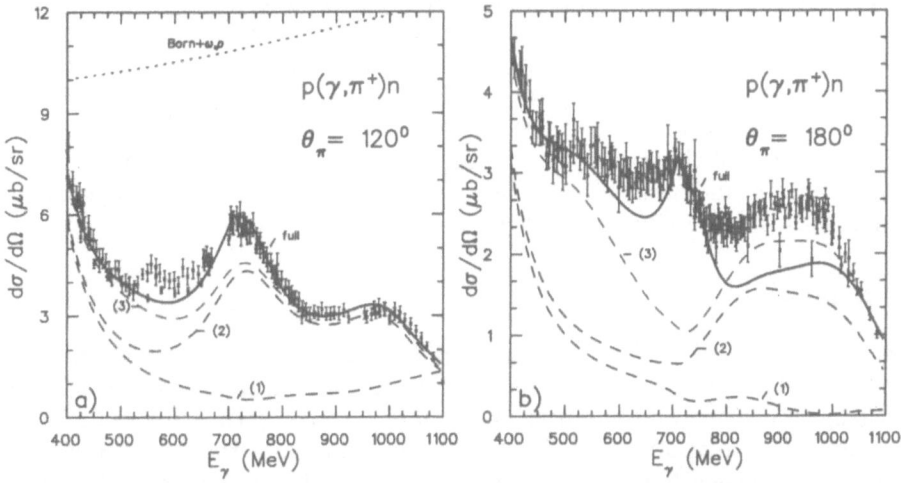

Figure 2. Differential cross sections for $p(\gamma, \pi^+)n$ at $\theta_\pi = 120^0$ (**a**) and $\theta_\pi = 180^0$ (**b**). The dotted curve is the *Born* $+ \omega, \rho$ contribution. The dashed curves are the results obtained with subsequent addition of the baryon resonances: (1) *Born*$+\omega, \rho +$ $P_{33}(1232)$; (2)=(1)+$D_{13} + F_{15} + D_{33}$ and (3)=(2)+$P_{11}(1440)$. The solid curves are the full calculations which also include the $S_{11}(1535)$ resonance. Experimental data from the VPI compilation [5].

As an example in Fig. 2 we demonstrate the evolution of the energy distribution in the $\pi^+ n$ channel at backward angles by subsequent addition of resonance contributions to the Born terms. First, we see that the large contribution from (non unitarized!) Born terms is strongly suppressed by the Δ resonance. The second interesting feature is related to the excitations of the $S_{11}(1535)$ and $D_{13}(1520)$ resonances. At $\theta_\pi = 120^0$ their contributions are comparable and it is difficult to separate them. However, at $\theta_\pi = 180^0$ the contribution of the $D_{13}(1520)$ resonance is very small, because the E_{2-} multipole does not contribute at this kinematics. This fact provides a possibility to study the contribution from the $S_{11}(1535)$ resonance and the related cusp effect due to the opening of the η channel, clearly separated from $D_{13}(1520)$ effects.

Virtual photons. The extension of the unitarization procedure to the case of pion electroproduction is straightforward for the Δ-resonance, because the Fermi-Watson theorem requires that the phases of the total $M_{1+}^{(3/2)}$, $E_{1+}^{(3/2)}$ and $S_{1+}^{(3/2)}$ multipoles should not depend on Q^2. Therefore, the position of the resonance peak does not depend on Q^2 either. In our model this important requirement is satisfied by an appropriate choice of the phase ϕ in Eq. (2).

The obtained Q^2 dependence of the $M_{1+}^{(3/2)}$ multipole describes very well the magnetic form factor G_M^*. We also give predictions for the Q^2 dependence of the $R_{EM} = E_{1+}/M_{1+}$ and $R_{SM} = S_{1+}/M_{1+}$ ratios in the isospin 3/2 channel at $0 \leqslant Q^2 \leqslant 3\,(GeV/c)^2$. In this range $-2.4\% \geqslant R_{EM} \geqslant -4.7\%$ and $-3.9\% \geqslant R_{SM} \geqslant -5.2\%$ [6].

For the $P_{11}(1440)$, $D_{13}(1520)$, $D_{33}(1700)$ and $F_{15}(1680)$ resonances we assumed that the total phase of the corresponding multipoles are independent of Q^2 at the resonance position. We have found that this requirement strongly modifies the Q^2 dependence of the helicity amplitudes for the P_{11} resonance, leading to a rapid fall-off with Q^2. However it does not change seriously the Q^2 dependence of the helicity amplitudes for the D_{13} and F_{15} resonances [6]. Finally our results, obtained for these resonances are in very good agreement with results of the analyses from DESY, NINA and Bonn.

4   Summary

We have developed a new operator for pion photo- and electroproduction for applications to reactions on nuclei which is simple, physically transparent, satisfies gauge invariance and unitarity. It provides a good description for individual multipoles as well as differential cross sections and polarization observables. We believe that the developed model can be used not only for nuclear applications, but also as a starting point to predict and analyze forthcoming data for pion photo- and electroproduction on proton and neutron targets.

References

1. R. L. Walker: Phys. Rev. **182**, 1729 (1969)

2. R. D. Moorehouse: In A. Donnachie and Shaw, *Electromagnetic Interactions of Hadrons*, N.Y. (1978)

3. M. G. Olsson: Nucl. Phys. **B78**, 55 (1974)

4. O. Hanstein, D. Drechsel, and L. Tiator: Nucl. Phys. **A632**, 561 (1998)

5. R. A. Arndt, I. I. Strakovsky and R. L. Workman: Phys. Rev. **C53**, 430 (1996)

6. D. Drechsel, O. Hanstein, S. S. Kamalov, L. Tiator, **nucl-th/9807001**

Few-Body Systems Suppl. 11, 173–176 (1999)

Few-
Body
Systems
© by Springer-Verlag 1999

Photoabsorption on Bound Nucleons in the Resonance Region

V. Muccifora

INFN-Laboratori Nazionali di Frascati C.P. 13, I-00044 Frascati, Italy

Abstract. The energy and nuclear mass dependence of the total hadronic cross section for real photons between 0.5 and 2.6 GeV have been measured at Bonn using the SAPHIR tagged photon beam. The measurement, performed on C, Al, Cu, Sn and Pb, has revealed also in the region 1-2 GeV a significant reduction of the photoabsorption strength on the bound nucleon compared to the free nucleon case. This reduction has been interpreted as a first signature of a low energy onset of the shadowing effect.

The properties of baryon resonances in nuclei have been studied at Frascati [1], Mainz [2], and Bonn [3], in photoabsorption experiments which showed significant medium effects. While the Δ-resonance is only slightly distorted, higher excited nucleon states N*, in the second and third resonance region, appear to be washed out. Furthermore, the absolute value of the cross section per nucleon is reduced, with respect to the free-nucleon case, above 0.6 GeV. The mechanism of this damping is not yet well understood (see Ref .[4] and references therein). In the resonance region (0.6-1.2 GeV) standard nuclear effects like Fermi-motion and Pauli-blocking are unable to reproduce the resonance disappearance. This seems to imply strong effects in the excitation, propagation and interaction of the N* resonances in the nuclear medium. Nevertheless this strong interaction between nucleon resonances and the nucleus do not predict any reduction of the strength above ~ 1 GeV region. On the contrary recent data from Ref. [3] show a significant reduction of the photoabsorption strength also above $k \sim 1$ GeV where resonance effects are expected to be small. This reduction could be a first signature of a low energy onset of the shadowing effect.

In order to clarify this hypothesis the total photoabsorption cross section has been measured on a broad range of mass number (C, Al, Cu, Sn and Pb) and in a wide energy region (connecting the resonance-region to the shadowing region). The measurement was carried out at Bonn using the SAPHIR tagged photon beam [5] of the ELSA accelerator in the photon energy range 0.5-2.6 GeV. The photoabsorption cross section on nuclei has been measured with the

174

photohadronic technique using an apparatus similar to the one previously used at Frascati [1].

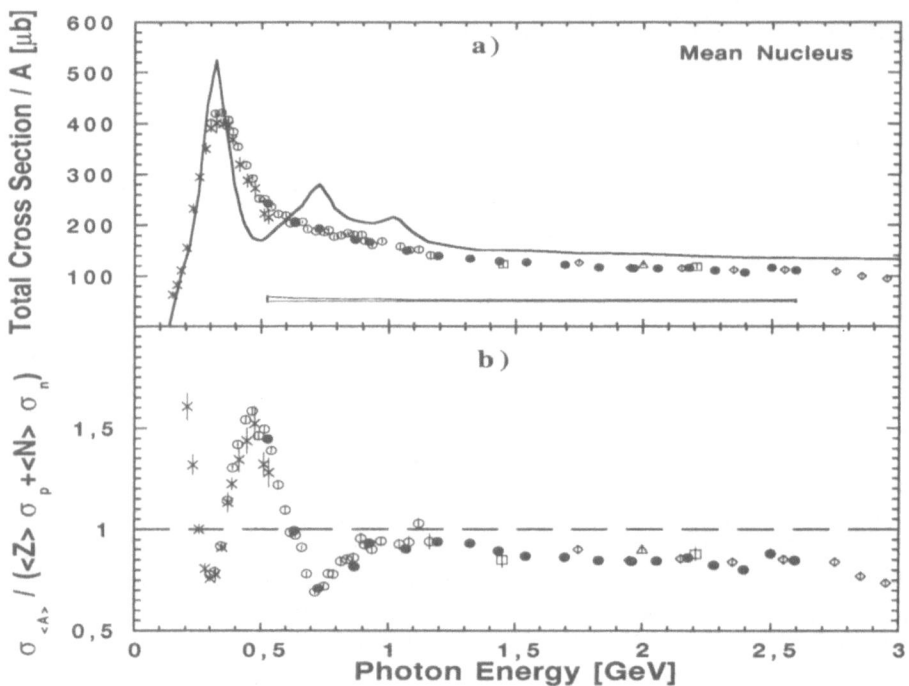

Figure 1. a) Total cross section b)Ratio bound/free nucleon cross section for the average nucleus (solid circles) compared with previous experiments (notations are the same of Ref. [3]). Also shown is the proton absorption cross section (solid line).

In Fig. (1a) is shown the total cross section, normalized to the mass number A, obtained from the average of the values for all the five nuclei studied. This can be considered as the cross section on an "average nucleus" with $Z/A = 0.469$ and an average density $\rho_A = 0.109$ nucleons fm^{-3}, i.e. an "aluminum-like" nucleus. The bars indicate the statistical errors, while the systematic ones are represented by the band in the bottom of the Fig. (1a). Total cross section values have been found in agreement within the experimental errors with both the low and high energy data available in the literature. They confirm with reduced statistical uncertainties the absence of the second and third resonance for the bound nucleon compared to the free nucleon case. However, the most striking result is the persistence of the absorption strength reduction above 1.2 GeV. To better evaluate this reduction, in Fig. (1b) the ratio between the total cross section on the average nucleus ($\sigma_{<A>}$), and that obtained from the free

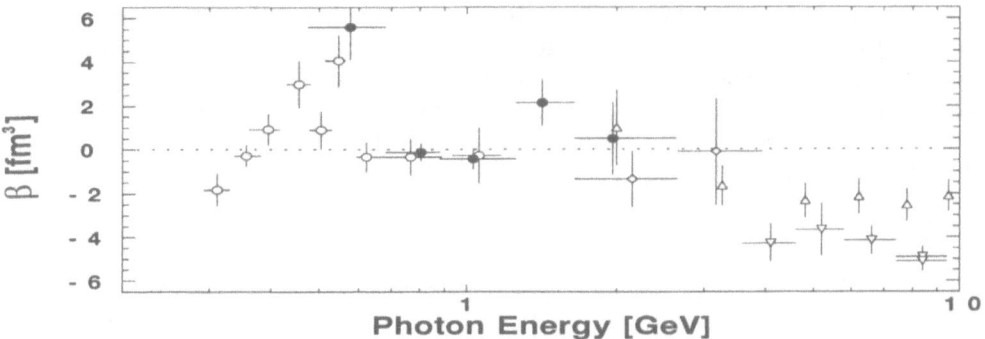

Figure 2. Linear coefficient β derived from our data on C, Al, Cu, Sn and Pb (solid circles), compared with the ones obtained from other experiments.

nucleons ($< Z > \sigma_p + < N > \sigma_n$), is shown.

The integral Σ_A of the measured cross sections has been calculated in five energy regions: Δ-tail, D_{13} and F_{15}-resonances; in the shadowing threshold region between $1.2 \div 1.7$ GeV; and in the shadowing region. In each region of mean energy $\overline{k}$, the Σ_A dependence on the nuclear density ρ_A has been parameterized in the form:

$$\frac{\Sigma_A(\overline{k})}{A} = \Sigma_0(\overline{k})[1 + \beta(\overline{k})\rho_A] \tag{1}$$

The $\beta(\overline{k})$ coefficients obtained in each region from the fits, are shown in Fig.2. The energy behavior of the $\beta(\overline{k})$ coefficient is similar to the one (see Fig.1b)) shown by the ratio bound/free nucleon cross section in the resonance regions (Δ-tail, D_{13} and F_{15}) and in the region above ~ 2 GeV. This is an indication of an increase with the nuclear density of the mechanism responsible for the observed medium effects. On the contrary in the shadowing threshold region the positive $\beta(\overline{k})$ value indicates that the strength reduction is more effective on light nuclei, thus suggesting a different mechanism for the photoabsorption strength reduction which is not dependent on the nuclear density alone.

The reduction of the strength in this region has been interpreted as a first signature of a low energy onset of the shadowing effect. The latter has been described in terms of properties (width, masses and cross sections σ_{VN}) of vector mesons in the nuclear medium. The result of a VMD calculation [6], performed by taking into account the width of the vector-meson masses and the σ_{VN} low energy behavior, is shown for the average nucleus in Fig. 3. For comparison the result of the calculation performed by using δ-function vector-meson masses is also given in the figure. A further improvement in the description of the data

176

is found by including the shift of the ρ-meson mass in the nuclear medium as shown in the figure. The analysis evidenced a significant decrease δm_ρ of the in medium ρ-meson mass: on the average nucleus the value $\delta m_\rho = 180 \pm 7 \pm 46$ MeV has been found. This value is well above the theoretical expectations [7], while is in agreement with the large ρ-meson mass shift recently observed via $^3He(\gamma, \pi^+\pi^-)X$ reaction [8].

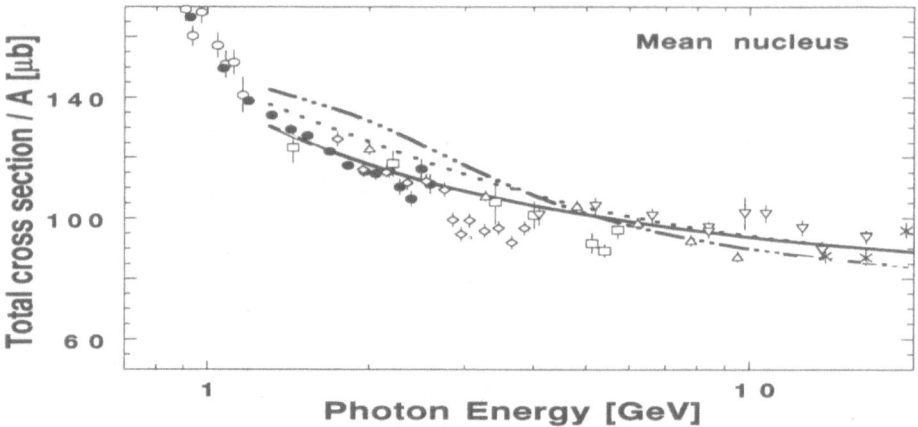

Figure 3. Results of a VMD calculation [6]. Dotted line (dot-dashed) is the result obtained by taking into account the width of the vector meson masses (δ-function vector meson masses) and the σ_{VN} low energy behaviour. Solid line is the result obtained by including also the vector-meson mass shift. Solid circles indicate the data from this experiment.

References

1. N. Bianchi et al.: Phys. Lett **B309**, 5 (1993); *ibid* **B325**, 333 (1994); Phys. Rev. **C 54**, 1688, (1996)

2. M. MacCormick et al.: Phys. Rev. **C 53**, 41 (1996); *ibid* **C 55**, 1033 (1997)

3. M. Mirazita et al.: Phys. Lett. **B 407**, 225 (1997)

4. V. Muccifora: Nucl. Phys. **A 623**, 116c (1997)

5. W.J. Schwille et al.: Nucl. Instr. and Meth. **A 344**, 470, (1994)

6. M. Mirazita: Ph.D. Thesis Milan University March 1998, Frascati Internal Report LNF-98/006; V. Muccifora et al.: to be submitted to Phys. Rev.C.

7. K. Saito, K. Tsushima and A.W. Thomas: Phys. Rev. **C 56**, 566 (1997)

8. G.J. Lolos et al.: Phys. Rev. Lett. **80**, 241 (1998); G.M. Huber, G.J. Lolos and Z. Papandreou: *ibid* **80** 5285 (1998)

Few-Body Systems Suppl. 11, 177–181 (1999)

Few-
Body
Systems
© by Springer-Verlag 1999

The Gerasimov-Drell-Hearn Sum Rule - Experimental Aspects

G. Anton [*] [†]

Physikalisches Institut der Universitaet Erlangen-Nuernberg, Erwin-Rommel
Str. 1 , D-91058 Erlangen, Germany

Abstract. The Gerasimov-Drell-Hearn sum rule relates the total cross section
of polarised photons on polarised nucleons with the anomalous magnetic moment of the nucleon. It is based on very general physics principles and provides
important constraints for models of the nucleon. The experimental test of the
sum rule is performed by the 'GDH-collaboration' at the electron accelerators
MAMI at Mainz and ELSA at Bonn. The data taking in Mainz has just begun.

1 Introduction

The Gerasimov-Drell-Hearn (GDH) sum rule relates the total cross section of
polarised photons on polarised nucleons with the anomalous magnetic moment
of the nucleon [2]. It is based on very general physics principles and provides
important constraints for models of the nucleon. The GDH prediction was formulated in the 1960's, but the experimtal test would have been too complicated
at that time. Due to the technical developments in recent years we have founded
a collaboration and have started to perform this experiment.

The GDH sum rule is embedded in the field of spin structure studies of the
nucleon, because the above mentioned polarized cross section is proportional to
the spin structure function g_1 at a four momentum transfer $Q^2 = 0$. While the
Bjorken and Ellis-Jaffe sum rule make predictions for the limit of large Q^2 and
the GDH sum rule makes a prediction for $Q^2 = 0$, the region of intermediate
momentum transfer values, where non perturbative phenomena occur, are of
increasing interest for model calculations.

---

[*] *on behalf of the GDH Collaboration*
[†] e-mail: Anton@Physik.Uni-Erlangen.De

2 Physics Motivation

The GDH sum rule concerns the interaction of circularly polarized photons with longitudinally polarized particles of arbitrary spin S [1]. The respective cross sections for parallel and antiparallel spin orientation σ_p resp. σ_a are predicted to obey the relation

$$\int_0^\infty d\nu \frac{\sigma_p - \sigma_a}{\nu} = \frac{4\pi^2 e^2}{m^2} S \cdot \kappa^2,$$ (1)

where ν is the photon energy, e is the elementary charge, m is the mass and κ is the anomalous magnetic moment of the particle.

For particles with spin 1/2 - like proton and neutron - this relation obtains the form

$$\int_0^\infty d\nu \frac{\sigma_{3/2} - \sigma_{1/2}}{\nu} = \frac{2\pi^2 \alpha}{m^2} \kappa^2.$$ (2)

Mostly this formula is referred to as the GDH sum rule. It is derived from very general principles applied on the Compton forward amplitude on the nucleon. The general ansatz for this amplitude can be written out between initial and final nucleon Pauli spinors χ_i and χ_f and initial and final real photon polarisation vectors e and e'

$$f(\nu) = \chi_f^* (f_1(\nu) e'^* . e + i f_2(\nu) \sigma . (e'^* \times e)) \chi_i.$$ (3)

For the two relative spin configurations we find $f_{3/2} = f_1 - f_2$ and $f_{1/2} = f_1 + f_2$. The three following fundamental principles are the base of the GDH sum rule:

1. Causality implies analyticity of f_2 which allows one to write a dispersion relation for the forward amplitude without subtraction :

   $$\mathrm{Re} f_2(\nu) = \frac{2\nu}{\pi} P \int_0^\infty d\nu' \frac{\mathrm{Imm}\, f_2(\nu')}{\nu'^2 - \nu^2}.$$ (4)

2. Flux conservation implies unitarity which is expressed in the optical theorem:

   $$\mathrm{Imm} f_2(\nu) = \frac{\nu}{8\pi} (\sigma_{1/2} - \sigma_{3/2}).$$ (5)

3. Relativity and gauge invariance are the only ingredients needed to prove the Low theorem [3]. It states that the first two terms in the expansion of the Compton scattering amplitude as a function of energy of a system of spin 1/2 are expressed in terms of the macroscopic properties of the system:

   $$f(\nu) = -\frac{\alpha}{m} e'^* . e - i \frac{\alpha}{2m^2} \kappa^2 \nu \sigma . (e'^* \times e).$$ (6)

   The constant represents the Thomson term which depends on the charge and the mass, the linear term depends also on the anomalous magnetic moment of the system.

By combining these 3 relations the Gerasimov-Drell-Hearn sum rule is obtained. The only questionable assumption is the no-subtraction hypothesis made in writing the dispersion relation of the Compton forward amplitude. This hypothesis relies on a certain asymptotic behaviour of the Compton forward amplitude f_2 at infinity (decreasing to 0). There are no theorical arguments to rule out this assumption, but due to its fundamental character this sum-rule deserves a verification. A good view on the theoretical implications of the sum rule is given in [4].

The prediction of the sum rule - i.e. the right hand side of equation (1) - is mainly given by the anomalous magnetic moment squared and can be calculated for the proton and for the neutron. Alternatively, the nucleon isoscalar and isovector magnetic moments resp. sum rules can be formulated. The left hand side of equation (1) has been estimated by looking into the existing single pion photoproduction data (cross sections and single polarisation observables) and the related multipole analyses; some assumptions on double pion production are also included [5]. From these estimates the isoscalar sum rule seems to be satisfied while the isovector sum rule shows a deviation. As the uncertainties of these estimates cannot be quantified, these observations can only be taken as a motivation to perform the GDH measurement for the proton and for the neutron. We will use a polarized deuteron target for the neutron measurement.

3   The GDH Experiment

The GDH experiment consist of the mesurement of the polarized total hadronic photoabsorbtion cross sections $\sigma_{3/2}$ and $\sigma_{1/2}$ as a function of the photon energy. We choose diffenrent techniques to measure this cross sections for low photon energies compared to high photon energies.

The GDH experiment for low photon energies (140 - 800 MeV) is performed at MAMI in Mainz using the DAPHNE detector [6] and special detector components for forward angles down to $\Theta = 1.5°$. At these energies mainly only 2 or 3 particles occur in the final state so that the identification of the reaction channel can be performed event by event. As the produced particles may have relatively low energies, a major task is the extrapolation of cross sections into kinematical regions where the particles do not emerge from the target and thus do not trigger the detector.

For high photon energies (500 - 3000 MeV) the GDH experiment is being set-up at ELSA in Bonn. Multi particle final states dominate the total cross section in this energy region. The energy of the produced particles is high enough to trigger the detector. Here, the problem is mainly to separate atomic background events from hadronic products, because both types of events are concentrated at forward angles due to the Lorentz boost at high energies. A special "GDH detector" has been built to meet these requirements.

A top view on the experimental set-up at ELSA (similar components are used in a similar arrangement for the GDH experiment at MAMI) is shown in Fig. 1. The beam of polarized electrons are extracted from ELSA. The transverse and vertical polarization of the electrons is rotated into the horizontal plane by a superconducting solenoid magnet and then rotated into the longitudinal direction by dipole magnets. The electrons are directed to a radiator foil from which Bremsstrahlung photons are radiated. The individual electrons are analysed in the Tagging spectrometer so that the energy and flux of the related photons is determined. Behind the Tagging spectrometer the polarization of the electrons is measured in all three space directions by a large acceptance two arm Moller spectrometer.

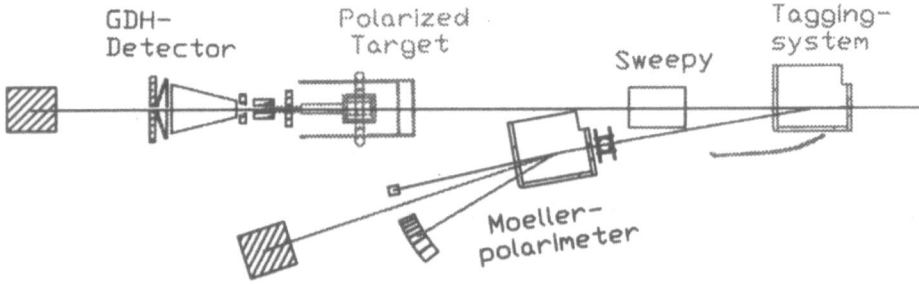

Figure 1. Top view on the GDH experiment at ELSA.

The photon beam is defined in its transverse size by a set of active collimators, it is cleaned from charged particles by a dipole magnet and at the end of the beam line the photons are detected individually in a beam dump detector.

The polarized protons are produced in a solid state frozen spin target with butanol as target material [7].

The hadronic cross section is measured by the GDH detector which has a complete azimuthal coverage and polar acceptance of $1.5° \leq \Theta \leq 170°$. It consists of a lead scintillator sandwich structure with practically 100 % efficiency for charged particles and 90 % for photons ($E_\gamma \geq 100$ MeV).

A gas cerenkov detector covering polar angles $\Theta = 0°$ to $15°$ gives a veto signal on atomic background events like Compton scattering or pair production. The whole set-up has been tested with unpolarized cross section measurments.

At MAMI in Mainz the GDH collaboration is taking data off a polarized proton target since May 1998. Very prelimenary results of a first analysis show the expected behaviour in the Delta-resonance region: the cross section $\sigma_{3/2}$ is larger than the $\sigma_{1/2}$. This is demonstrated in Fig. 2 for the channel $\gamma p \to p\pi°$.

After the low energy measurements in Mainz the polarized target will be brought to Bonn and the GDH experiments will be continued for the high photon enegies.

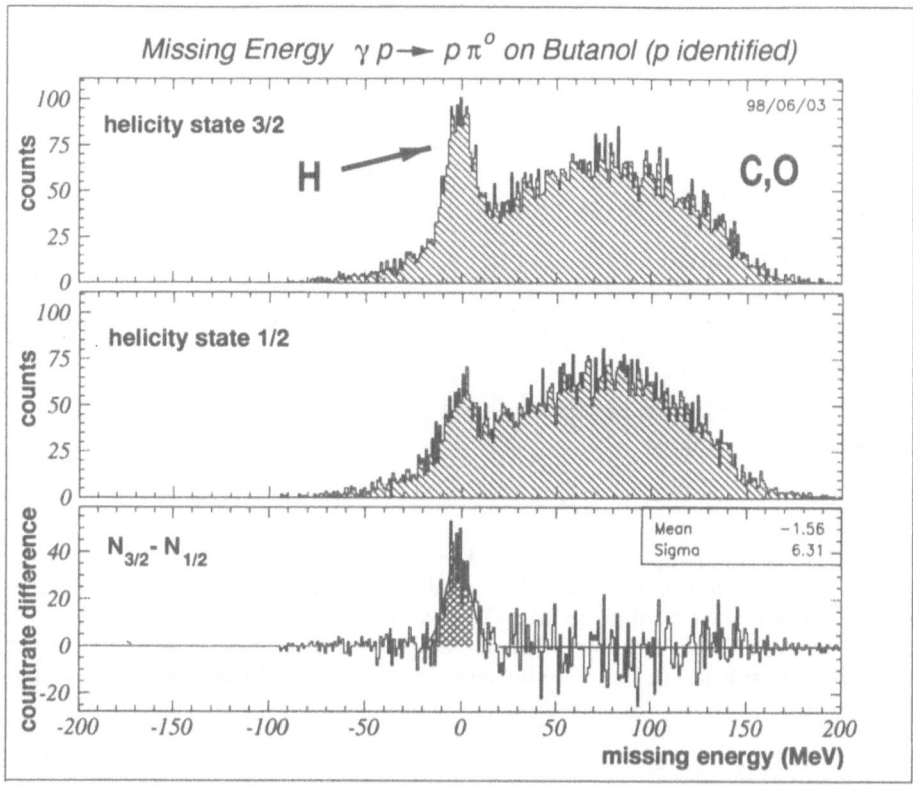

Figure 2. "Missing energy" spectrum under the assumption of a $\gamma p \to p\pi^o$ reaction for the helicity states 3/2 and 1/2 and for the difference of both.

References

1. S.B. Gerasimov: Sov. J. Nucl. Phys. **2**, 430 (1966); M. Hosada: Prog. Theor. Phys. **36**, 426 (1966)

2. S.D. Drell and A.C. Hearn: Phys. Rev. Lett. **16**, 908 (1966); S.B. Gerasimov: Yad. Fiz. **2**, 598 (1965); Sov. J. Nucl. Phys. **2**, 430 (1966)

3. F. Low: Phys. Rev. **96**, 1428 (1954); M. Gell-Mann and M.L. Goldberger: Phys. Rev. **D1**, 1319 (1970)

4. R. Pantförder, H. Rollnik, W. Pfeil: Eur. Phys. J **C1** , 585 (1998)

5. I. Karliner: Phys. Rev. **D7**, 2717 (1973); R.L. Workman, R.A. Arndt: Phys. Rev. **D45**, 297 (1992)

6. G. Audit et al.: Nucl. Instr. Meth. **A301**, 473 (1991)

7. H. Dutz et al.: Nucl. Inst. Meth. **A356**, 111 (1995)

Few-Body Systems Suppl. 11, 182–192 (1999)

Few-
Body
Systems
© by Springer-Verlag 1999

Dispersion Approach to Pion Photoproduction and GDH Sum Rule

D. Drechsel

Johannes Gutenberg-Universität, Institut für Kernphysik, Becherweg 45, 55099 Mainz, Germany

Abstract. Pion photoproduction has been analyzed in the framework of dispersion relations at constant t. The results for S-wave pion production at threshold and P-wave production in the Δ (1232) resonance region are presented and compared to the data and theoretical expectations. The predictions for the GDH sum rule are shown to be very sensitive to the threshold S-wave amplitude.

1   Introduction

Ever since Yukawa introduced the pion in 1935, the pion has been the centerpiece of our understanding of the strong interactions. When charged and neutral pions were finally discovered in the late 1940's, Kemmer's isospin invariant theory of nuclear forces was already well accepted. Though in the meantime QCD has taught us that quarks and gluons are the building blocks of matter, the elementary particle "pion" has not dissolved into these partons. Instead it has been revived in the form of a Goldstone boson, as a necessary consequence of the (partial) conservation of the axial current, due to the small (current) masses of the light quarks. The extent to which the axial current is not conserved is essentially ruled by the decay of the pions: by the decay of a charged pion, axial current leaks from the hadronic into the leptonic world, and by the neutral pion decay into the electromagnetic field.

The photoproduction of pions has been studied in the 60's and 70's at various places. Due to bad statistics and large systematical errors, the accuracy of the data was generally moderate. This has changed over the past decade with modern high current, high duty-factor electron accelerators or laser backscattering facilities. By now statistical errors of the order of 1 % have been achieved, and coincidence experiments (will) make it possible to separate the various structure functions appearing with polarization degrees of freedom.

The threshold production of pions is ruled by low energy theorems (LET), which were established by Kroll and Ruderman as early as 1954 (see Ref. [1] for a review). Recently Bernard, Kaiser and Meißner [2] have shown that the LET has to be modified by loop corrections in the framework of chiral perturbation theory (ChPT). While these developments have provided a good theoretical understanding in the threshold region, model calculations at the higher energies are badly failing in comparison with the accurate experimental data. As an example, all quark model calculations miss the strength of the Δ (1232) resonance by typically 30 %. In order to analyse and interpret the existing precision data, we therefore have studied pion photoproduction in the framework of dispersion relations [3, 4, 5]. Such an approach makes it possible to probe the internal consistency of the data and, by means of continuation of the analytical functions involved, to predict the results in neighbouring kinematical regions.

In section 2 I briefly summarize the formalism of pion photoproduction. An introduction to dispersion theory at fixed t is given in section 3, together with a description of our numerical methods and approximations. I then discuss pion threshold production in section 4, and the excitation of the Δ (1232) resonance in section 5. The connection between Compton scattering and pion photoproduction by sum rules is reviewed in section 6, followed by our predictions for the Gerasimov-Drell-Hearn (GDH) sum rule in section 7. I conclude with a brief summary in section 8.

2 Formalism

We consider the scattering matrix for the reaction $\gamma + N \rightarrow \pi + N'$, with γ a real photon, N and N' nucleons in the initial and final states respectively, and π a pion. Due to the spins of the involved particles, there are four independent amplitudes for transversely polarized photons, which can be written in terms of relativistic invariants, CGLN amplitudes, or helicity amplitudes (see Ref. [6] for a recent review). These amplitudes can be decomposed into a series of electric multipoles $E_{l\pm}(W)$ and magnetic multipoles $M_{l\pm}(W)$, which are functions of only one variable, the total energy in the *cm* frame, W, or the corresponding *lab* energy of the photon, E_γ. The multipoles are denoted by the orbital angular momentum of the pion-nucleon system, l, the $\pm$ sign indicating whether angular momentum and spin are parallel or antiparallel ($J = l \pm \frac{1}{2}$).

In addition the amplitudes are decomposed in isospace,

$$\mathcal{M}_l^{(\alpha)} = \frac{1}{2}[\tau_\alpha, \tau_0]\mathcal{M}_l^{(-)} + \frac{1}{2}\{\tau_\alpha, \tau_0\}\mathcal{M}_l^{(+)} + \tau_\alpha \mathcal{M}_l^{(0)} . \tag{1}$$

The first two amplitudes on the right hand side can also be combined to

$$\mathcal{M}_l^{(\frac{3}{2})} = \mathcal{M}_l^{(+)} - \mathcal{M}_l^{(-)}, \; \mathcal{M}_l^{(\frac{1}{2})} = \mathcal{M}_l^{(+)} + 2\mathcal{M}_l^{(-)} , \tag{2}$$

the upper index denoting the isospin of the final state. The 4 physical amplitudes are then given in terms of linear combinations of the 3 isospin amplitudes [6]. We note, however, that the isospin symmetry is broken by the mass differences between the nucleons (n, p) and pions $(\pi^{\pm}, \pi^0)$ and by explicit Coulomb effects, in particular near threshold.

3 Dispersion relations at fixed t

With the usual assumptions of causality, analyticity, unitarity and crossing symmetry, we may write a dispersion relation at fixed t for the multipoles leading to a system of integral equations coupled by kernels $K_{ll'}$. These kernels are known but rather complicated functions of the energy. The convergence of the multipole series has been proven up to $E_{\gamma} < 450$ MeV. At this energy the so-called Lehmann ellipse touches the spectral region in the Mandelstam plane. Since little is known about this region, it might well be that the multipole series gives meaningful results even above 450 MeV.

Defining the phases Φ_l of the multipoles, $\mathcal{M}_l = |\mathcal{M}_l|e^{i\Phi_l}$, and the function $h_l = \sin\Phi_l e^{i\Phi_l}$, we obtain

$$
\begin{aligned}
\mathcal{M}_l(W) &= \mathcal{M}_l^{\text{Born}}(W) + \frac{1}{\pi}\int_{\text{thr}}^{\infty}\frac{h_l^*(W')\mathcal{M}_l(W')}{W' - W - i\varepsilon}dW' \\
&+ \frac{1}{\pi}\sum_{l'}\int_{\text{thr}}^{\infty}K_{ll'}(W', W)h_{l'}^*(W')\mathcal{M}_{l'}(W')dW',
\end{aligned}
\tag{3}
$$

where the singular part of the diagonal kernel K_{ll} has been treated explicitly. The result is a system of integral equations with inhomogenous terms (Fredholm equations). Solving first the characteristic equations ($l = l'$), the Fredholm equations can then be solved directly or by iteration, which we prefer.

In our explicit numerical evaluation of Eq. (3) we encounter the following problems:

1) As is well known, we can add arbitrary multiples of the homogeneous solutions to obtain the final result. In the case of 6 of the partial waves this seems to be necessary to obtain a good fit to the data, which accounts for 6 parameters in our description.

2) Up to two-pion threshold the phase Φ_l is simply the scattering phase δ_l of the pion-nucleon system (Fermi-Watson theorem). Above two-pion threshold we use an ansatz for Φ_l in terms of the scattering phase δ_l and the inelasticity η_l. This ansatz is unitary but not unique, and we are guided by early investigations in that field [7] and the existing data.

3) Above $E_{\gamma} = 2$ GeV the existing information is scarce. Therefore we replace the integral from 2 GeV to infinity by t-channel exchange of ρ and ω mesons. Since some part of the t-channel contribution is already contained in the dispersion integral up to 2 GeV (duality!), only a certain fraction will have to be added, which introduces 4 free parameters (the vector and tensor couplings of ω and ρ).

Altogether we describe the coupled S, P and D_{13} waves by 10 free parameters, while the higher partial waves are approximated by the Born terms. If necessary, these Born terms are unitarized with the pion-nucleon scattering phase.

All observables are now uniquely defined in terms of the multipoles discussed in the preceding section. They include the differential cross section $d\sigma/d\Omega$, the single polarization asymmetries Σ (polarized photon beam), T (polarized target) and P (recoil polarization), and double polarization variables.

We have fitted the free parameters of our model to recent Mainz data for π^0 and π^+ production in the range of 160-420 MeV for both $d\sigma/d\Omega$ and Σ, to recent and older Bonn data for π^+ production including T, to old Frascati data for $\gamma n \to p\pi^-$ and recent TRIUMF data for radiative pion capture, $p\pi^- \to \gamma n$. A complete list of references to these experimental investigations is given in Ref. [5]. Altogether our basis includes 1377 data points. With only 10 parameters in our global fit, we obtain $\chi^2 = 2.8$ per data point.

4 Threshold pion photoproduction

As has been pointed out before, our data basis does not include the threshold region. At these energies we have to expect isospin symmetry breaking due to the pion mass splitting, while our dispersion calculation is based on the isospin symmetry of the pion-nucleon phase shifts. Therefore, the threshold S-wave amplitudes described below are a prediction, in the sense that the cross sections above $E_\gamma = 160$ MeV determine these threshold values quite accurately by continuation of the analytical functions constructed in dispersion theory. Table 4.1 compares our threshold values for the charged pion channels to the "classical" low energy theorem [1] (LET), ChPT [8] and experiment. Note that ChPT contains the lowest order loop corrections, while "LET" is based on tree graphs only. Due to the coupling between the channels, the real part of E_{0+} obtains large contributions from the imaginary parts of the higher multipoles via the dispersion integrals of Eq. (3). Altogether these contributions nearly cancel the large contribution of the Born terms, which correspond to the result of pseudoscalar coupling, leading to a total threshold value [4]

Table 4.1. The S-wave amplitude E_{0+} at threshold in units of $10^{-3}/m_\pi$.

	$\gamma p \to \pi^+ n$	$\gamma n \to \pi^- p$	$\gamma p \to \pi^0 p$
"LET" [1]	27.5	-32.0	-2.4
ChPT [8]	$28.2 \pm .6$	$-32.7 \pm .6$	-1.16
this work	28.4	-31.9	-1.22
experiment	$28.3 \pm .2$ [9]	$-31.8 \pm .2$ [9]	$-1.31 \pm .08$ [10, 11]

$$\mathrm{Re}E_{0+}^{\mathrm{thr}}(p\pi^0) = -7.63 + 4.15 - 0.41 + 2.32 + 0.29 + 0.07 = -1.22,$$
$$\mathrm{Re}E_{0+}^{\mathrm{thr}}(n\pi^0) = -5.23 + 4.15 - 0.41 + 3.68 - 0.93 - 0.05 = 1.19, \quad (4)$$

where the individual contributions on the *rhs* are, in that order, the Born term, $M_{1+}, E_{1+}, E_{0+}, M_{1-}$ and higher multipoles.

While the differential cross section receives its forward-backward asymmetry essentially from the combination $\mathrm{Re}\{E_{0+}^*(M_{1+}+3E_{1+}-M_{1-})\}$, Σ is dominated by $\mathrm{Re}\{M_{1+}^*(E_{1+}+M_{1-})\}$ and T by $\mathrm{Im}\{E_{0+}^*(E_{1+}-M_{1+})\}$. Since E_{1+} is small, the threshold value of Σ is surprisingly sensitive to the multipole M_{1-} resonating at the Roper resonance $N^*(1440)$. The observable T, on the other side, measures the phase of pion-nucleon S-wave scattering at threshold relative to the phase of the Δ (1232) multipole.

5 Pion production in the Δ (1232) region

The search for a deformation of the "elementary" particles is a longstanding issue. Such a deformation is evidence for a strong tensor force between the constituents, originating in the case of the nucleon from the residual force of gluon exchange. Depending on one's favourite model, such effects can be described by D-state admixture in the quark wave function, tensor correlations between the pion cloud and the quark bag, or by exchange currents accompanying the exchange of mesons between the quarks. Unfortunately, it would require a target with a spin of at least 3/2 (e.g. Δ matter) to observe a static deformation. An alternative is to measure the transition quadrupole moment between the nucleon and the Δ, i.e. the amplitude E_{1+}, which is sensitive to model parameters responsible for a possible deformation of the hadrons.

Table 5.1. Results of the speed plot technique for the pole position of the Δ (1232) and the complex residues of the multipoles at the pole [3].

	M_R (MeV)	Γ_R (MeV)	$r(10^{-3}/m_\pi)$	Φ (deg)
M_{1+}	1212 ± 1	99 ± 2	21.16	-27.5
E_{1+}	1211 ± 1	102 ± 2	1.23	-154.7

The experimental quantity of interest is the ratio $R_{EM} = E_{1+}/M_{1+}$ in the region of the Δ. The comparison of our dispersion analysis [3] with the analysis of the Mainz data by Beck et al. [12] shows nice agreement with the multipole M_{1+} but some differences for the quadrupole transition E_{1+} at energies above the resonance. Unfortunately, there appears to be a discrepancy of up to 15 % between the absolute cross sections measured at Mainz (upon which our analysis is largely based) and the LEGS data [13] at Brookhaven. As far as ratios of cross sections are concerned, the data seem to be in general agreement. As an example, the ratio of the LEGS cross sections for photons with polarization parallel and orthogonal to the scattering plane, is nicely described by our analysis, though these data have not been included in our fit. In view of the importance of the quantity R_{EM} for our understanding of the internal structure of the nucleon, the existing discrepancies between the data deserve further studies. Based on "our" data selection we obtain a value $R_{EM} = (-2.5 \pm 0.1)$ %, which is close to the value derived by Beck et al. [12].

Due to the analyticity of the functions involved, dispersion theory is particularly apt to study the amplitudes in the complex energy plane. Therefore, we have analyzed the amplitudes by comparing the "speed"

$$SP\left[\mathcal{M}_{1+}(W)\right] = |d\mathcal{M}_{1+}(W)/dW| \; ,$$

with the speed of an ideal resonance,

$$\mathcal{M}^{\mathrm{res}}(W) = re^{i\Phi}\Gamma_R/(M_R - W - i\Gamma_R/2) \; ,$$

with $re^{i\Phi}$ the complex residue at resonance, $W = W_R - i\Gamma_R/2$. As may be seen in Table 5.1, the values for the pole position are in perfect agreement with the analysis of pion-nucleon scattering. Essentially based on investigations of Höhler et al., the PDG [14] lists $M_R = (1210 \pm 1)$ MeV and $\Gamma_R = (100 \pm 2)$ MeV.

6 Compton scattering and sum rules

The Compton amplitude is described by 6 independent structure functions depending on the photon energy ω and the scattering angle θ, $A_i = A_i(\omega, \theta)$. In the case of forward scattering ($\theta = 0°$), only the structure functions A_1 and A_3 survive. Their low-energy expansion is

$$A_1(\omega, \theta = 0) \quad = \quad f(\omega) = -\frac{e^2}{m} + 4\pi(\alpha + \beta)\omega^2 + \; ... \; , \tag{5}$$

$$A_3(\omega, \theta = 0) \quad = \quad g(\omega) = -\frac{e^2\kappa^2}{2m^2}\omega + 4\pi(\gamma_1 - \gamma_2 - 2\gamma_4)\omega^3 + \; ... \; , \tag{6}$$

where α and β are the familiar elctric and magnetic polarizabilities respectively, and $\gamma = \gamma_1 - \gamma_2 - 2\gamma_4$ is the "forward" spin polarizability.

On the basis of analyticity/causality and crossing symmetry, we can derive dispersion relations for these amplitudes. Expressing then the imaginary part of the functions by cross sections via the optical theorem and expanding in ω, we obtain

$$f(\omega) = -\frac{e^2}{m} + \left[\frac{2}{\pi}\int\frac{\sigma_{\mathrm{tot}}(\omega')}{\omega'^2}d\omega'\right]\omega^2 + \; ... \; , \tag{7}$$

$$g(\omega) = \left[\frac{2}{\pi}\int\frac{\Delta\sigma(\omega')}{\omega'}d\omega'\right]\omega + \left[\frac{2}{\pi}\int\frac{\Delta\sigma(\omega')}{\omega'^3}d\omega'\right]\omega^3 \; ... \; , \tag{8}$$

i.e. the coefficients of the power series expansion in ω can be constructed by integrals over the cross sections σ_{tot} and $\Delta\sigma$. The latter may be obtained by a double-polarization experiment, absorbing circularly polarized photon on the nucleon, with polarization in the direction of the photon or opposite to it, leading to intermediate states with spin projection 1/2 or 3/2. In terms of the absorption into these states, we have $\sigma_{\text{tot}}(\omega) = \sigma_{1/2}(\omega) + \sigma_{3/2}(\omega)$ and $\Delta\sigma(\omega) = \sigma_{1/2}(\omega) - \sigma_{3/2}(\omega)$. Of course, the expansion in ω is only possible below the onset of the branch cut at pion threshold, i.e. for $\omega < \omega_{\text{thr}}$, beyond which value the amplitudes f and g become complex functions.

From Eqs. (5) and (8), we obtain a series of relations, Baldin's sum rule [15]

$$\alpha + \beta = \frac{1}{2\pi^2} \int \frac{\sigma_{\text{tot}}(\omega')}{\omega'} d\omega' \ , \tag{9}$$

the Gerasimov-Drell-Hearn sum rule (GDH) [16]

$$-\frac{\kappa^2 e^2}{2m^2} = \frac{1}{\pi} \int \frac{\Delta\sigma(\omega')}{\omega'} d\omega' \ , \tag{10}$$

a sum rule for the "forward" spin-polarizability,

$$\gamma = \gamma_1 - \gamma_2 - 2\gamma_4 = \frac{1}{4\pi^2} \int \frac{\Delta\sigma(\omega')}{\omega'^3} d\omega' \ , \tag{11}$$

and similar expressions for the higher moments. In particular we have evaluated the one-pion contribution to the four spin polarizabilities γ_1 to γ_4 [17] and to the GDH integral [18].

Concerning Baldin's sum rule there is all reason to believe that it converges. Above the resonance region, for $\omega \gtrsim 1$ GeV, σ_{tot} settles at a constant value of about 110 μb with a slow logarithmic rise at the highest energies between 100 und 300 GeV. The GDH integrand, on the contrary, is a rapidly varying function with both positive and negative contributions. Though convergence is generally assumed on the basis of Regge models, a divergence of the integral cannot be ruled out.

7 The GDH sum rule

The Constituent Quark Model (CQM) developed by Isgur, Karl and others in the 70's has been quite successful in describing the excitation spectrum of the nucleon and the transition probabilities to its excited states. However, a correct calculation of the GDH in that framework requires a very careful consideration of relativistic corrections. This becomes evident already in the case of the simple nonrelativistic CQM in an oscillator basis, without the hyperfine interaction. With the quark masses $m_q = m/3$ and no anomalous magnetic moments of the quarks, $\kappa_q = 0$, we have $\kappa_p = 2$ and $\kappa_n = -2$, hence $\kappa^2/4 = 1$ for both nucleons. Normalizing then the GDH prediction to that value, i.e. to unity, we obtain for the contribution of the excited states the GDH integral for the proton [19].

Table 7.1. Predictions from various models and data analyses for the GDH integral for proton (I_p), neutron (I_n), and the difference $I_p - I_n$ in units of μb. With the exception Ref. [23], the two-pion background has been taken from Ref. [21].

	I_p	I_n	$I_p - I_n$
GDH integral	-204.5	-232.8	28.3
Karliner [21]	-261	-183	- 78
Workman and Arndt [22]	-260	-157	-103
Burkert and Li [23]	-223		
Sandorfi et al. [24]	-289	-160	-129
this work	-261	-180	- 81

$$I = 1 - \frac{5}{4} \sum_{n=1}^{\infty} \frac{1}{n!} (n\zeta)^{2n} e^{-(n\zeta)^2} , \tag{12}$$

where the 1 on the *rhs* is the contribution of the NΔ transition, and the sum runs over the contributions of the higher oscillator shells, i.e. $n = 1$ for the $1\hbar\omega_0$ states etc. With $\zeta^{-2} = 3m_q^2 < r^2 > \approx (0.57)^{-1}$, corresponding to a bag radius of about 0.5 fm, the Δ contribution is practically completely compensated by the contributions of the higher shells, leading to an overall "prediction" I=0.04. Since the contributions of the higher shells are $O(m_q^{-2n})$ for the nth shell, it is obvious that they originate from terms beyond the validity of a nonrelativistic model. But even if we introduce the usual spin-orbit and Darwin-Foldy terms $O(m_q^{-2})$, we can only expect to correct the next-but-leading term, but not the contributions of higher orders. The prediction would be still bad, I=0.44, except if magically all inaccuracies in the higher orders would cancel.

With the hyperfine interaction switched on, the ground state configuration of the nucleon contains admixtures of the 2nd oscillator shell, and the correct result of the GDH can only be reproduced if relativistic two-body exchange currents are taken into account [20]. Such currents are of order $(m_q m_N)^{-1}$, and they depend on the charges and momenta of both the constituents and the composite system.

Due to the strong energy dependence of $\Delta\sigma$, the GDH is extremely sensitive to small changes of the multipole content as may be seen in Table 7.1. Comparing, e.g., the recent analysis of Sandorfi et al. [24] with our work [18], we find the origin of the difference in an underestimation of the threshold S-wave multipole of SAID [25] for π^+ photoproduction, E_{0+}^{thr} (SAID) $= 24.9 \cdot 10^{-3}/m_\pi$, while our value [3], the prediction of ChPT [26] and the old threshold data [9] are in the range of $(28.3 \pm 0.1)10^{-3}/m_\pi$. Because of the opposite signs of the E_{0+} and M_{1+} contributions, relatively small differences translate into a substantial effect for the GDH prediction, the 38 % overprediction (in absolute values) of [24] is reduced to 28 % [18].

The remaining discrepancy of 56.5 μb between our analysis and the GDH is of the same order as the two-pion contribution, estimated to be 65 μb in a resonance model [21]. However, it is not obvious to which extent the two-pion continuum should follow the resonance structure of the single-pion amplitudes.

The most striking discrepancy concerning the GDH is the proton– neutron difference (last column of Table 7.1). The overestimate for the proton and a similar underestimate for the neutron lead to a proton-neutron difference of -129 μb, while the GDH predicts 28 μb. However, one should keep in mind that the GDH for the neutron is measured by using ^3He or ^2H targets. The GDH for ^3He can be obtained from Eq. (10) by inserting the appropriate values for mass, charge and anomalous magnetic moment. The result is $I(^3He) = -496$ μb, which is more than twice the value of the "neutron target", $I(n) = -233$ μb. Of course, the bulk part of the difference of -263 μb will be contained in nuclear excitations of ^3He at energies far below pion threshold, but how should one account for meson exchange currents near pion threshold or two-body break-up by isobaric currents in the Δ resonance region? In these cases the driving term is certainly related to the E_{0+} or M_{1+} strength respectively, producing a pion from a nucleon followed by reabsorption by another target nucleon. In the case of the deuteron, Arenhövel et al. [27] have carefully studied the interplay between nuclear and nucleonic degrees of freedom. Naively one would expect that the deuteron, a weakly bound object, should have a GDH integral of $I(p) + I(n) = -438$ μb. However, due to its spin structure, the deuteron has $I(^2H) = -0.65$ μb. The nuclear and nucleonic contributions cancel to 3 decimals! In the energy region up to $E_\gamma =550$ MeV, Arenhövel et al. [27] find a total of 413 μb from two-body break-up and about -94 μb resulting from mesonic and isobaric currents. Since the GDH for the neutron has been evaluated from pion photoproduction multipoles, the reabsorption of the pion due to mesonic and isobaric currents is certainly not taken care of in the existing analysis. If we attribute half of the -94 μb originating from two-body currents to the neutron, and add them to our result in table 7.1, the discrepancy with the GDH has magically disappeared! While these arguments are certainly no proof of the GDH, they indicate that the GDH strength of a bound nucleon is not described by quasi-free pion production only, but that there are non negligible contributions from other kinematical corners.

8   Summary

Dispersion relations at constant t provide a framework for analysing the multipole content of pion photoproduction. The differential cross sections at threshold are predicted in good agreement with recent data and the results of ChPT. Our predictions are very sensitive to the spectral functions at the higher energies, in particular the polarization observables are strongly influenced by the Roper multipole M_{1-}.

In the region of the Δ (1232), the analytical functions provided by dispersion theory reproduce the position of the Δ pole in the complex energy plane even in the case of the weak E_{1+} multipole. The ratio $R = E_{1+}/M_{1+}$ at resonance is predicted to be (-2.5 ± 0.2) %.

The spin structure of the nucleon seen in deep inelastic scattering (DIS) is connected with the physics in the resonance region by the generalized GDH integral and the Burkhardt-Cottingham sum rule [28]. At the real photon point, the GDH predicts this integral in terms of the anomalous magnetic moment of the nucleon. The existing discrepancies for both neutron and proton value, and in particular for the difference of these two values, could be due to shortcomings of the present analysis, such as the contributions of two-pion and higher mass states and, in the case of the neutron, binding effects.

The GDH is now being measured at MAMI and ELSA with circularly polarized photons and longitudinally polarized targets [29]. This will be the first direct measurement of the sum rule, which is expected to yield the contributions of the individual decay channels at the same time. Further experiments have been proposed and partly scheduled at Jefferson Lab, in order to study the gradual transition from the coherent processes seen by real photons to the regime of incoherent DIS off the constituents of the nucleon [30, 31]. The outcome of these experiments will be invaluable for our understanding of low-energy QCD in general and of the spin structure of the nucleon in particular.

References

1. P. de Baenst: Nucl. Phys. **B24**, 633 (1970)

2. V. Bernard et al.: Phys. Lett. **B268**, 219 (1991)

3. O. Hanstein et al.: Phys. Lett. **B385**, 45 (1996)

4. O. Hanstein et al.: Phys. Lett. **B399**, 13 (1997)

5. O. Hanstein et al.: Nucl. Phys. **A632**, 561 (1998)

6. D. Drechsel and L. Tiator: J. Phys. **G18**, 449 (1992)

7. W. Pfeil and D. Schwela: Nucl. Phys. **B45**, 379 (1972)

8. V. Bernard et al.: Phys. Lett. **B378**, 337 (1996); **B383**, 116 (1996)

9. M.I. Adamovich: Proc. P.N. Lebedev Phys. Inst. **71**, 119 (1996)

10. M. Fuchs et al.: Phys. Lett. **B368**, 20 (1996)

11. J.C. Bergstrom et al.: Phys. Rev. **C53**, 1052 (1996)

12. R. Beck et al.: Phys. Rev. Lett. **78**, 606 (1997)

13. G. Blanpied et al.: Phys. Rev. Lett. **79**, 4337 (1997)

14. Review of Particle Physics: Phys. Rev. **D54**, Part I (1996)

15. A.M. Baldin: Nucl. Phys. **18**, 310 (1960)

16. S.B. Gerasimov: Yad. Fiz **2**, 839 (1965); S.D. Drell and A.C. Hearn: Phys. Rev. Lett. **16**, 908 (1966)

17. D. Drechsel et al.: Phys. Lett. **B429**, 248 (1998)

18. D. Drechsel and G. Krein: submitted to Phys. Rev., MKPH-T-98-9

19. D. Drechsel and M.M. Giannini: Few-Body Systems **15**, 99 (1993)

20. M. de Sanctis et al.: Few-Body Systems **16**, 143 (1993)

21. I. Karliner: Phys. Rev. **D7**, 2717 (1973)

22. R.L. Workman and R.A. Arndt: Phys. Rev. **D45**, 1789 (1992)

23. V. Burkert and Z. Li: Phys. Rev. **D47**, 46 (1993)

24. A.M. Sandorfi et al.: Phys. Rev. **D50**, R6681 (1994)

25. VPI-SAID program: *"The Scattering Analysis Interactive Dial-in program"*, data available via telnet to VT-INTE.PHYS.VT.EDU, solution SP97K

26. V. Bernard, N. Kaiser, and U.-G. Meißner: Nucl. Phys. **B383**, 442 (1992)

27. H. Arenhövel et al.: Phys. Lett. **B407**, 1 (1997)

28. H. Burkhardt, W.N. Cottingham: Ann. Phys. (N.Y.) **56**, 543 (1970)

29. J. Arends et al. (GDH Collaboration): ELSA and MAMI Proposal A2/2-95

30. V.D. Burkert et al.: CEBAF Proposal 91-023

31. S.E. Kuhn et al.: CEBAF Proposal 93-009

Few-Body Systems Suppl. 11, 193–196 (1999)

Few-
Body
Systems
© by Springer-Verlag 1999

DIS Contribution to the GDH Sum Rule

N. Bianchi
for the HERMES Collaboration

Istituto Nazionale di Fisica Nucleare, Laboratori Nazionali di Frascati,
00044 Frascati, Italy

Abstract. We present the HERMES preliminary data of virtual photon absorption cross section difference $[\sigma_{1/2} - \sigma_{3/2}]$ for the proton and neutron in deep inelastic scattering of 27.6 GeV longitudinally polarised positrons from polarised ^{1}H and ^{3}He internal gas targets. The data were collected in the region above the nucleon resonances. The DIS contribution to the generalised Gerasimov-Drell-Hearn (GDH) integral for the nucleons is shown.

The GDH sum rule [1] was derived starting from a general dispersive relation for the forward Compton scattering amplitudes which follows from causality, crossing symmetry and unitarity principles

$$\int_0^\infty \left[\sigma_{1/2}(\nu) - \sigma_{3/2}(\nu)\right]\frac{d\nu}{\nu} = -\frac{2\pi^2\alpha}{M^2}\kappa^2, \tag{1}$$

where $\sigma_{1/2}$ and $\sigma_{3/2}$ are the photo-absorption cross sections for total helicities 1/2 and 3/2, ν is the photon energy, M is the nucleon mass and κ is the anomalous contribution to the nucleon magnetic moment.

This sum rule has never been tested experimentally. Evaluations from single pion photo-production amplitudes in the resonance region suggested a possible violation of the sum rule, mainly in the isovector channel. For this reason, several real photon experiments are currently planned. (See G.Anton's contribution to this workshop).

The integral defined in Eq. (1) can be generalised to the absorption of virtual photons with energy ν and squared four-momentum $-Q^2$

$$I(Q^2) = \int_{Q^2/2M}^\infty \left[\sigma_{1/2}(\nu, Q^2) - \sigma_{3/2}(\nu, Q^2)\right]\frac{d\nu}{\nu}, \tag{2}$$

$$= \frac{8\pi^2\alpha}{M} \int_0^1 \frac{g_1(x, Q^2) - \gamma^2 g_2(x, Q^2)}{K}\frac{dx}{x},$$

where g_1 and g_2 are the polarised structure functions of the nucleon, x is the Bjorken variable and K is the flux factor of virtual photons, defined as $\nu\sqrt{1+\gamma^2}$, with $\gamma^2 = Q^2/\nu^2$.

The interest in the generalised GDH sum rule was raised by Anselmino *et al.* [2] which related this integral to the Ellis-Jaffe integral $\Gamma_1 = \int_0^1 g_1(x)dx$. In fact, for $\gamma \ll 1$ and with the approximation that g_1 depends only on x, the right side of Eq. (2) reduces to $I_1(Q^2) = 16\pi^2\alpha\Gamma_1/Q^2$. A strong variation of $I(Q^2)$, including a change of sign for the proton, is required in order to connect the Γ_1 measurement in the DIS region to the real photon limit. Several phenomenological models were developed in order to predict the low-Q^2 behaviour of the integrals $I(Q^2)$ and $I_1(Q^2)$ [3]. It is worth stressing that in the low-Q^2 region $I_1(Q^2)$ is only an approximation of $I(Q^2)$ due to the effects of the flux factor K and of the second polarised structure function g_2. Furthermore, while the GDH sum rule for $Q^2=0$ is almost saturated by the resonance excitation processes and the Ellis-Jaffe integral and Bjorken sum rule for high-Q^2 are largely dominated by DIS processes, in the low-Q^2 region the contributions of resonance and DIS regions should be of the same order of magnitude.

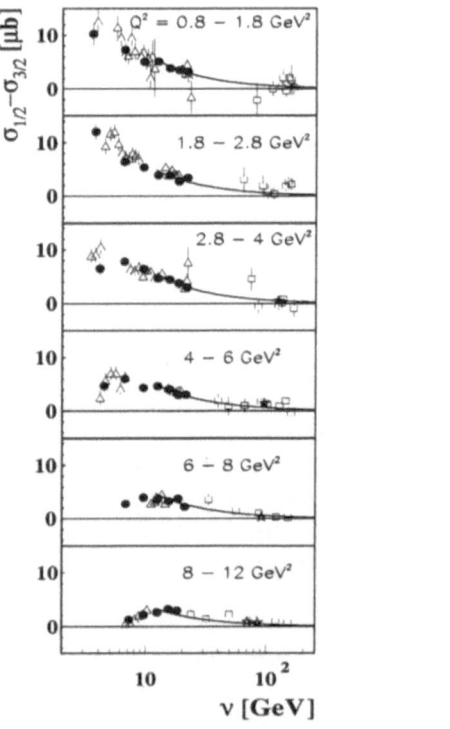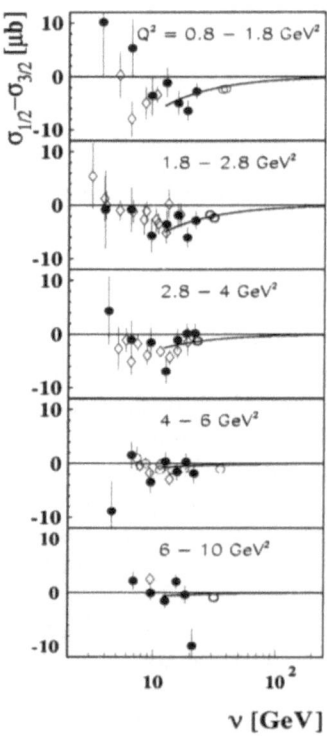

Figure 1. Cross section differences measured in different Q^2 bins for the proton (a) and the neutron (b). Close circles are our preliminary data. Open symbols are values derived from other experiments[6].

The preliminary data of a measurement of $[\sigma_{1/2} - \sigma_{3/2}]$ for the neutron and proton in the DIS region are presented. The data were collected during the 1995 and 1997 years of operation of the HERMES experiment. The experimental details can be found in Ref. [4] and in references therein.

The difference $[\sigma_{1/2} - \sigma_{3/2}]$ has been evaluated from the experimental virtual photon asymmetry A_1 and the unpolarised structure function F_1 derived from the literature[5]: $[\sigma_{1/2} - \sigma_{3/2}] = (8\pi^2\alpha/M)\cdot(A_1F_1/K)$. The preliminary cross section differences $[\sigma_{1/2} - \sigma_{3/2}]$ extracted from the measured values of A_1^p and A_1^n are presented in Fig. 1. Our preliminary results are compared to the values of $[\sigma_{1/2} - \sigma_{3/2}]$ that we have extracted from the published data for A_1^p from ^{1}H target and for A_1^n for ^{3}He target [6]. In Fig. 1 are also shown fits to our data, assuming the $a_1(1260)$ Regge trajectory with intercept $\alpha = 0$. The fit extrapolations seem to slightly underestimate the absolute values of most of the data at higher energy. Recent predictions of the QCD evolution of polarised parton densities show a larger contribution from the high energy region.

In each Q^2-bin we evaluated the GDH integral over the range $\nu_0 \leq \nu \leq 23.4$ GeV, being $\nu_0 = (W_0^2 - M^2 + Q^2)/2M$ where $W_0 = 1.8$ (2.0) GeV for the proton (neutron). The results shown in Fig. 2 represent the first direct determination of the DIS contribution $I^{DIS}(Q^2)$ to $I(Q^2)$.

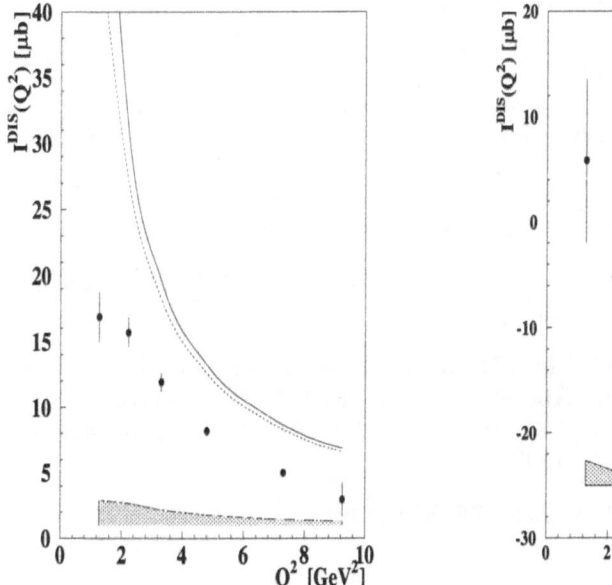
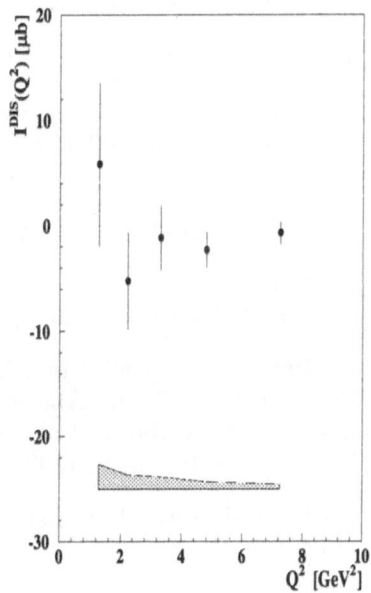

Figure 2. $I^{DIS}(Q^2)$ as measured from preliminary data in the range $\nu_0 \leq \nu \leq 23.4$ GeV for the proton (a) and for the neutron (b). The bands are our estimates of the systematic errors. The two curves are two estimates of the complete integral $I(Q^2)$ for the proton, as described in the text.

To estimate the order of magnitude of the full integral $I(Q^2)$ for the proton we have used two parameterisations of the polarised structure functions in the right side of Eq.(2). These estimates, in which the nucleon resonances excitation is not included and the second structure function $g_2(x)$ is calculated following the Wandzura-Wilczek expression are presented in Fig. 2. While at high Q^2 the difference between our data and the curves (about 4μb) can be mainly ascribed to the unmeasured high energy region, at low Q^2 also the contribution of the unmeasured low energy resonance region is important.

Up to now no measurement of the resonance region contribution to $I(Q^2)$ has been performed; the E143 collaboration measured the resonance part ($W \leq 2$ GeV) of $\Gamma_1(Q^2)$ for two values of Q^2 [7]. From their results at $Q^2 = 1.2$ GeV^2 it follows $I_1^p = 15\pm1\mu$b and $I_1^n = 6\pm5\mu$b for the resonance region contribution. These values are of the same order of magnitude of our determination of the DIS part of I at the lowest Q^2. Therefore our results together with the precise measurements of the resonance region contribution planned at TJNAF and with a more refined estimate of the high-energy extrapolation, will provide the Q^2 evolution of the entire GDH integral.

References

1. S.B. Gerasimov: Sov. J. Nucl. Phys. **2**, 430 (1966); S.D. Drell and A.C. Hearn: Phys. Rev. Lett. **16**, 908 (1966)

2. M. Anselmino, B.L. Ioffe and E. Leader: Sov. J. Nucl. Phys. **49**, 136 (1989)

3. N.Bianchi: *Proc. Workshop on Electron Nucleus Scattering*, O.Benhar and A.Fabrocini editors, Elba, July 1-5, 1996, p.363 and reference therein

4. K. Ackerstaff et al.: Phys. Lett. **B 404**, 383 (1997) A.Airapetian et al.: DESY 98-072 and Phys. Lett. **B**, to be submitted

5. M. Arneodo et al.: Phys. Lett. **B364**, 107 (1995); L.W. Whitlow et al.: Phys. Lett. **B250**, 193 (1990)

6. J.Ashman et al.: Nucl. Phys. **B328**, 1 (1989); B. Adeva et al.: Phys. Lett. **B412**, 414 (1997); K. Abe et al.: SLAC-PUB-7753, submitted to Phys. Rev. **D**; P.L. Anthony et al.: Phys. Rev. **D54**, 6620 (1996); K. Abe et al.: Phys. Rev. Lett. **79**, 26 (1997)

7. K.Abe et al.: Phys. Rev. Lett. **78**, 815 (1997)

Few-Body Systems Suppl. 11, 197–206 (1999)

Few-
Body
Systems
© by Springer-Verlag 1999

η Meson Photoproduction

J.-P. Didelez*

IN2P3, Institut de Physique Nucléaire, 91406 Orsay, France

Abstract. I review the recent (γ, η) data in the $S_{11}(1535)$ region and emphasize their impact on our understanding of the η meson photoproduction process.

1   Introduction

The underlying substructure of the nucleon gives rise to a rich spectrum of excited states, many of which have been associated with the baryon resonances seen in πN scattering [1] and π-photoproduction. These baryon resonances and their associated $\gamma N N^*$ electromagnetic couplings form bench-marks for models of hadron structure. Quark models have been quite successful in accounting for the masses of most observed resonances, although in some cases the electromagnetic couplings have presented more of a challenge [2].

Multipole analyses of meson photoproduction are necessary to extract resonance properties. These require a minimum of eight observables to avoid ambiguities: the cross section, the three single-polarization observables (polarized beam, target and nucleon-recoil) and four appropriately chosen double-polarization observables [3]. Such complete information is not available and the extraction of resonance parameters is necessarily model dependent. For the photoproduction of the π meson, the experimental situation is extensive [4], while for the η meson, only differential cross section data [5] and four polarization points, corresponding to the recoil proton polarization [6], all measured before 1974 using untagged photon beams, were available. The more recently published results concerning the differential and total cross sections, are devoted either to the study of the threshold behavior [7] or to the investigation at energies above the $S_{11}(1535)$ region [8]. For the Σ beam asymmetry observable, a few points exist above 1.39 GeV [9], but no measurement has been done near threshold.

Recently, a series of new data were measured from thresholds to above 1 GeV. They concern the proton [10] and neutron [11] differential cross sections,

*E-mail address: didelez@ipno.in2p3.fr

the single-polarization observables T (polarized target) [12] and Σ (linearly polarized beam) [13]. Those data, in particular the precise target and beam single-polarization observables, provide a new dimension to the previously existing data base for the η meson photoproduction and will put strong constraints on theoretical models. Together with the threshold differential cross sections on the proton [7], it should now be possible to fix the contribution of individual resonances as well as to understand non resonant mechanisms (Born terms including vector meson exchange) [14, 15]. Accordingly, approaches as those of Ref. [8] and [16], where the determination of the $S_{11}(1535)$ electromagnetic coupling strengths from the available η photoproduction data was attempted, will provide more reliable results. The electromagnetic photon couplings are a sensitive test of the resonances wave functions that quark-based models produce [17]. Recent models are able to reproduce quite well the mass spectrum and predict "missing" resonances which could have escaped experimental investigation, because they are not strongly coupled to the πN scattering channel. It has been pointed out that those missing resonances could show up in photoexcitation followed by a decay channel other than πN [18]. Spin observables in η photoproduction are therefore, a good place to look for possible candidates.

2 PHOENICS

In addition to proton differential cross sections remeasured with a hydrogen target [10], new break-up and coherent data for η meson photoproduction on the deuteron, using a deuterium target and tagged Bremsstrahlung photons up to 1 GeV, were produced at Bonn [11]. The differential cross sections for the coherent process, were measured between threshold and 800 MeV. The break-up channels, which provide a direct measurement of the neutron to proton differential cross sections ratios, were measured till above 1 GeV.

For the excitation of isospin 1/2 resonances, both isospin components of the photon I = 0 (isoscalar) and I = 1 (isovector) can contribute, accordingly the amplitude splits in two terms A_s and A_v, for the isoscalar and isovector excitations respectively. The separation of those components requires measurements on the proton and on the neutron. For an ηN final state one can write:

$$A_p(\gamma p \rightarrow p\eta) = A_s + A_v \tag{1}$$
$$A_n(\gamma n \rightarrow n\eta) = A_s - A_v \tag{2}$$

Two complementary approaches were followed. First, considering the target deuteron as a system of only two losely bound nucleons with identical distributions for proton and neutron and detecting in the final state the η and the leading (fast) outgoing nucleon, one can assume that the elementary reaction took place on the detected nucleon, while the undetected one behaved as a "spectator". For the outgoing protons or neutrons, the corresponding counting rate ratios are related to the relevant "free" nucleon cross sections ratios: σ_n/σ_p. Those ratios are related to the ratios A_n/A_p as defined in Eqs. (1) and (2). They provide a connection to A_s and A_v.

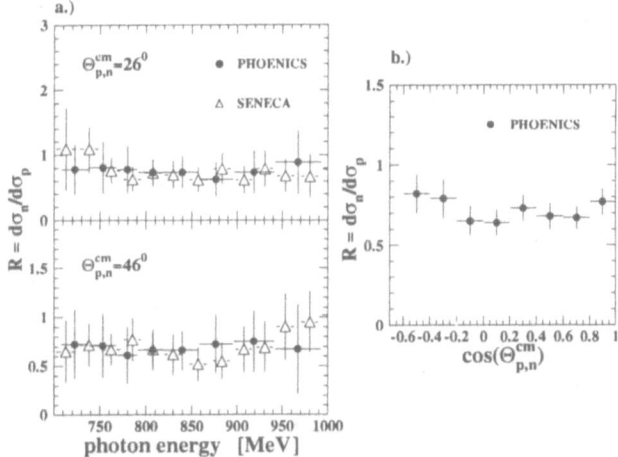

Figure 1. a.) Ratios $R = d\sigma(\gamma, \eta n)p/d\sigma(\gamma, \eta p)n$ of the cross sections corresponding to the "quasifree" reactions, as a function of incident photon energies. b.) Angular dependence of the ratio R. The corresponding differential cross sections have been integrated from 720 to 860 MeV. The error bars take into account statistical and systematical uncertainties..

Second, since the target deuteron is an isoscalar object, the coherent reaction ($\gamma d \rightarrow \eta d$) proceeds through A_s. Therefore, large coherent production cross sections would mean that the photoproduction of η mesons is mainly an isoscalar process, as suggested by the first results reported in 1969 by Anderson and Prepost [19]. Recent experimental investigations [20, 21] provide only upper limits, three times smaller than those of Ref. [19]. The major breakthrough of the Bonn work is to provide the first measurements of those coherent differential cross sections. In connection with the ratios σ_n/σ_p, they lead to a ratio A_s/A_v of the order of 10%, close to the value expected from the $S_{11}(1535)$ photon couplings extracted from pion photoproduction data [22] and consistent with quark model predictions [17, 23].

All the measurements were done with the PHOENICS [24] tagged Bremsstrahlung photon beam up to 1.2 GeV at the electron stretcher ring ELSA in Bonn [25], using the neutral meson spectrometer SPES0 [26] to detect the 2γ and $3\pi^0$ η-decay channels. The recoil baryons were identified with a variety of large scintillating detector systems, comprising the traditional neutron and charged hadrons PHOENICS detectors [27] and the SENECA neutron detector [28] array for increased neutron detection efficiency at forward angles. To increase the system efficiency for coherent deuteron detection, the AMADEUS [29] detector was also used in separate runs. The basic idea of this system is to identify simultaneously the η in SPES0 by its invariant mass and the leading outgoing baryon (p, n or d) in the forward scintillators by time of flight and energy loss. For two body reactions, this procedure gives a straightforward identification of the (η, p) and (η, d) final states. The break-up channel requires

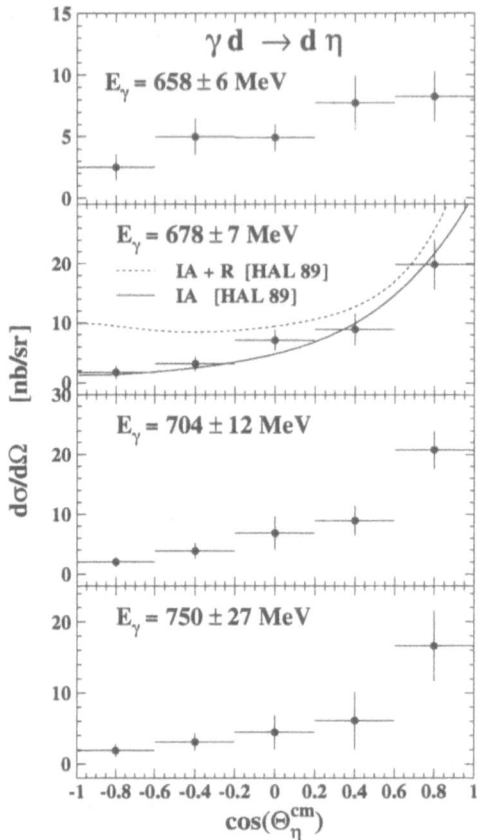

Figure 2. Differential cross sections for the reaction $\gamma d \rightarrow \eta d$ at four energy bins between $650 MeV$ and $780 MeV$. For $678 MeV$ the results are compared to the theoretical predictions of Halderson and Rosenthal [30]. The curves were scaled down by a factor five to fit the experimental points.

a more complicated procedure to isolate the quasifree process. Details are given in Ref. [11]. Fig. 1 gives the results for the break-up channels. The σ_n/σ_p ratios do not depend very much on energy, but there is a marked angular dependence: for the neutron, the η is more forward peaked. In the energy region between 720 and 860 MeV, where the $S_{11}(1535)$ resonance peaks, the average value of the σ_n/σ_p experimental ratios is 0.68 ± 0.06.

For the coherent process, Fig. 2 displays the angular distributions at four different energies, between 650 and 780 MeV. The curve which fits the data at 678 MeV, was obtained by scaling down those of Ref. [30] by a factor five. It is interesting to note that a satisfactory fit to the angular distributions is obtained with only the Impulse Approximation contribution (full curve), without much need for the rescattering terms corresponding to π and η meson exchange (dotted curve). Similar conclusions would arise from a comparison of the data with the older calculation [31] or the more recent ones [15, 32]; in all approaches, the Impulse Approximation generates good fits to the angular distributions.

Compared to the free proton case, the coherent deuteron cross section will scale approximately as $(2A_s/A_p)^2 = [2A_s/(A_s + A_v)]^2$. The authors of Ref. [30] had to use an A_s/A_p ratio of 0.6 to fit the points of Ref. [19]. In the framework of their model, this changes to an A_s/A_p ratio of 0.27 to fit the new coherent deuteron data. On the other hand, knowing that A_s is smaller than A_v, we can use the relation:

$$A_s/A_p = (1/2) \cdot (1 - \sqrt{\sigma_n/\sigma_p}) \tag{3}$$

and the experimental ratios σ_n/σ_p, averaging to 0.68 ± 0.06 between 720 and 860 MeV where the $S_{11}(1535)$ peaks, to determine A_s/A_p in a model independent way, from the break-up data. One finds $A_s/A_p = 0.09 \pm 0.02$. Such a small isoscalar amplitude would bring the impulse approximation curve of Ref. [30] a factor of 9 below the data. New theoretical investigations of the coherent production mechanism are necessary to understand why the experimental cross sections are still so large compared to first order theoretical predictions. In particular, the role of the Roper resonance should be reconsidered [31]. A completely new approach to the structure of the $S_{11}(1535)$ resonance itself [33], could also be invoked to solve this puzzle.

Using the PHOENICS-SPES0 experimental setup and the Bonn frozen spin Butanol target BOFROST [34], for which 85% polarization can be achieved by the dynamic method, the single-polarization observable T, corresponding to the target proton polarized in a plane perpendicular to the propagation direction of the incident photon, could be measured [12]. For this particular experiment, the holding field (0.35 T) was produced by a small superconducting coil located inside the in-beam dilution refrigerator, around the target cell [35]. In the Butanol molecule (C_4H_9OH), 10 out of the 42 present protons are polarized, generating a dilution factor of 24%. The unpolarized contribution of the (C_4O) nuclei is increased by those of the $^3He/^4He$ cooling mixture in which the target material is immersed with a 55% filling factor and of the numerous separation windows, further reducing the effective dilution factor. Fortunately, due to the kinematic correlation of the (η, p) final state, this unpolarized background can be considerably suppressed. The resulting T asymmetries are small, reaching maximum values of the order 0.2. They are structured, with a clear node appearing at 90 degree (η CM angle), for E_γ less than 750 MeV. See Ref. [12] for more details.

3   GRAAL

New data for the beam asymmetry Σ, i.e. the analyzing power of a photoproduction reaction induced by linearly polarized photons, have been measured for η meson photoproduction on the proton, using the GRAAL tagged, laser backscattered, linearly polarized novel photon beam up to 1.1 GeV.

The backscattering takes place on the stored electrons of the ESRF (European Synchrotron Radiation Facility) in Grenoble. The linear polarization of the beam is close to 100% at the maximum photon energy. The LAGRANγE (Large Acceptance GRaal Apparatus for Nuclear γ Experiments) detector for neutral and charged particles covers 99.95% of 4π. With the tagger providing the energy of the incoming γ and all particles detected in the final state, the reaction is overdetermined and the events from η photoproduction are easily identified using kinematical constraints. This allows much better identification of the (ηp) final state than previously accomplished by the detection of a single particle [7] or a poorly detected η [11, 12] and provides the potentiality to generate high quality data. See Ref. [13, 36] for more details.

For a linearly polarized photon, the differential cross section for η photoproduction is:

$$\frac{d\sigma}{d\Omega} = \frac{d\sigma_{unp}}{d\Omega}[1 - P_\gamma \Sigma cos(2\varphi)] \tag{4}$$

where P_γ is the polarization degree of the photon and φ the angle between the reaction plane and the polarization axis. The cylindrical symmetry of the GRAAL detector allows this $cos(2\varphi)$ dependence to be observed directly in the azimuthal distribution of events. The results are shown in Figures 3. The asymmetries are large and positive. Below 900 MeV, the angular distribution is fairly symmetric around 90^0, but at the highest γ incident energies, there is a marked forward peaking, with large asymmetries, reaching values as high as 70%.

4 Interpretation

Several predictions for the Σ asymmetries can be derived from the existing models that fit the differential cross sections. The authors of Ref. [15] use a coupled channel analysis based on $\gamma\pi$, $\pi\pi$, and $\pi\eta$ data to predict the $\gamma\eta$ channel. They include in their model, only the first N^* resonances: Roper $P_{11}(1440)$, $S_{11}(1535)$ and $D_{13}(1520)$. The corresponding predictions are shown by the dotted curves. Although positive asymmetries result from this approach, the agreement is poor, except at 808 MeV. In Ref. [37], an effective lagrangian formalism, that contains Born terms and both vector meson and nucleon resonance contributions [$P_{11}(1440)$, $S_{11}(1535)$, $D_{13}(1520)$ and $D_{15}(1675)$], has been developed and fitted to recent $\gamma\eta$ cross section data [7, 10]. The corresponding predictions for the Σ observable follow the dashed curves. The agreement is quite good at energies below 900 MeV, but the large asymmetries observed at higher energies are not reproduced.

In an attempt to reproduce the large forward asymmetries, the "nodal approach" developed by the authors of Ref. [38] was used. In this framework, the angular structure of spin observables provides a powerful tool to find out which angular momentum quantum numbers, corresponding to excited intermediate resonances, must play a role in pseudoscalar meson photoproduction processes

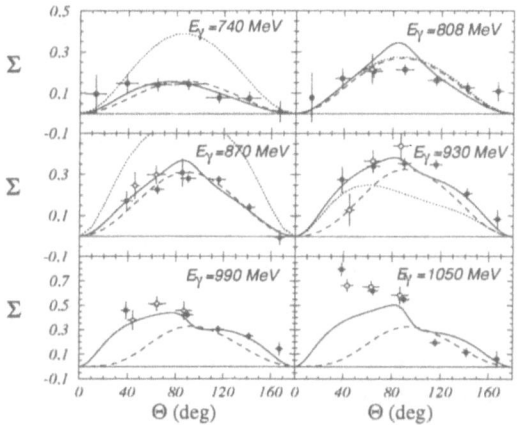

Figure 3. Σ observable, at different incident γ energies, corresponding to the photoproduction of the η meson on the proton as a function of the η angle Θ in the CM. The full circles are the results when the 2 or 6 decay photons are all detected in the BGO ball. The open circles are the results for one photon in the BGO and the other one in the shower detector. The error bars include statistical and systematic errors. The horizontal bars indicate the angular resolution. Predictions from already published approaches are: dotted curves, Ref. [15]; dashed curves, Ref. [37]. Full curves, fit to our data resulting from the nodal approach, Ref. [41].

[39]. The relevant profile function $\Sigma(\theta)$ is a Legendre class 2 polynomial. It has the general form

$$\Sigma(\theta) = sin^2\theta \Sigma_{l=0}^{n} a_l cos^l \theta \qquad (5)$$

where the polynomial coefficients a_l can be expressed in terms of the basic multipole amplitudes. Knowing their dependence on the resonance angular quantum numbers, one can find out which ones are necessary to reproduce the experimental angular pattern and guess which corresponding resonances must play a role.

Such an analysis, limited to resonances with intrinsic angular momenta $L \leq 2$, has been carried out for the target asymmetry observable T [12] and has shown that all "4 stars" S, P and D resonances must be taken into account [40]. A similar nodal analysis including our Σ data points, shows that S_{11}, P_{13}, D_{13} and D_{15} resonances are needed, with no evidence for P_{11} [41]. The fit obtained with the inclusion in the dynamical model [42] of the established $S_{11}(1535)$, $D_{13}(1520)$, $P_{13}(1720)$ and $D_{15}(1675)$ resonances and a "missing" P_{13} (M = 1880MeV, Γ = 150MeV) resonance is shown by the full curve in Figure 3. There is a significant improvement towards fitting the large forward asymmetries at the highest energies. It is worth noting that it is the inclusion of a missing P_{13} resonance, predicted by quark model calculations [18], but not yet observed experimentally, which generates significant forward peaking as observed experimentally in the Σ data [41]. The inclusion of this resonance, also produces the best fit to the T observable [40]. Reduced χ^2s for three configurations ($S_{11}D_{13}$; $S_{11}P_{13}D_{13}D_{15}$ and $S_{11}P_{13}P_{13}D_{13}D_{15}$), fitting both Σ and T asymmetry data, are 2.7; 2.2 and 2.0 respectively [42]. The missing P_{13} resonance is predicted to have a weak coupling to the πN channel but a significant coupling to the ηN decay channel and could therefore show-up in photoproduction data.

However, in spite of the contribution of a significant number of low multipole ($L \leq 2$) resonances, the large forward asymmetries, cannot be reproduced. Large forward asymmetries were also observed on earlier data above 1.39 GeV [9]. Therefore, it is likely that the contribution of higher multipoles is needed. In particular, the $F_{15}(1680)$ resonance which is strongly excited in the initial channel, should play an important role. More comprehensive theoretical investigations, including higher multipole resonances, are expected using the available formalisms quoted above [16, 37]. Such an analysis, limited to $E_\gamma \leq 800$ MeV has already nicely fixed the relative contributions of the $S_{11}(1535)$ and $D_{13}(1520)$ resonances [43]. Within the framework of a recent quark-based approach [44], the new data have been investigated. It was found that the cross sections and the Σ asymmetry data require significant contributions from the $D_{13}(1520)$ and the $F_{15}(1680)$ resonances, in addition to those from the dominant $S_{11}(1535)$. The contributions from the $S_{11}(1650)$, $D_{13}(1700)$ and $D_{15}(1675)$ resonances are consistently small, but sensitive to the T asymmetry. Those results were found to be consistent with the quark model predictions [45].

5 Acknowledgments

The data reviewed in this talk have been mainly published in 3 papers [11, 12, 13] for which I am one author among 70 others from France, Germany, Italy, Russia and USA. I thank all of them for a fruitful collaboration. I gratefully acknowledge C. Bennhold, S. Kamalov, Z. Li, N. Mukhopadhyay, B. Saghai, L. Tiator for their prompt analysis of the data, for providing the results of their calculations and for many enlightening discussions.

References

1. Review of Particle Physics: Phys. Rev. **D54**, 561 (1996)

2. M. Giannini: Rep. Prog. Phys. **54**, 453 (1991), and references therein

3. W. Chiang and F. Tabakin: Phys. Rev. **C55**, 2054 (1997)

4. SAID data base, quoted in R. A. Arndt et al.: Phys. Rev. **C52**, 2120 (1995)

5. P.S.L. Booth et al.: Nucl. Phys. **B71**, 211 (1974), and references therein

6. C.A. Heusch et al.: Phys. Rev. Lett. **25**, 1381 (1970)

7. B. Krusche et al.: Phys. Rev. Lett. **74**, 3736 (1995); S.A. Dytman et al.: Phys. Rev. **C51**, 2710 (1995); J.W. Price et al.: Phys. Rev. **C51**, R2283 (1995)

8. S. Homma et al.: J. Phys. Soc. Japn. **57**, 828 (1988)

9. G.A. Vartapetyan and S.E. Piliposyan: Sov. J. Nucl. Phys. **32**, 804 (1980)

10. M. Breuer et al.: 7th Int. Conf. on *"Nuclear Reaction Mechanisms*, Varenna, Italy, June 1994, Uni. Studi Milano, Suppl. $n°$ 100, p. 584. E. Gadioli (ed.) (1994)

11. P. Hoffmann-Rothe et al.: Phys. Rev. Lett. **78**, 4697 (1997)

12. A. Bock et al.: Phys. Rev. Lett. **81**, 534 (1998)

13. J. Ajaka et al.: Phys. Rev. Lett. **81**, 1797 (1998)

14. H. R. Hicks et al.: Phys. Rev. **D7**, 2614 (1973); F. Tabakin, S. A. Dytman and A. S. Rosenthal: Proc. of the Workshop on *"Excited Baryons*, Troy (USA), World Scientific, Singapore, 1989

15. C. Bennhold et al.: Nucl. Phys. **A530**, 625 (1991); L. Tiator et al.: Nucl. Phys. **A580**, 455 (1994)

16. M. Benmerrouche and N. C. Mukhopadhyay: Phys. Rev. Lett. **67**, 1070 (1991); M. Benmerrouche et al.: Phys. Rev. **D51**, 3237 (1995); N. C. Mukhopadhyay, J-P Zhang and M. Benmerrouche, Phys. Lett. **364**, 1 (1995)

17. R. Koniuk and N. Isgur: Phys. Rev. **D21**, 1868 (1980); S. Capstick: Phys. Rev. **D46**, 2864 (1992)

18. S. Capstick and W. Roberts: Phys. Rev. **D47**, 1994 (1993); Phys. Rev. **D49**, 4570 (1994)

19. R.L. Anderson: R. Prepost: Phys. Rev. Lett. **23** 46 (1969)

20. W. Beulertz: PhD Thesis, Bonn University (1994)

21. B. Krusche et al.: Phys. Lett. **B358**, 40 (1995)

22. Review of Particle Properties: Phys. Rev. **D54**, 579 (1996)

23. R.P. Feynman et al.: Phys. Rev. **D3**, 2706 (1971); Z. Li and F. E. Close, Phys. Rev. **D42**, 2207 (1990); M. Warns et al.: Phys. Rev.**D42**, 2215 (1990); R. Bijker et al.: Ann. of Phys. **236**, 69 (1994); Zhenping Li, Phys. Rev. **D52**, 4961 (1995)

24. P. Detemple et al. Nucl. Instr. Meth **A321**, 479 (1992)

25. K.H. Althoff et al.: Internal Report, Bonn-IR-87-30 (1987)

206

26. G. Rappenecker et al.: Nucl. Phys. **A590**, 763 (1995)

27. K. Buchler et al.: Nucl. Phys. **A570**, 580 (1994)

28. G. v. Edel et al.: Nucl. Instr. and Meth., **A32**, 224 (1993)

29. G. Anton et al.: Nucl. Inst. and Meth., **A306**, 89 (1991)

30. D. Halderson and A. S. Rosenthal: Nucl. Phys. **A501**, 856 (1989)

31. N. Hoshi et al.: Nucl. Phys. **A324**, 234 (1979)

32. E. Breitmoser and H. Arenövel: Nucl. Phys. **A612**, 321 (1997)

33. N. Kaiser: These Proceedings

34. H. Dutz et al.: Nucl. Inst. and Meth. **A340**, 272 (1994)

35. R. Gehring: Ph.D. thesis, University of Bonn, (1996)

36. J. Ajaka et al.: These Proceedings

37. G. Knöchlein, D. Drechsel and L. Tiator: Z. Phys. **A352**, 327 (1995)

38. C. G. Fasano, F. Tabakin and B. Saghai: Phys. Rev. **C46**, 2430 (1992); B. Saghai and F. Tabakin: Phys. Rev. **C53**, 66 (1996)

39. B. Saghai and F. Tabakin: Phys. Rev. **C55**, 917 (1997); F. Tabakin and B. Saghai, Proc. of the 4th CEBAF Worksh. on "*N* Physics*", Ed. T.S.H. Lee and W. Roberts, Seattle, Sept 1996. World Scientific

40. B. Saghai et al.: 6th Conference on the "*Intersections of Particle and Nuclear Physics*", Big Sky, Montana, (1997), edited by T.W. Donnelly (AIP Conference Proceedings), Vol. 412, p. 567

41. B. Saghai: Private Communication

42. B. Saghai et al.: in preparation

43. N. C. Mukhopadhyay: These Proceedings

44. Z. Li: Phys. Rev. **D52**, 4862 (1995)

45. Z. Li and B. Saghai, Nucl. Phys. **A644**, 345 (1998)

Few-Body Systems Suppl. 11, 207–215 (1999)

Few-
Body
Systems
© by Springer-Verlag 1999

The Role of $N^*(1520)$ Resonance in Eta Photoproduction in the Light of New Polarization Data *

Nilmani Mathur † and Nimai C. Mukhopadhyay**

Department of Physics, Applied Physics and Astronomy, Rensselaer Polytechnic Institute, Troy, New York 12180-3590

Abstract. We examine the η photoproduction from nucleons in the cm energy region around 1530 MeV and study the role of $N^*(1520)$ resonance with the help of recently obtained polarized target and photon asymmetry data. We show here that these polarization data, in conjunction with the differential cross-section data, yield model-insensitive constraints on the electrostrong parameters for the excitation and decay of the $N^*(1520)$ resonance.

1 Introduction

Studies of the electromagnetic [1] and weak [2] transition amplitudes to various resonance states of the nucleon (N) as a function of the square of the four-momentum transfer, q^2, is a powerful way to explore the chromodynamic structure of the nucleon. The real photon point, for which $q^2 = 0$, is at one end of the domain of non-perturbative QCD, which continues until some large q^2, as yet unknown, which marks the onset of perturbative QCD (pQCD). For the N to N^* transitions, with the spin of the N^* being $\frac{3}{2}$, there are two helicity amplitudes, $A_{\frac{1}{2}}$ and $A_{\frac{3}{2}}$, for real photon excitations. In contrast to the pQCD domain, where counting rules yield $A_{\frac{1}{2}} >> A_{\frac{3}{2}}$ [3], the non-perturbative region is characterized by a large helicity violation [1]. Here we shall study this in the N to $N^*(1520)$ real photon transition, via the reaction

$$\gamma + p \quad \longrightarrow \quad p + \eta, \tag{1}$$

with the photon lab energy from the eta photoproduction threshold of 707 MeV, up to about 900 MeV.

---

*Invited talk given by NCM

† E-mail address: mathun@rpi.edu

** E-mail address : mukhon@rpi.edu

Though a well-established resonance from the analyses of the pion-nucleon scattering and pion photoproduction, the electromagnetic properties of $N^*(1520)$, are yet to be studied experimentally via complementary reactions such as (1) and are not fully understood in the framework of the QCD-inspired models. Our interest in the reaction (1) is also enhanced by the recent availability of high quality data from different photon factories: differential cross-section data from the Mainz microtron [4], the polarized target asymmetry (PTA) from the upgraded electron facility at Bonn [5] and a precise data set, just released, on the polarized photon asymmetry (PPA) from the European laser light source, GRAAL [6]. We show below that a combination of these observables provide a *powerful constraint* on relatively small effects from the excitation of the $N^*(1520)$ resonance and its decay, *amplified by* the interference with the dominant contribution of the $N^*(1535)$ resonance. There are also subtle issues arising from the nodal structures [7] of these observables. Finally, we discuss implications of these data on the non-perturbative QCD violation of the helicity conservation in the electromagnetic process $N \rightarrow N^*(1520)$, and in particular, on the testing of various QCD-inspired models [8] of hadrons.

Absent lattice gauge theoretic estimates, our theoretical knowledge of helicity amplitudes for the baryon resonance excitation comes from the QCD-inspired models [8]. Current level of their uncertainties for the helicity amplitudes for the $N \rightarrow N^*(1520)$ excitation is as follows : the $A_{\frac{1}{2}}$ amplitude for the proton target ranges from -13 to -51, in units of $10^{-3}\text{GeV}^{1/2}$, in various versions [8] of the constituent quark model, while the $A_{\frac{3}{2}}$ amplitude is predicted to be in the range 117 to 173 in the same units. The 1996 PDG [9] values of these amplitudes from pion photoproduction quote relatively small errors, but they do not include theoretical uncertainties of the position, branching ratios and width of the resonance. The relatively large uncertainties in the partial width Γ_η (~ 0.14 MeV) of the resonance $N^*(1520)$ to decay into ηN channel and the total width, Γ, known to be between 110 to 135 MeV, result in errors in extracting A_i from the photoproduction data, *much bigger* than the errors quoted by the PDG. An attempt to get these amplitudes from the $(\gamma, \pi\pi)$ reaction has yielded the $A_{\frac{3}{2}}$ amplitude only within a factor of two [10], with no meaningful constraint on $A_{\frac{1}{2}}$.

We shall extract from the process (1) the parameters $\xi_i (i = \frac{1}{2}, \frac{3}{2})$ defined as [11] :

$$\xi_i = \sqrt{\chi \Gamma_\eta} A_i / \Gamma, \tag{2}$$

where χ is a kinematic parameter, $Mk/(qM_R)$, k and q are the photon and eta meson momenta in the ηN cm frame, M and M_R are the nucleon and the resonance masses. One of our findings is that the new cross-section and polarization data of the reaction (1), *taken together*, give us precise estimates of the quantities $\xi_{1/2}$ and $\xi_{3/2}$ for $N^*(1520)$ for the first time from the reaction (1). Thus the quantity $\xi_{3/2}/\xi_{1/2}$ yields an estimate of the ratio of electromagnetic helicity amplitudes $A_{3/2}/A_{1/2}$ *essentially independent of uncertainties of the strong interaction parameters*, which drop out in the ratio.

Our theoretical tool for analysis of the reaction (1) is the effective Lagrangian approach, which consists, in the tree approximation of the s- and u- channel nucleon and resonance Born terms and the t-channel vector meson (ρ and ω) exchanges [1, 11]. Dominant contributions for eta photoproduction around $W \sim 1.3$ GeV are well-studied, consisting of the nucleon Born terms and s-channel excitation of $N^*(1535) \frac{1}{2}^-, T = \frac{1}{2}$ resonance [11, 12]. Our goal here is to get at the relatively small contributions from the excitation of $N^*(1520), \frac{3}{2}^-, T = \frac{1}{2}$ resonance. We cannot do that from the differential cross-section data alone, even though there is a hint [1, 4, 12, 13] of its presence from these data. It is a *combination* of these differential cross-section data with the recently gathered data [5, 6] on polarization observables that allows us to put powerful constraints on $N^*(1520)$ amplitudes. The PPA turns out to be rather insensitive to $A_{1/2}$, thereby giving us a better fix on $A_{3/2}$, while the differential cross-section and the PTA help us to constrain the $A_{1/2}$ amplitude.

2 Our Model : Effective Lagrangian Approach

We shall now briefly discuss the general structure of the interaction Lagrangian for the $\frac{3}{2}^-, T = \frac{1}{2}$ resonance excitation. The strong and electromagnetic pieces are [11]:

$$L_{\eta NR} = \frac{g_R}{\mu} \bar{R}^\mu \theta_{\mu\nu}(Z) \gamma_5 N \partial^\nu \eta + h.c., \tag{3}$$

$$L^1_{\gamma NR} = \frac{ie}{2M} \bar{R}^\mu \theta_{\mu\nu}(Y) \gamma_\lambda (G^1_s + \tau_3 G^1_v) N F^{\lambda\nu} + h.c., \tag{4}$$

$$L^2_{\gamma NR} = -\frac{e}{4M^2} \bar{R}^\mu \theta_{\mu\nu}(X) (G^2_s + \tau_3 G^2_v)(\partial_\lambda N) F^{\nu\lambda} + h.c., \tag{5}$$

where the tensor $\theta_{\mu\nu}(A)$ is defined as follows [14]:

$$\theta_{\mu\nu}(A) = g_{\mu\nu} - [\frac{1}{2}(1 + 2A)]\gamma_\mu \gamma_\nu. \tag{6}$$

Parameter A is not *a priori* known and it must be determined from the fits to the data on the reaction (1). R^μ is the vector-spinor field for the spin-$\frac{3}{2}$ resonance; the resonant three-point couplings for the proton target, g_R, $G^i_P = G^i_s + G^i_v$ are all to be determined from the fits to the data of the reaction (1); $F^{\mu\nu}$ is the electromagnetic field tensor representing the external real photon field. In the broadest fit we have attempted, we have *nine* effective parameters: in the non-resonant part of the amplitude, these are the eta-nucleon coupling and two vector meson couplings; in the resonant sector we have one helicity amplitude for $N^*(1535)$ and two helicity amplitudes $A_{\frac{1}{2}}$, $A_{\frac{3}{2}}$ for $N^*(1520)$ excitation and three "off-shell" parameters, X, Y, Z[14].

We start with the expressions for the observables of our interest in terms of the helicity amplitudes $H_i(i = 1, 2, 3, 4)$ and write them in terms of multipole amplitudes up to d-waves in the ηN channel [1], exhibiting only terms involving the dominant E_{0+} multipole. Thus, the differential cross-section $\frac{d\sigma}{d\Omega}$, the PPA

$[\Sigma]$ and the PTA $[T]$ are given by [1, 16]

$$
\frac{d\sigma}{d\Omega} = \frac{|q|}{2\,|k|} \sum_{i=1}^{i=4} |H_i|^2
$$

$$
= \frac{|q|}{|k|} \left[E_{0+}^2 - \mathrm{Re}\{E_{0+}^*(E_{2-} - 3M_{2-} + 3M_{2+} + 6E_{2+})\} \right.
$$

$$
+ 2\cos\theta \, \mathrm{Re}\{E_{0+}^*(3E_{1+} + M_{1+} - M_{1-})\}
$$

$$
\left. + 3\cos^2\theta \, \mathrm{Re}\{E_{0+}^*(E_{2-} - 3M_{2-} + 6E_{2+} + 3M_{2+})\} \right], \quad (7)
$$

$$
\frac{d\sigma}{d\Omega}\Sigma = \frac{|q|}{|k|} \mathrm{Re}\{H_1 H_4^* - H_2 H_3^*\}
$$

$$
= -\frac{|q|}{|k|} 3\sin^2\theta \, \mathrm{Re}\left[E_{0+}^*(M_{2+} - E_{2+} - M_{2-} - E_{2-}) \right], \quad (8)
$$

$$
\frac{d\sigma}{d\Omega}T = \frac{|q|}{|k|} \mathrm{Im}\{H_1 H_2^* + H_3 H_4^*\}
$$

$$
= \frac{|q|}{|k|} 3\sin\theta \, \mathrm{Im}\left[E_{0+}^*(E_{1+} - M_{1+}) \right.
$$

$$
\left. + E_{0+}^*(4E_{2+} - 4M_{2+} - M_{2-} - E_{2-})\cos\theta \right]. \quad (9)
$$

Above we are omitting the interference terms between p- and d-wave multipoles for brevity, although they are included in our calculation. They are crucial to understand sensitivity to the $A_{1/2}$ helicity amplitude in our chosen observables. This subtle interference effect of the p-wave multipoles is ignored in a recent analysis [13] of Tiator $et\ al.$.

In the "second" resonance region, around $W \sim 1.3$ GeV, of interest here, the dominant multipole for eta photoproduction is E_{0+}, and its primary contribution is from the excitation of the $N^*(1535)$ resonance [12]. Both the E_{0+} and M_{1-} multipoles also receive contributions from the spin-$\frac{1}{2}$ sector of the $N^*(1520)$, often referred to as the off-shell sector [14] of the spin-$\frac{3}{2}$ resonance. This is controlled by the parameters X, Y and Z introduced earlier in Eqs. (3)-(5). The multipoles E_{2-} and M_{2-}, in which $N^*(1520)$ is resonant, are relatively small in the energy region of our interest, but are retained for an important reason. Their effects are enhanced by the interference with the large E_{0+} multipole (Eqs.7-9), in contrast to pion photoproduction where no single multipole stands out.

The differential cross-section $\frac{d\sigma}{d\Omega}$ of the reaction (1), recently determined at the Mainz Microtron [4], is very flat near the eta photoproduction threshold characteristic [1] of the dominance of the E_{0+} multipole and $N^*(1535)$ excitation. As the photon energy increases, the differential cross-section begins to deviate from near isotropy and shows angular dependence (Figs.1, first column). This has been interpreted [11, 13] as a complicated effect of a combination of nucleon Born terms and the role of the $N^*(1520)$ excitation. However, the best fit of the Mainz data alone *misses* the sign of the PTA (Figs.1, second column) and *cannot* reproduce the magnitude of the PPA [6] (Figs.1, third column). It is only a *combination of these three data sets*, encompassing broad energy range, [4-6] that results in the acceptable fits to all these diverse data, a sample of which is represented by the solid lines. But in the low energy region if we force a fit to the data, we cannot use that fit (dashedlines) to describe the higher energy data sets. We should recall that isobar model fits of the low-energy PTA data have been so far unsuccessful in reproducing these data [13]. It is not clear whether the predictions of the different models are wrong here or there is some incompatibility between the low energy PTA data and other observables.

This brings us to the subject of nodal structure of the observables as fingerprints of resonance contributions [7]. The nodal structure anticipated by Saghai and Tabakin [7] for *pure* s- and d-wave resonances, follows from a $sin\theta \cdot cos\theta$ distribution for the PTA. However, the *s*-wave interfering with the *p* and *d*-wave multipoles, predicted in our effective Lagrangian approach, spoils this simple expectation. The data at higher energies support the latter, with no node appearing in the PTA (Figs.1, second column). Such deviations are indicative of subtle roles of background contributions.

We have to deal with the problem of broad range of parameters for resonances in the PDG compilation. We adopt the following strategy to do our fits. For a particular set of $N^*(1520)$ parameters, we vary properties of $N^*(1535)$, such as mass, width etc., within the permitted PDG 1996 boundaries. We then change the parameter set of $N^*(1520)$ and repeat the procedure. In this way we cover many possible parameter sets of $N^*(1520)$ and $N^*(1535)$. Shown in Table 2.1 are the parameter sets for which we got the χ^2 per degree of freedom around 1.3, our lowest χ^2 level. The fitted values of the parameters for the $N^*(1520)$ resonance correlate strongly with the properties of $N^*(1535)$, such as its position, total width and the eta-nucleon partial width, as shown in Table 2.1. However, *the electrostrong parameters for the $N^*(1520)$, $\xi_{3/2}^D$ and $\xi_{1/2}^D$ are relatively stable*. By taking a broad band of allowed parameters around the lowest value of the χ^2 per degree of freedom, (~ 1.30), we can ascertain the following set of $N^*(1520)$ parameters :

$$\xi_{\frac{3}{2}}^D = 0.165 \pm 0.015 \pm 0.035, \tag{10}$$

$$\xi_{\frac{1}{2}}^D = -0.065 \pm 0.010 \pm 0.015. \tag{11}$$

where the first error is statistical and the second one is systematic. The systematic error reported here is due to uncertainties in positions, branching ratios and decay widths of the resonances (Table 2.1), while the statistical error is obtained from the fitting program MINUIT [15].

In the ratio of these two parameters, strong interaction uncertainties drop and we can determine, in a nearly *model independent* manner, a ratio of the helicity amplitudes $A_{3/2}$ to $A_{1/2}$ for $N^*(1520)$:

$$A_{3/2}/A_{1/2} = -2.5 \pm 0.4 \pm 0.4. \tag{12}$$

This ratio has been reported by the 1996 PDG as -6.9 ± 2.6 [8], from pionic processes. *Ours is the first determination of this quantity from eta photoproduction.* The importance of the difference between our value and the PDG one will become apparent below, when we compare it with model estimates. This ratio could only be determined here by exploiting both the differential cross-section [4] and polarization observables [5, 6] together.

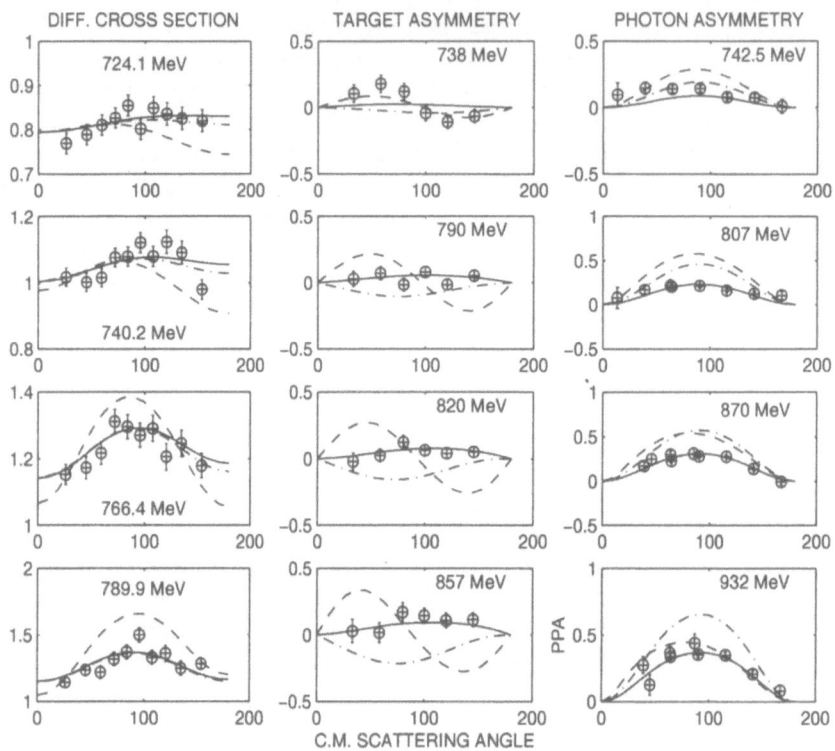

Figure 1. Differential cross-section and polarization observables in our effective Lagrangian fit with different choices of energy segments. Solid : Fitting Mainz [4], Bonn [5] and GRAAL [6] data upto 790, 895 and 932 MeV respectively; dash-dot: fitting Mainz data alone; dash: fitting a low-energy truncation of the Mainz and Bonn data, upto 750 MeV.

We now discuss the significance of the ratio (12) from the point of view of the structure of the nucleon and $N^*(1520)$. It should be zero in the pQCD regime [3]. We are clearly dealing here with non-perturbative physics that is altering the helicity structure dramatically in the $q^2 \to 0$ limit. This helicity structure is very model-dependent. Many topical models [8] of baryon structure in the literature attempt to address this. A sample of their predictions is given in Table 2.2. We note that the non-relativistic quark model of Isgur and Koniuk [8] yields a value -5.6, while Li and Close [8], taking into account effects of color hyperfine interaction in the quark transition operators, are estimating its value in between -2.5 to -4.5. Our phenomenologically extracted value is in excellent agreement with the prediction (-2.5) of Bijker *et al.*[8]. Their model deals with a dynamical $U(7)$ symmetry of the nucleon structure that has a oblate top spectrum. For the $N^*(1520)$ excitation at $q^2 \to 0$, we have confirmed the helicity inequality $|A_{1/2}| < |A_{3/2}|$, from the data, in contrast to the inequality $|A_{1/2}| \gg |A_{3/2}|$ expected at high q^2.

Table 2.1. The electrostrong parameters as determined for $N^*(1535)$ and $N^*(1520)$ for different sets of resonance position M_R, total width Γ and ηN branching ratio. The sets are : a1 = 1544, 212, 0.45; a2 = 1535, 185, 0.45; b1 = 1515, 135, 0.0012; b2 = 1530, 135, 0.0012; b3 = 1530, 110, 0.0012, b4 = 1520, 120, 0.0012.

Parameter Set	χ^2 per d.f.	$\xi^S_{1/2}$	$\xi^D_{1/2}$	$\xi^D_{3/2}$	R
$a1, b1$	1.349	2.221	-0.053 ± 0.009	0.145 ± 0.017	-2.73 ± 0.56
$a1, b2$	1.308	2.218	-0.075 ± 0.011	0.186 ± 0.021	-2.48 ± 0.46
$a1, b3$	1.310	2.219	-0.080 ± 0.011	0.198 ± 0.021	-2.47 ± 0.43
$a1, b4$	1.329	2.221	-0.063 ± 0.009	0.167 ± 0.019	-2.65 ± 0.48
$a2, b1$	1.329	2.308	-0.051 ± 0.005	0.132 ± 0.011	-2.59 ± 0.33
$a2, b2$	1.301	2.307	-0.074 ± 0.007	0.172 ± 0.015	-2.32 ± 0.30
$a2, b3$	1.295	2.307	-0.074 ± 0.007	0.181 ± 0.015	-2.44 ± 0.31
$a2, b4$	1.308	2.307	-0.059 ± 0.006	0.152 ± 0.013	-2.58 ± 0.34

Table 2.2. The ratio $R = \xi^D_{3/2}/\xi^D_{1/2}$ as predicted in various models [8] for the $N^*(1520)$ excitation and decay into ηN, and as determined in a model-independent manner from this work.

Isgur-Koniuk	Capstick	Li-Close	Bijker *et al.*	Inferred from PDG 96	This work
-5.56	-8.93	-2.49 to-4.86	-2.5	-6.9 ± 2.6	-2.5 $\pm 0.4 \pm 0.4$

In summary, the recent experimental advances in the study of photoproduction of eta mesons in the second resonance region ($W \sim 1.3$ GeV) has immediate theoretical pay-off for the knowledge of electromagnetic amplitudes that excite $N^*(1520)$. Even though this resonance is a relatively minor player in this reaction, a combination of the differential cross-section and polarization data, coming out of the recent experiments at the photon facilities, have allowed us to infer in a nearly model-independent way the value of the ratio $A_{3/2}/A_{1/2}$, which is predicted to be negative in the QCD-inspired models, but is strongly model-dependent. The magnitude of this ratio is selective among these topical models, and is very different from being zero, expected in the pQCD regime. New theoretical work is needed to explore this ratio on the lattice. On the experimental side, electroproduction of pseudoscalar mesons, at facilities like the CEBAF, will throw new light on the q^2 developments of these helicity amplitudes and their longitudinal partner.

Acknowledgement. We thank R.M. Davidson, M. Benmerrouche and J.-F. Zhang for many discussions and numerical helps. We are grateful to A. Bock, J.-P. Didilez, B. Krusche, and L. Sözüer for communicating to us preliminary results from eta experiments at Bonn, GRAAL and Mainz. We also thank J. A. Gomez-Tejedor, E. Oset, B. Saghai, F. Tabakin and R. Workman for clarifying related theoretical issues. Our research is supported by the U. S. Department of Energy.

References

1. R. M. Davidson, N. C. Mukhopadhyay and R. Wittman: Phys. Rev. **D43**, 71 (1991); R. M. Davidson and N. C. Mukhopadhyay: Phys. Rev. Lett. **60**, 748 (1988); M. Benmerrouche and N. C. Mukhopadhyay: Phys. Rev. Lett. **67**, 1070 (1991); C. Bennhold and H. Tanabe: Nucl. Phys. **A530**, 455 (1991); M. Benmerrouche, N. C. Mukhopadhyay and J. -F. Zhang: Phys. Rev. **D51**, 3237 (1995); G. Knöchlein, D. Drechsel and L. Tiator: Z. Phys. **A352**, 327 (1995); B. Krusche et al.: Phys. Lett. **B397**, 171 (1997)

2. J. Liu, N. C. Mukhopadhyay and L. Zhang: Phys. Rev. **C52**, 1630 (1995); N. C. Mukhopadhyay et al.: Nucl. Phys. **A633**, 481 (1998)

3. C. E. Carlson: Phys. Rev. **D34**, 2704 (1986)

4. B. Krusche et al.: Phys. Rev. Lett. **75**, 3736 (1995); Phys. Lett. **B358**, 40 (1995)

5. A. Bock et al.: Phys. Rev. Lett. **81**, 537 (1998)

6. J. Ajaka et al.: Phys. Rev. Lett. **81**, 1797 (1998); J. -P. Didilez: priv. comm. (1998)

7. B. Saghai and F. Tabakin: Phys. Rev. C55, 917 (1997) ; F. Tabakin and B. Saghai: Proceedings of the *Forth CEBAF/INT Workshop on N* Physics*,

Editors: T.-S. H. Lee and W. Roberts, (World Scientific, 1997); and priv. comm. (1998).

8. R. Koniuk and N. Isgur: Phys. Rev. **D21**, 1868 (1980); S. Capstick: Phys. Rev. **D46**, 2864 (1992); F. E. Close and Z. Li: Phys. Rev. **D42**, 2194 (1990); *ibid* **D42**, 2207 (1990); Z. Li: Phys. Rev. **D52**, 4961 (1995); S. Capstick and W. Roberts: Phys. Rev. **D49**, 4570 (1994); R. Bijker et. al.: Ann. Phys. 236, 69 (1994)

9. Particle Data Group, R.M. Bernett et al.: Phys. Rev. **D54**, 1 (1996); Particle Data Group, L. Montanet et al.: Phys. Rev. **D50**, 1173 (1994)

10. J. A. Gomez Tejedor, F. Cano and E. Oset, Phys. Lett. **B379**, 39 (1996); and priv. comm. (1998).

11. N. C. Mukhopadhyay, J. -F. Zhang and M. Benmerrouche: Phys. Rev. Lett. **75**, 3022 (1995)

12. M. Benmerrouche, N. C. Mukhopadhyay and J. -F. Zhang: Phys. Rev. Lett. **77**, 4716 (1996)

13. C. Deutsch-Sauermann, B. Friman and W. Norenberg: Phys. Lett. **B409**, 51 (1997); L. Tiator, G. Knöchlein and C. Bennhold: nucl-th/9802064

14. M. Benmerrouche, R. M. Davidson and N. C. Mukhopadhyay: Phys. Rev. **C39**, 2339 (1989)

15. F. James and M. Roos: Comp. Phys. Comm. **10**, 343 (1975)

16. N. C. Mukhopadhyay et al.: Phys. Lett. **B410**, 73 (1997)

Few-Body Systems Suppl. 11, 216–221 (1999)

Few-
Body
Systems
© by Springer-Verlag 1999

Beam Polarisation Asymmetries in Meson Photoproduction at Graal

J. Ajaka[1], V. Bellini[2], J.P. Bocquet[3], M. Breuer[4], M. Capogni[1], M. Castoldi[5], L. Ciciani[1], A. D'Angelo[4], J.P. Didelez[1], R. Di Salvo[1], M.A. Duval[1], C. Gaulard[6], F. Ghio[7], B. Girolami[7], M. Guidal[1], E. Hourany[1], A. Lapik[8], I. Kilvington[9], V. Kouznetsov[8], P. Levi Sandri[6], A. Lleres[3], D. Moricciani[4], V. Nedorezov[8], L. Nicoletti[10], C. Perrin[3], D. Rebreyend[3], F. Renard[3], N.V. Rudnev[8], C. Schaerf[10,4], A Turinge[11] A. Zucchiatti[5]

[1] IN2P3, Institut de Physique Nucléaire, F-91406 Orsay, France

[2] INFN, Laboratori Nazionali del Sud and Università di Catania I-95123 Catania, Italy

[3] IN2P3, Institut des Sciences Nucléaires, F-38026 Grenoble, France

[4] INFN, sezione di Roma II, I-00133 Roma, Italy

[5] INFN, Sezione di Genova, I-16146 Genova, Italy

[6] INFN, Laboratori Nazionali di Frascati, I-00044 Frascati, Italy

[7] INFN, sezione di Roma I and Istituto Superiore di Sanità, I-00161 Roma, Italy

[8] Institute for Nuclear Research, RU-117312 Moscow, Russia

[9] European Synchrotron Radiation Facility, F-38026 Grenoble, France

[10] Universitá di Roma "Tor Vergata", I-00133 Rome, Italy

[11] I. Kurchatov Institute of Atomic Energy, RU-123182 Moscow, Russia

Meson photoproduction on the nucleon has been a major source of information on the N and Δ resonances and the transition matrix elements from the nucleon ground state to its excited states.

Polarisation observables have enhanced the sensitivity to the contributions of the less pronounced resonances since they depend mostly on the interference terms among the four helicity amplitudes [1, 2]. In particular the beam asymmetry for linearly polarised photons is given by:

$$\Sigma \sim \mathrm{Re}(H_1 H_4^* - H_2 H_3^*). \tag{1}$$

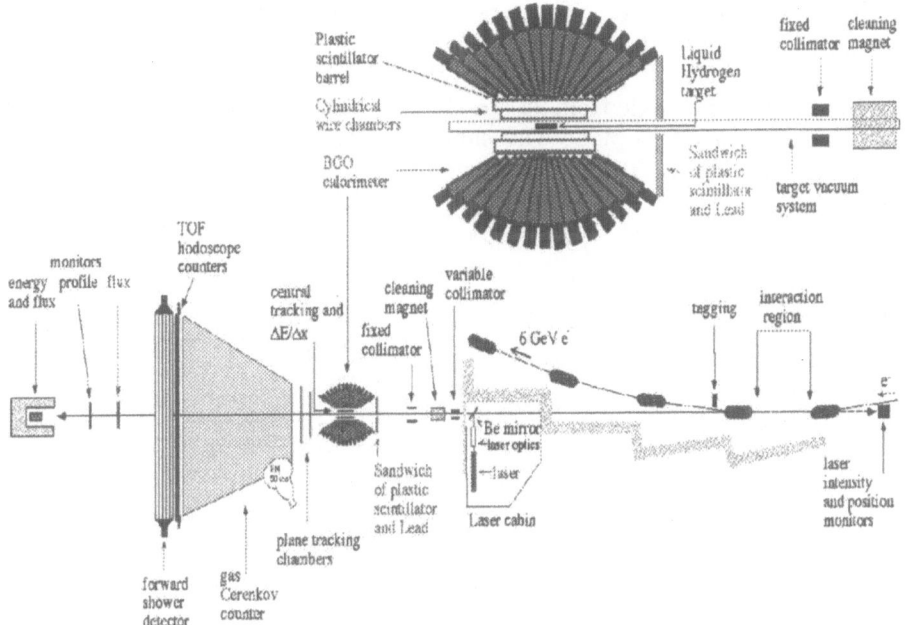

Figure 1. The Graal experimental layout (not in scale). The gas Cerenkov counter is still under construction.

The Graal facility provides a polarised and tagged photon beam by the backward Compton scattering of laser light on the high energy electrons circulating in the ESRF storage ring [3]. Using the UV line (350 nm) of an Ar-Ion laser we have produced a gamma-ray beam with an energy from 550 to 1470 MeV. Its polarisation is 0.98 at the maximum photon energy and the energy resolution has been measured to be 16 MeV (FWHM). Using the green line of the same laser we have measured the beam polarisation asymmetries in the photoproduction of η [4] π^o and π^+ in the energy region 550-1100 MeV.

The experimental apparatus is indicated in Fig. 1. Electrons, scattered by the photons, are momentum analysed by the first dipole of the ring lattice and their position, relative to the circulating electron beam, is measured by the tagger. The tagging detector is located inside a rectangular box with one side parallel to the electron beam at a distance of 10 mm from it. Inside this vacuum box are located a Densimet shielding box and inside it 10 plastic scintillators and a solid state Silicon microstrip detector with 128 channels and a pitch of 0.3 mm. The position of the microstrip traversed by the scattered electron, gives the energy lost by it and therefore the energy of the gamma-ray.

The gamma-ray beam traverses the Beryllium mirror, used to reflect the laser beam, exits the accelerator vacuum system and is collimated in air by four remotely controlled Lead blocks. Then it enters the vacuum system of

218

the liquid Hydrogen target. The target is surrounded by two cylindrical wire chambers, a barrel made of 32 strips of plastic scintillator parallel to the beam axis and the BGO ball made of 480 crystalls of BGO scintillator [5, 6]. The BGO ball covers an angle from 25° to 155°. Particles moving at angles smaller than 25° encounter two plane wire chambers, two walls of plastic scintillator bars 3 cm thin and a shower wall made by a sandwich of four layers of Lead and plastic scintillators 4 cm thick. Two disks of plastic scintillator separated by a disk of Lead close the backward cone.

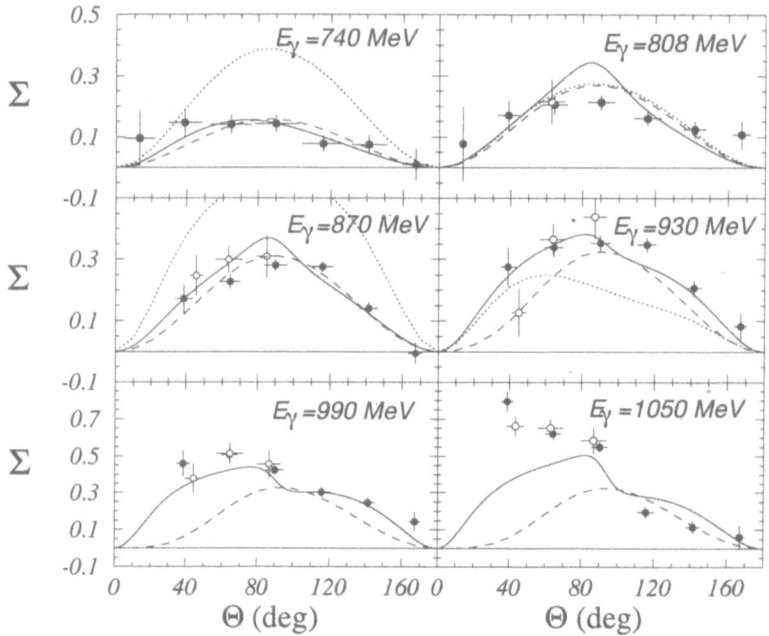

Figure 2. Beam Asymmetry Σ for η photoproduction. The full circles are the results when the photons are detected in the BGO. The open circles when one of the photons is in the shower detector. Dotted curve is from reference [10], dashed curve from reference [11], full curves from refence [8].

π^o events are detected when the two decay photons enter the BGO ball and the proton is in the BGO or the forward walls. For the η events we have added the condition that one photon is in the BGO and the other in the shower wall. π^+ events are detected when the neutron is in the shower wall and the pion in the BGO.

The symmetry of the apparatus, around the beam axis, provides directly the azimuthal distribution of the events and therefore the beam polarisation asymmetry. However, to correct for instrumental asymmetries, the data are collected with two orthogonal directions of the beam polarisation: horizontal and vertical. For each interval of gamma-ray energy and meson centre of mass angle the parameter B and the product ΣP are derived from the experimental data with a best fit procedure:

$$\frac{N_{hor}}{N_{hor} + N_{ver}} = B\left(1 + \Sigma P \cos(2\varphi)\right). \tag{2}$$

N_{hor} and N_{ver} are respectively the number of events collected with the beam

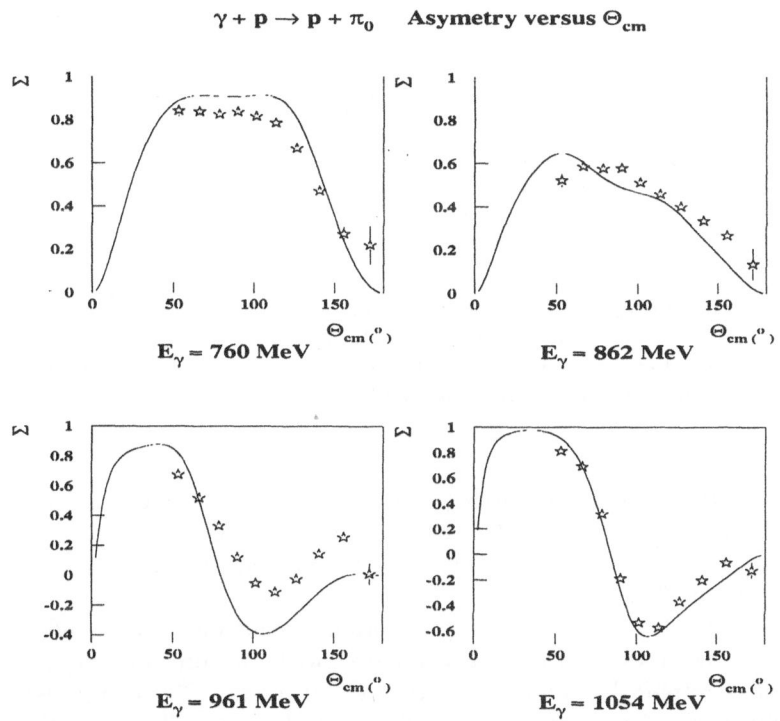

Figure 3. Sample of data on Beam Asymmetry Σ for π° photoproduction, full curve is from reference [2].

polarisation in the Horizontal and Vertical planes normalized to the respective numbers of gamma-rays impinging on the target; B is a normalization constant close to one; P is the polarisation of the beam evaluated with a Monte Carlo from the known polarisation of the laser beam, its geometry and the characteristics of the electron beam in the interaction region. It is a function of the gamma-ray energy; Σ is the beam asymmetry.

The asymmetries for η photoproduction are indicated in Fig. 2. The most interesting results are the large asymmetries at small angles and high energies. They have been interpreted [7, 9] as the contribution of four established resonances and the "missing" P13(1880).

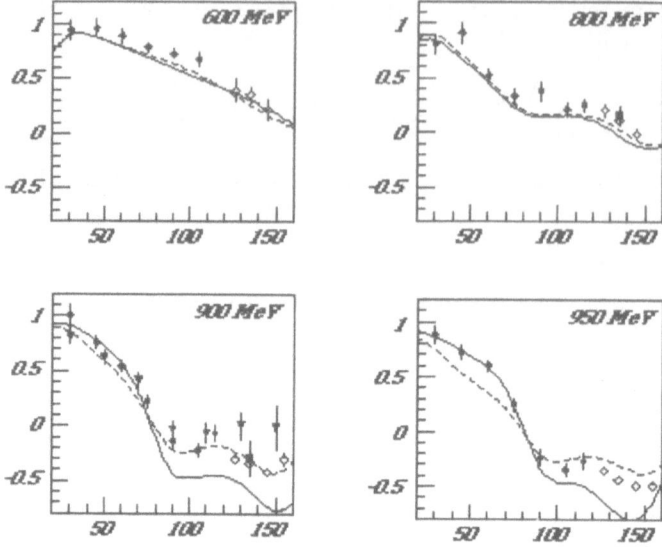

Figure 4. Preliminary data on Beam Asymmetry Σ for π^+ photoproduction as function of $\vartheta_{\pi^+}^{c.m.}$. Open circles this experiment, dark circle from Daresbury [13], dark triangles and squares from SLAC [14]. The solid line is from partial wave analysis [15-19]. The dashed line is from reference [20]

A preliminary set of measured asymmetries for π^o are indicated in Fig. 3. In Fig. 4 some selected curves are compared with the SAID model [2, 12]. The agreement seems poor at the energy of 961 MeV corresponding to a centre of mass energy of 1638 MeV.

The asymmetries for π^+ photoproduction at backward pion angle are indicated in Fig. 4 where they are compared with the previous experimental results of Daresbury [13] and several theoretical models. The comparison with the previous experiment seems good. The partial wave analysis of VPI misses the experimental points between 848 and 950 MeV.

References

1. B. Saghai and F. Tabakin: Phys. Rev. **C55**, 917, (1997); Phys. Rev. **C53**, 66, (1996); F. Fasano, F. Tabakin, and B. Saghai: Phys. Rev. **C46**, 2430, (1992)

2. R. Arndt et al.: Phys. Rev. **C42**, 1853, (1990)

3. Graal Collaboration: Nucl. Phys. **A622**, 110c, (1997)

4. J. Ajaka et al.: Phys. Rev. Lett. **81**, 1797 (1998)

5. F. Ghio et al.: Nucl. Inst. and Meth. **A404**, 71 (1998) A. Zucchiatti et al.: Nucl. Inst. and Meth. **A317**, 492 (1992)

6. P. Levi Sandri et al.: Nucl. Inst. and Meth. **A370**, 396 (1996).

7. B. Saghai et al.: Proceedings of the *6th Conference on the Intersections of Particle and Nuclear Physics* ed. T. W. Donnelly, AIP Conference Proceedings 412, New York, 1997

8. B. Saghai: Private Communication

9. B. Saghai et al.: in preparation

10. C. Bennhold et al.: Nucl. Phys. **A530**, 625 (1991); L. Tiator et al.: Nucl. Phys. **A580**, 455 (1994)

11. G. Knöchlein, D. Drechsel and L. Tiator: Z. Phys. **A352**, 327 (1995)

12. R. Arndt et al.: Phys. Rev. **C47**, 2759 (1993); Phys. Rev. **C52**, 2759 (1995)

13. P. Busley et al.: Nucl. Phys. **B154**, 205 (1979)

14. G. Knies et al.: Phys. Rev. **D10**, 2778 (1974); R. Zdarko and E. Dolly: Nuovo Cimento **10A**, 10 (1972)

15. R. Arndt et al.: Phys. Rev. **C42**, 1853 (1990)

16. R. Arndt et al.: Phys. Rev. **C42**, 1864 (1990)

17. Zhujun Li et al.: Phys. Rev. **C47**, 2759 (1993)

18. R. Arndt, I. Strakovsky and R. Workman: Phys. Rev. **C53**, 430 (1995)

19. R. Arndt, I. Strakovsky and R. Workman: Phys. Rev. **C56**, 577 (1997)

20. D. Drechsel, O. Hanstein, S.S. Kamalov and L. Tiator: Nucl-th/9807001, and S.S. Kamalov: Private communication

Few-Body Systems Suppl. 11, 222–224 (1999)

η Physics and Multi-Resonance Coupled-Channel Analysis

A. Švarc *

Rudjer Bošković Institute, Bijenička c.54, 10000 Zagreb, Croatia

Abstract. The final version of the multi-resonance, coupled-channel partial wave analysis with the significant improvement of the πN elastic S_{11} partial wave is presented. The need for the new, only theoretically predicted P_{11} resonance is pointed out. The value of the ηN S-wave scattering length is finally unambiguously established.

A three-channel, multi-resonance, unitary model, based on the formulation developed in Ref. [1], has been applied in Refs. [2, 3] to perform a partial-wave analysis (PWA) using the Karlsruhe-Helsinki PWA (KH80) [4] as input for the πN elastic scattering, and the weighted total and differential cross section data for the $\pi N \to \eta N$ reaction [5]. The partial-wave amplitudes for the $\pi N \to \eta N$ and $\eta N \to \eta N$ transitions are the predictions of the model. The complete input data set, with the assumption that the number of resonances per partial wave is given by PDG [6] could not be fitted with only three P_{11} resonances simultanuously obtaining a consistency with the former three-channel, multi-resonance unitary analysis of Manley and Salesky [7]. Therefore, the fourth P_{11} resonance had to be added [2, 3]. It turned out that the need for additional resonances was theoretically predicted by Capstick et al. [8]. The additional resonance of our model fairly coincides with one of the resonances given in [8], therefore, we claim that the serious search for the fourth P_{11} resonance should be initiated in other processes.

Using the obtained ηN elastic scattering partial-wave amplitudes, the ηN S-wave scattering length $a_{\eta N}$ has been extracted. A number of very different values for the real part was predicted in earlier articles[1]. Almost identical result was obtained in an independent K-matrix analysis [10, 11]. Therefore, we dare concluding that the value of the ηN S-wave scattering length is at last established.

---

*E-mail address: svarc@rudjer.irb.hr
[1] see the compilation of results in Ref. [9]

In Fig.1, we display the comparison between the quark model theoretical predictions of Capstick el al. [8], and the new P_{11} resonance of this model needed for the complete coupled-channel multi-resonance analysis of Ref. [3].

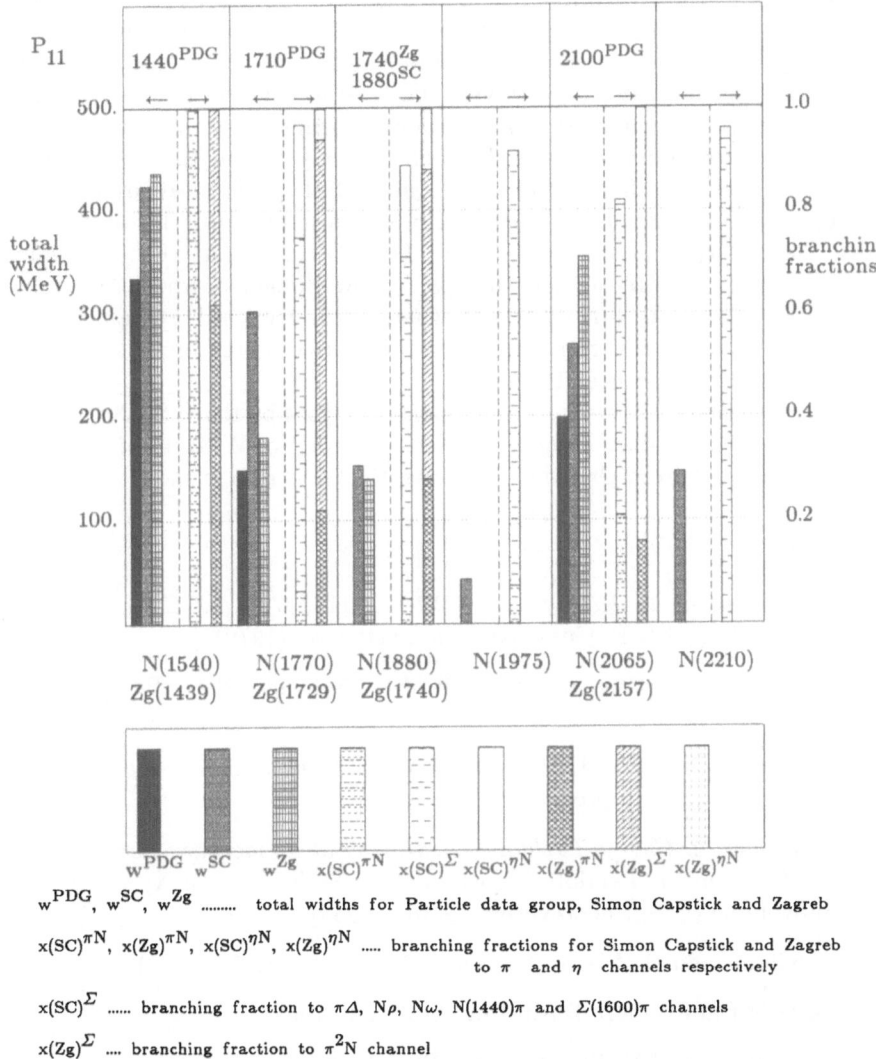

w^{PDG}, w^{SC}, w^{Zg} .... total widths for Particle data group, Simon Capstick and Zagreb

$x(SC)^{\pi N}$, $x(Zg)^{\pi N}$, $x(SC)^{\eta N}$, $x(Zg)^{\eta N}$ ..... branching fractions for Simon Capstick and Zagreb to π and η channels respectively

$x(SC)^{\Sigma}$ ...... branching fraction to $\pi\Delta$, $N\rho$, $N\omega$, $N(1440)\pi$ and $\Sigma(1600)\pi$ channels

$x(Zg)^{\Sigma}$ .... branching fraction to $\pi^2 N$ channel

Figure 1. The comparison of the P_{11} resonances of PDG [6], S. Capstick et al. [8], and Zagreb group [3]. The explanations are given in the figure.

All parameters of the fourth P_{11} resonance of both models roughly correspond. A detailed comparison and the additional verification is needed, but we do see a strong indication that the fourth P_{11} resonance does exist. It is interesting to remark that the parallel multi-channel analysis, presented at this conference by the Pittsburgh group [12], does not have any indication of the presence of the fourth P_{11} resonance. The reason lies in the fact that this analysis is using a simplifying approach that only single-channel T-matrices are fitted, and not the experimental data. That gives them the possibility to perform the fit partial wave by partial wave reducing significantly the number of fitting parameters, but eliminates the possibility for the transfer of flux from the partial in one channel to the partial wave in another channel. Such an simplification of the Cutkosky method is a serious assumption, and can hide a lot of important effects. For instance, it did not reveal the need for the fourth P_{11} resonance.

The determination of the ηN scattering length within a multi-channel, multi-resonance, unitary model depends strongly uppon the πN input data in S_{11} channel. The final extracted value of Ref. [3] is $a_{\eta N}(\text{fm}) = (0.717 \pm 0.030) + i \cdot (0.263 \pm 0.025)$. Taking into account the similarity of our results with the results reported in Ref. [10, 11], we claim that the final value of the ηN S-wave scattering length $a_{\eta N}$ is now pinned down to the value of $\overline{a_{\eta N}}(\text{fm}) = (0.72\pm?) + i \cdot (0.26\pm?)$ where error bars still remain to be determined. However, this result can be reliably used as the input to investigate η-nucleus interactions near threshold.

Acknowledgment. This work is supported in part by the U.S. Department of Energy, Nuclear Physics Division, under contract No. W-31-109-ENG-38. The financial support from Croatia-US grants JF 221 is also acknowledged.

References

1. R.E. Cutkosky et al.: Phys. Rev. **D20**, 2804 (1979); *ibid* **D20**, 2839 (1979); R.L. Kelly and R.E. Cutkosky: *ibid* **D20** 2782 (1979)

2. M. Batinić, I. Šlaus, A. Švarc, B.M.K. Nefkens: Phys. Rev. **C51**, 2310 (1995)

3. M. Batinić, I. Dadić, I. Šlaus, A. Švarc, B.M.K. Nefkens, and T.-S.H. Lee: Physica Scripta, in press

4. G. Höhler: in *Charge Exchange Scattering of Elementary Particles*, edited by H. Schopper, Landolt-Börnstein, New Series, Group X, Vol 9, Part 2, Subvolume b, (Springer-Verlag, Berlin, 1983)

5. M. Clajus and B.M.K. Nefkens, πN Newsletter **7**, 76 (1992)

6. Particle Data Group: Phys. Rev. **D45**, S1 (1992)

7. D.M. Manley, E.M. Salesy: Phys. Rev. **D45**, 4002 (1992)

8. S. Capstick: These Proceedings, and references therein

9. M. Batinić and A. Švarc: Few Body Syst. **20**, 69 (1996)

10. A.M. Green and S. Wycech: Phys. Rev. **C55**, R2167 (1997)

11. S. Wycech: private communication

12. S. Dytman: contribution to this conference, and references therein

Few-Body Systems Suppl. 11, 225–232 (1999)

Few-
Body
Systems
© by Springer-Verlag 1999

K Meson Photo- and Electro- Production with CLAS at Jefferson Lab

Reinhard A. Schumacher*

Department of Physics, Carnegie Mellon University, Pittsburgh, PA 15213, U.S.A.

Abstract.
In this talk we give an overview of experiments involving strange particles using the CLAS spectrometer at Jefferson Lab. Also, a first "pre-production" look at strange particle detection in CLAS is presented based on recently-completed photoproduction and electroproduction runs. Experimental issues of particle identification and missing mass resolution are discussed. The results show that the outlined program of experiments is well underway to producing first physics results.

1 The CLAS Program of Strangeness Physics

The strange particle physics program at CLAS has been established for several years, with no new additions to the experimental program since it was reviewed some time ago [1]. In Table 1 we list the strange-particle experiments at Jefferson Lab which are being done at CLAS. The experiments on a proton target obtained their first data by mid-1998, so that at present there is an intense analysis effort underway. In this talk it is therefore not possible to present final results of any sort, but rather to outline again the physics motivation for the experiments, and to discuss the experimental aspects of identifying strange particle production at CLAS.

Table 1 lists in condensed form the measurements that each of seven experiments seeks to make. For a brief discussion of the physics goals of the experiments, please refer to Ref [1]. Since the present analysis is concentrating on elementary production, it is worth noting that numerous recent theoretical efforts have improved our understanding of strangeness photo- and electroproduction, including approaches using chiral perturbation theory [3], isobar models [4, 5, 6, 7], isobars with significant input from quark models [8, 9], and the Regge approach [10].

---

*E-mail address: schumacher+@cmu.edu

Table 1. Experiments at CLAS involving associated strangeness production, as of summer 1998 [2].

Experiment Title and Main Goals	Experiment	Beam
Electromagnetic Production of Hyperons - $K^+\Lambda$, $K^+\Sigma^0$, $K^0\Sigma^+$ differential cross sections - hyperon recoil polarizations - beam-recoil double polarization observables - extraction of resonance production parameters	89-004 R. Schumacher	real photon $.91 < E\gamma < 2.5$ GeV
Study of Kaon Photo-production on Deuterium - elementary amplitudes on the neutron - hyperon-nucleon final state interactions - Λ - Σ channel coupling effects - S = -1 dibaryon search	89-045 B. Mecking	real photons $.91 < E\gamma < 1.5$ GeV
Radiative Decays of the Low-Lying Hyperons - decay of Λ(1520), Λ(1405), and Σ(1385) to ground state Λ - tests of quark models of hyperon structure	89-024 G. Mutchler	real photons $E\gamma < 3.2$ GeV
Quasi-free Strangeness Production in Nuclei - ^{3}He ^{4}He and ^{12}C targets - test one-body hypothesis for (γ,K) in nuclei - probe kaon-nucleus final state interactions	91-014 C. Hyde-Wright	real photons $1.6 < E\gamma < 2.0$ GeV
Measurement of the Structure Functions for Kaon Electroproduction - $1.0 < Q^2 < 2.5$ GeV^2, W < 2.2 GeV - separate σ_T, σ_L, σ_{LT}, σ_{TT} for the Λ(1115) to the Λ(1520) - t dependence of σ_L - missing N* resonance search in strange final states	93-030 M. Mestayer K. Hicks	electron 2.4 and 4.0 GeV
Measurements of the Electroproduction of the Λ(gnd), Λ^*(1520), and f_0(975) via the K^+K^-p and $K^+\pi^-p$ Final States - s, t, and u channel contributions to Λ production	89-043 L. Dennis H. Funsten	electron 2.4 and 4.0 GeV
Measurement of K^0 Electroproduction - sensitivity to the Q^2 dependence of the $\gamma K^* K$ transition form factor	95-003 R. Schumacher K. Dhuga	electron 2.4 and 4.0 GeV

2 Recent Progress in Strangeness Production at CLAS

Analysis has begun of CLAS data stemming from the electroproduction runs in December 1997 and February 1998, and from the photoproduction run in April and May 1998. An early goal has been to verify that the strangeness production signal can be isolated, and that the aims of the strangeness production experiments can be met. To this end, we have done a "pre-production" analysis of both the photoproduction and electroproduction data sets to investigate the particle (kaon) identification quality and the missing mass resolution we can obtain. Some of the results of this preliminary analysis are shown here. It must be emphasized that all of these data are very preliminary, with no acceptance corrections, and without having achieved the final resolutions in particle ID, momentum, etc. All the distributions are also un-normalized, which is to say that we are not yet showing actual cross-sections.

2.1 Photoproduction

We have analyzed 1.1 million events from CLAS, where the trigger requirement was that at least one charged particle be present in coincidence with a tagged photon. This amounts to about 45 minutes worth of data, or less than 2% of the data set accumulated. The endpoint energy was 2.534 GeV, and the photon tagging range was from 0.2 to 0.95 of this energy. For the preliminary data shown, a cutoff at 2.0 GeV was applied in the photon energy.

The photon beam was circularly polarized by virtue of the 35% longitudinal polarization of the electron beam. This led to an energy-dependent photon circular polarization of, e.g., 26% at 1.45 GeV. We hope eventually to exploit this feature of the data to extract the double polarization (beam-recoil) observables $C_{x'}$ and $C_{z'}$ (c.f. Ref [4]).

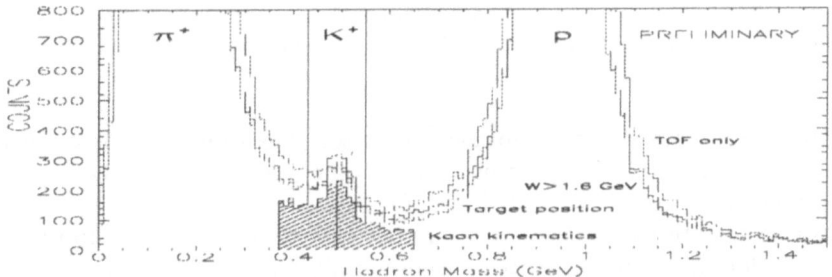

Figure 1. Particle mass spectrum for photoproduction using time-of-flight, and cutting on target position, $W > 1.6$ GeV, and kinematically allowed kaon angles.

Fig. 1 shows a particle ID mass distribution obtained for photoproduction. It is based mainly on time-of-flight (TOF) from the target to the counters that surround CLAS over a pathlength of 4 to 5 meters. Pions of roughly 2 GeV/c momentum are the main background under the kaons. Much effort has already gone into calibration of the 288 counters in this TOF array, but more improvements can be expected. In addition to TOF, rough kaon kinematics and vertex cuts were made to select kaons. Within the mass region of the kaons, the signal to noise ratio is about 1:1.5. Sidebands on either side of the kaon peak were defined for subsequent subtraction of background.

After selecting events in the kaon region, the resulting missing mass recoil spectrum for photoproduction shown in Fig. 2 was obtained. The mass resolution for the Λ and Σ^o hyperons is about 25 MeV FWHM. In the sideband-subtracted lower panel there is evidence of some known broad higher-mass hyperon states.

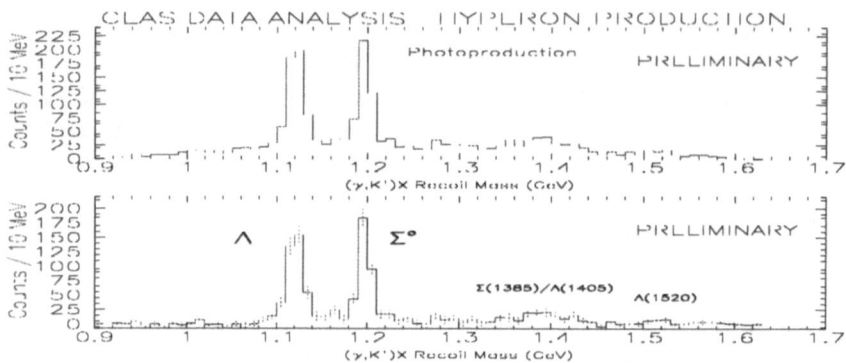

Figure 2. Missing mass spectrum for the reaction $\gamma + p \to K^+ + X$ for $W > 1.6$ GeV and $E_\gamma < 2.0$ GeV. Top panel is for hadrons that passed kaon selection cuts; lower panel includes sideband subtraction of events from hadrons that failed the kaon selection cuts.

Because the data are not acceptance corrected it is difficult to say much about the underlying physics at this stage of analysis. A study which does have meaning is the comparison of Λ and Σ^o data, since acceptance and kaon decay corrections are likely to be very similar. In Fig. 3 we show preliminary distributions for the W dependence of hyperon photoproduction, where W is the invariant energy of the γp system (*i.e.* $\sqrt{s}$). We choose to display W since this facilitates comparison with electroproduction data and with models of s-channel baryon resonance production. These distributions are proportional to the total cross sections. One sees that the rise with W of Σ^o production seems much slower that that for Λ production; this is consistent with recent results from SAPHIR for the total cross sections [11]. The differing slopes are understood as due to the differing resonance structure for the production of the two hyperons. The decrease of the distributions on the high energy side is controlled by CLAS's acceptance.

We have also examined the c.m. kaon angular distributions and the hyperon polarization distributions. 37% of all hyperons seen in the recoil spectra have an associated detected proton which is kinematically consistent with the decay of the hyperon. We can use these events to study the hyperon polarizations; the polarizations for Λ's and Σ's are different, but they are not shown here

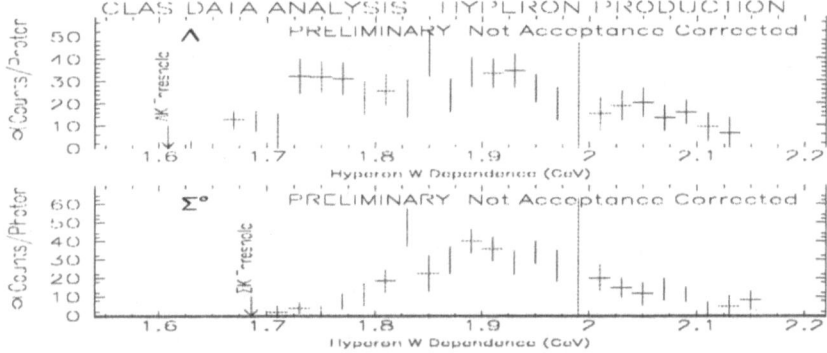

Figure 3. W dependence of hyperon photoproduction for Λ's (top) and Σ^{o}'s (bottom). The error bars are statistical only, and the normalization is proportional to the total number of photons detected in a given W bin.

since the acceptance corrections we need to apply are quite large.

The strange particle yield was 0.5% of all events with a c.m. energy above the threshold at W = 1.61 GeV. This is consistent with expectations from the known total cross sections. Also, the ratio of the total yield of Λ's to Σ^{o}'s for $E_{\gamma} < 2.0$ GeV was close to 1:1.

2.2 Electroproduction

In electroproduction, we have analyzed about 1.3 million events from CLAS, where the trigger requirement was that an electron be detected in the electromagnetic calorimeter; any other tracks present at the same time, due to hadrons, were then recorded. This corresponds to about 4.2 hours worth of data, or 3.5% of the data set accumulated. The electron beam energy was 2.245 GeV, and the beam was not polarized. Electroproduction data were also accumulated at 4.045 GeV, but the analysis of that set is not discussed here.

Fig. 4 shows the particle ID spectrum obtained for electroproduction. Again, it is based mainly on time-of-flight from the target to the TOF counters that surround CLAS. In this case, more meaningful cuts could be made on the vertex at the target, since for electroproduction the target was smaller (4 cm. rather than 17 cm long) and there were always at least two tracks: the electron and the kaon. Within the mass region of the kaons, the signal to noise ratio is nearly 4:1. Again, sidebands on either side of the kaon peak were defined for subsequent subtraction of background.

After selecting events in the kaon mass region, the resulting missing mass spectrum for electroproduction shown in Fig. 5 was obtained. The mass reso-

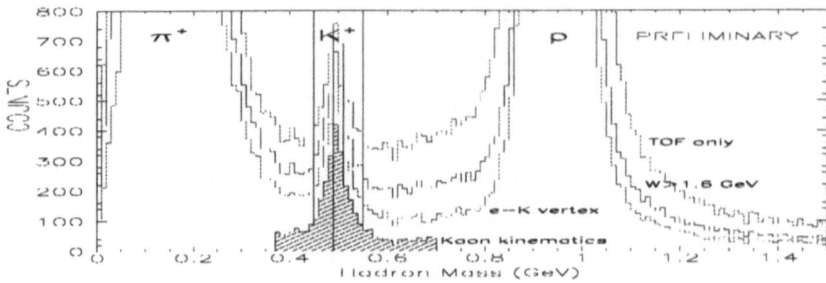

Figure 4. Particle identification mass spectrum for electroproduction based on time-of-flight, vertex constraint, and kaon kinematics, for $W > 1.6$ GeV.

lution for the Λ and Σ^o hyperons is about 35 MeV FWHM. In the sideband-subtracted lower panel there is weak evidence of known broad higher-mass hyperon states.

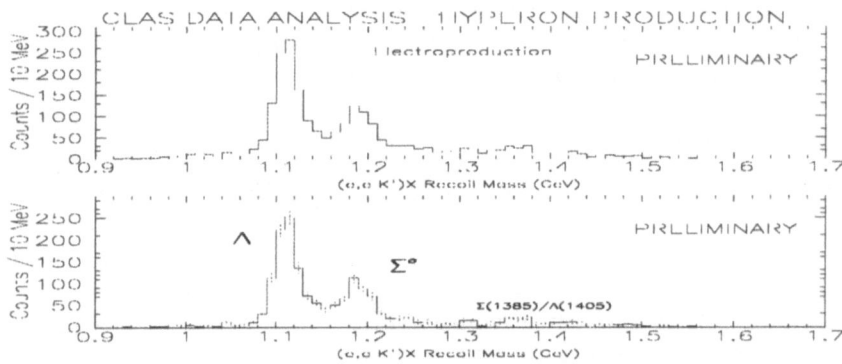

Figure 5. Missing mass spectrum for the reaction $e + p \rightarrow e' + K^+ + X$ for $W > 1.6$ GeV. The top panel is for hadrons that passed kaon selection cuts; lower panel includes sideband subtraction of events from hadrons that failed the kaon selection cuts.

In Fig. 6 we show preliminary distributions for the W dependence of hyperon electroproduction averaged over the range $0.4 < Q^2 < 1.25$ $(\text{GeV/c})^2$. These distributions are proportional to the total cross sections. This figure can be compared to the photoproduction case shown in Fig. 3. One sees again that the rise with W of Σ^o production seems much slower that that for Λ production, a reflection of differing resonance structure in electroproduction of the two hyperons. The fall of the distributions on the high energy side are controlled by CLAS's acceptance. The staircase-like histogram is the arbitrarily-scaled non-strange W dependence, which is being cut off by the CLAS acceptance in the same way as the strange acceptance.

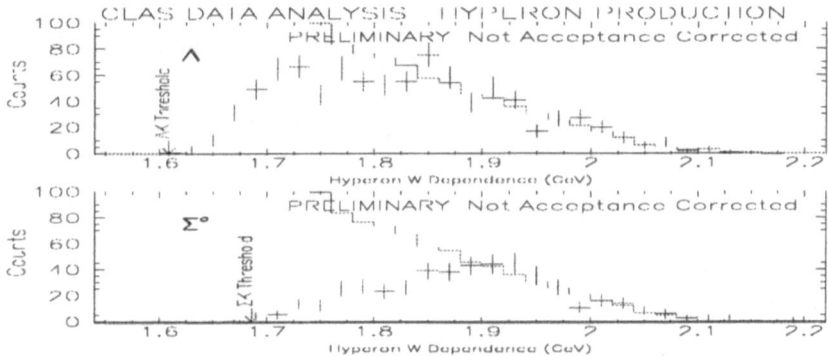

Figure 6. W dependence of hyperon electroproduction for Λ's (top) and Σ^o's (bottom). The error bars are statistical only.

We have also examined the c.m. kaon angular distributions and the hyperon polarization distributions. 25% of all hyperons seen in the recoil spectra have an associated detected proton which kinematically consistent with the decay of the hyperon. We see that there is more Λ yield at forward angles relative to the Σ^o's than in photoproduction. It is also the case that the ratio of yields of Λ's to Σ^o's is now 1.8:1 rather than 1:1. Both of these observations may be consistent with the fact that Λ production away from $Q^2 = 0$ is dominated by the longitudinal piece of the electron interaction, a piece which is absent in photoproduction.

3 Summary / Conclusions

In conclusion, this study has demonstrated that the CLAS strangeness physics program is underway. The particle identification of kaons is already adequate to do a physics analysis, and will improve further. This pre-production look at the photo- and electro- production data, which examined only a few percent of the data in hand at this time, has shown that the analysis tasks are tractable and that physics results are coming. It should be noted that this analysis work was somewhat independent of the main production analysis being implemented by the collaboration for reducing the bulk of the data. As such, these results should only be construed as showing that the data will likely yield valid results. Acceptance corrections will be very important, and are underway within the collaboration. While no new physics conclusions are possible based on this first look at the new CLAS data, we are optimistic that a wealth of new results are on the way.

References

1. Reinhard A. Schumacher: Nucl. Phys. **A585**, 63c (1995)

2. Copies of proposals are available upon request from the Jefferson Lab Users' Liaison Office, at 757-269-7586.

3. S. Steininger and U. Meissner: Phys. Lett. **B391**, 446 (1997)

4. J.C. David. C. Fayard, G.H. Lamot, and B. Saghai: Phys. Rev. **C53**, 2613 (1996)

5. R.A. Adelseck and B. Saghai: Phys. Rev. **C42**, 108 (1990); Phys. Rev. **C45**, 2030 (1992)

6. R.A. Williams, C.-R. Ji, and S.R. Cotanch, Phys. Rev. **C46**, 1617 (1992)

7. T. Mart and C. Bennhold: Few-Body Systems Suppl. **9**, 369 (1995)

8. Zhenping Li: Phys. Rev. **C52**, 1648 (1995); Phys. Rev. **C54**, R2171 (1996)

9. V. Keiner: Z. Phys. **A352**, 215 (1995)

10. M. Guidal, Ph.D. thesis, Saclay, DAPNIA-SPhN-96-03T (1996)

11. J. Barth et al.: Proceedings of *Intersections Between Particle and Nuclear Physics*, AIP Conf **412**, 849, T.W. Donnelly, Ed. (1997). See also M. Bockhorst et al.: Z. Phys. **C63**, 37 (1994).

Few-Body Systems Suppl. 11, 233–240 (1999)

Few-
Body
Systems
© by Springer-Verlag 1999

Study of the Strange Hadrons Electromagnetic Form Factors via the Process $e + p \rightarrow e' + K^+ + \Lambda$

B. Saghai[1], P. Oswald[1,2], C. Fayard[2], G. H. Lamot[2], R. A. Williams[3]

[1] Service de Physique Nucléaire, DSM/DAPNIA, CEA/Saclay, F-91191 Gif-sur-Yvette Cedex, France

[2] Institut de Physique Nucléaire de Lyon, IN2P3-CNRS, Université Claude Bernard, F-69622 Villeurbanne Cedex, France

[3] JLAB, Newport News, Virginia 23606, USA

Abstract. Results of recent theoretical studies on the electromagnetic form factors of strange baryons and kaons are embodied in a strangeness electromagnetic production model based on an effective Lagrangian approach. The sensitivity of different observables that are the subject of current or planned measurements at JLAB, are investigated with respect to the phenomenological ingredients of the model.

1 Introduction

The form factors of the nucleons have been extensively studied both experimentally and theoretically. However, there are no data on the hyperons form factors and theoretical investigations on this subject are scarce. In our effort to shed light on the non-perturbative features of hadron structure, the study of such form factors constitutes a fundamental topic.

Electroproduction processes offer an appropriate tool to probe the structure of hadrons by virtual photons. In phenomenological models, this structure is introduced via form factors appropriate to each intervening hadron. One of the major issues in strangeness electroproduction is thus the study [1] of hyperons and strange mesons form factors.

2 Strange hadrons electromagnetic form factors

Among the strange particles of interest in this work, only the form factor of the K^+ has been studied. However, the available experimental data [2] are restricted to the low momentum transfer domain ($-Q^2 < 0.15$ GeV^2), and therefore do not allow us to select a model valid in the whole kinematical

234

region considered here. For kaons (K^+, K^*, and $K1$), we therefore investigate [3, 4] three models based on *i)* a simple monopole parameterization [5], *ii)* a VDM approach [6], and *iii)* a more recent relativistic constituent quark model formalism [7].

Recently, an extended vector meson dominance model (EVDM) approach has been developed [8] for the baryon octet electromagnetic form factors. This model reproduces very well the proton and neutron form factors data and makes predictions for the other members of the octet. In Fig. 1 the results of the above model for the two form factors, F_1 and F_2, for the neutron, Λ and Σ° are depicted.

The above form factors are introduced in a model, summarized below, describing the associated strangeness electromagnetic production on the proton.

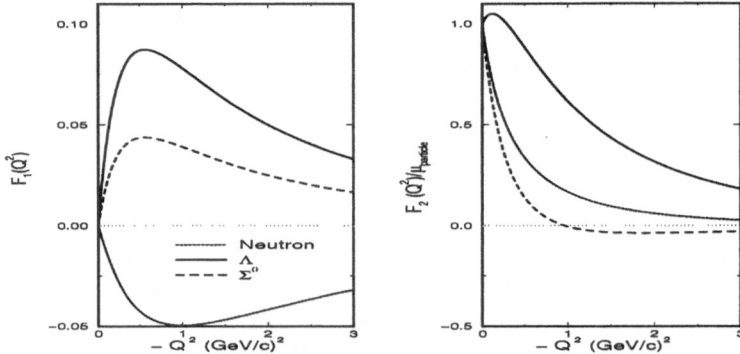

Figure 1. Form factors of neutral hadrons [8] used in the present work.

3 Strangeness electromagnetic production on the proton

Our starting point is the Saclay-Lyon approach [4] implemented with the off-shell effects treatment [9] required for spin-3/2 resonances. The formalism is based on an isobaric approach using the Feynman diagrammatic technique in first order perturbation theory, where each diagram corresponds to the exchange of *one* particle or resonance (tree approximation). The corresponding diagrams are the so-called extended Born terms for the exchange of the proton (p), kaon (K), and hyperons ($Y \equiv \Lambda$, Σ°), and the resonant terms for the exchange of the N^*, K^*, and Y^* resonances.

Application of the Feynman rules yields the invariant amplitudes $\mathcal{A}_j$, $j=1,6$. To highlight the form factor dependence of these amplitudes, we single out two of them with the usual notations (see, e.g., Ref. [4])

$$\mathcal{A}_2^{Born} = \frac{eg_{K\Lambda N}}{(s - M_p^2)(t - M_K^2)}\left[(F^K + F_1^p) + (F^K - F_1^p)\frac{p_\gamma \cdot p_K + p_\gamma \cdot p_p}{p_\gamma \cdot p_\Lambda}\right] \quad (1)$$

$$\mathcal{A}_5^{Born} = \frac{eg_{K\Lambda N}}{s - M_p^2}\frac{\kappa_p}{2M_p}F_2^p - \frac{eg_{K\Lambda N}}{u - M_\Lambda^2}\frac{\kappa_\Lambda}{2M_\Lambda}F_2^\Lambda - \frac{eg_{K\Sigma^\circ N}}{u - M_{\Sigma^\circ}^2}\frac{\kappa_{\Sigma^\circ \Lambda}}{2M_p}F_2^{\Sigma^\circ}. \quad (2)$$

As we will see in the next Section, the free parameters of this approach are determined by fitting all the available data for the photoproduction ($E_\gamma^{lab} \leq 2.5$ GeV) and electroproduction of the kaons on the proton, as well as the associated radiative capture branching ratio $R_{\gamma\Lambda} = \frac{\Gamma(K^-p \to \gamma\Lambda)}{\Gamma(K^-p \to \text{all})}$ at threshold, namely, the following reactions:

$$\gamma + p \quad \to \quad K^+ + \Lambda, \tag{3}$$

$$e + p \quad \to \quad e' + K^+ + \Lambda, \tag{4}$$

$$K^- + p \quad \to \quad \gamma + \Lambda. \tag{5}$$

At the present time, the most complete model (called model C in Ref. [9]) contains, besides the extended Born terms, the following exchanged resonances:

- s-channel: $P_{13}(1720)$
- u-channel: $\Lambda(1405), \Lambda(1670), \Lambda(1810), \Lambda(1890), \Sigma(1660)$,
- t-channel: $K^*(892), K1(1270)$.

The relevant coupling constants are treated as free parameters. Moreover, the off-shell effects treatment requires [9] three free parameters for each of the spin-3/2 resonances $P_{13}(1720)$ and $\Lambda(1890)$.

4 Results and discussion

To study the hyperons electromagnetic form factors, we concentrate on the electroproduction process, for which the differential cross section is given by the general well known expression

$$\frac{d\sigma}{d\Omega_K} = d\sigma_U + \epsilon_L \, d\sigma_L + \epsilon \, d\sigma_P \sin^2\theta \cos2\phi + \sqrt{2\epsilon_L(1+\epsilon)} \, d\sigma_I \sin\theta \cos\phi. \tag{6}$$

Table 4.1. Form factors discussed here. In the first column, the three letters in each set refer to the hyperon, kaon, and kaonic resonances form factors, respectively, as described in columns 2 to 4. The second column concerns the hyperons form factors approximated by those of the neutron (N) [10] or coming from a vector meson dominance model (VDM) [8]. The third and fourth columns refer to kaon and kaonic resonances, respectively, and read as: monopole [5], vector meson dominance model (VDM) [6], and quark model (QM) [7].

Set	Hyperons	K	K* & K1
NVM	Neutron	VDM	Monopole
VVM	VDM	VDM	Monopole
VVV	VDM	VDM	VDM
VQM	VDM	QM	Monopole

Table 4.1 summarizes the combinations of the form factors used for the hyperons, kaon, and kaonic resonances in columns 2 to 4, respectively.

The first row corresponds to the combination used to obtain the original model C [4, 9]. There, the neutral hyperons form factors were approximated by those of the neutron obtained within an EVDM approach [10]; this latter work was also used for the proton form factors. The sensitivity of the original electromagnetic strangeness production model to different combinations of kaonic form factors can be found in Ref. [1].

In the second row, the form factors for Λ and Σ hyperons come from Ref. [8]. The third row corresponds to the use of a VDM approach (instead of monopole) for the kaonic resonances, while in the fourth row we use the results of a quark model (instead of VDM) for the kaon form factor.

All these form factors configurations, within model C, have been used to

Table 4.2. Exchanged particles, coupling constants (CC), and off-shell (OS) parameters (X, Y, Z) for $K\Lambda$ channels for the different sets of form factors summarized in Table 4.1. The reduced χ^2's are given in the last row.

Exchanged particles	CC & OS	NVM	VVM	VVV	VQM
Λ	$g_{K\Lambda N}/\sqrt{4\pi}$	-3.22±0.01	-3.16±0.01	-3.16±0.03	-3.16±0.85
Σ	$g_{K\Sigma N}/\sqrt{4\pi}$	0.86±0.02	0.79±0.06	0.92±0.04	0.89±0.01
K^{*+}	$G_V/4\pi$	0.02±0.01	-0.01±0.01	0.08±0.08	0.89±0.01
	$G_T/4\pi$	0.17±0.01	0.15±0.01	0.21±0.01	0.17±0.01
$K1$	$G_{V1}/4\pi$	-0.15±0.01	-0.20±0.01	-0.12±0.01	-0.16±0.01
	$G_{T1}/4\pi$	-0.39±0.03	-0.39±0.04	-0.11±0.01	-0.51±0.02
N(1720)	$G^1_{N7}/4\pi$	-0.04±0.01	-0.04±0.01	-0.04±0.01	-0.04±0.04
	$G^2_{N7}/4\pi$	-0.10±0.01	-0.12±0.02	-0.17±0.01	-0.11±0.01
	X	-1.03±0.08	-2.25±0.57	3.69±0.21	-0.79±0.07
	Y	8.18±0.15	7.93±0.57	9.57±0.40	8.31±0.08
	Z	0.01±0.01	-0.14±0.02	0.09±0.02	-0.08±0.01
Λ (1405)	$G_{L1}/\sqrt{4\pi}$	-0.28±0.01	-0.27±0.05	-0.28±0.03	-0.28±0.01
$\Lambda(1670)$	$G_{L3}/\sqrt{4\pi}$	1.26±0.01	1.26±0.05	1.32±0.04	1.27±0.01
$\Lambda(1810)$	$G_{L5}/\sqrt{4\pi}$	-1.78±0.02	-1.97±0.12	-1.75±0.08	-1.81±0.02
$\Lambda(1890)$	$G^1_{L5}/\sqrt{4\pi}$	0.01±0.01	0.02±0.01	-0.04±0.013	0.01±0.01
	$G^2_{L5}/\sqrt{4\pi}$	0.01±0.05	-0.05±0.03	0.20±0.01	0.16±0.01
	X	-0.02± 3.92	-9.88± 7.88	-2.30±0.09	-1.36±0.03
	Y	0.23± 9.20	-8.51± 1.16	0.66±0.32	1.85±0.06
	Z	0.23± 9.00	-0.35±0.03	1.07±0.07	0.68±0.02
$\Sigma(1660)$	$G_{S1}/\sqrt{4\pi}$	-5.36±0.02	-5.30±0.11	-5.45±0.09	-5.37±0.02
χ^2		1.69	1.64	1.95	1.63

fit the available data on the reactions (3) to (5). The free parameters, as determined by the minimization procedures, are given in Table 4.2. The total number of the fitted data points is 309 and the number of free parameters is 20.

The values shown in Table 4.2 deserve a few comments. The χ^2s turn out to be comparable except in the case of all form factors coming from the VDM approaches (fifth column in Table 4.2), where the χ^2 is about 20% larger than for the three others. However, the present data base is not accurate enough [11] to make such a discrepancy significant. The extracted values for the main coupling constants $g_{K\Lambda N}$ and $g_{K\Sigma N}$ are consistent with the broken SU(3)-symmetry values [11]. The other free parameters show in general quite small variations according to the sets used. Given these observations, we believe that the resonance content of the model is realistic enough for the relative effects investigated in this work.

Figure 2 shows the unpolarized component of the differential cross section $d\sigma_{UL} = d\sigma_U + \epsilon_L\, d\sigma_L$ in Eq. (6), as a function of the momentum transfer. The comparison between data and the curves obtained with different form factors combinations does not allow us to favor any of the sets in a sharp manner, in lines with the reduced χ^2 values.

In Figure 3, the four components of the differential cross section [Eq. (6)] are depicted as a function of the outgoing kaon c.m. angle. Examination of the longitudinal and transverse terms indicates a general feature: these terms are larger when based on the hyperons form factors of Ref. [8] as compared to the

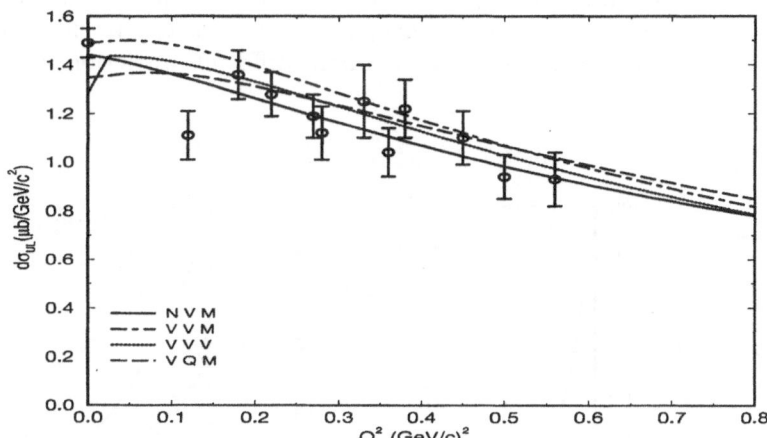

Figure 2. Differential cross section $d\sigma_{UL}$ as a function of momentum transfer (Q^2) for the reaction $ep \to e'K^+\Lambda$, for $s = 5.02$ GeV^2, $t = -0.15$ GeV^2, $\epsilon = 0.72$. Curves are for the combinations of the relevant hadrons electromagnetic form factors (given in Table 4.1). Data are from Ref. [12].

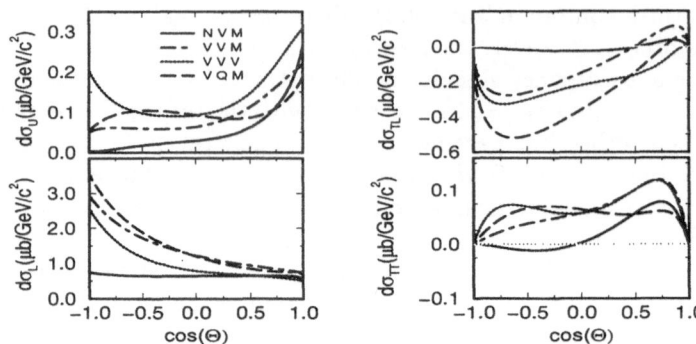

Figure 3. Same as Fig. 2, but for the components of the differential cross sections: $d\sigma_U$, $d\sigma_L$, $d\sigma_{TL} \equiv d\sigma_I \sin\theta$ and $d\sigma_{TT} \equiv d\sigma_P \sin^2\theta$, see Eq. (6) at $s = 5$ GeV^2, $\epsilon = 0.72$ $Q^2 = 1$ (GeV/c)^2.

case where they are approximated by the neutron form factors. This behavior can be understood by the large differences in the relevant form factors shown in Fig. 1. As expected, the longitudinal part is very sensitive to the different sets of the form factors and discrepancies among different predictions increase with

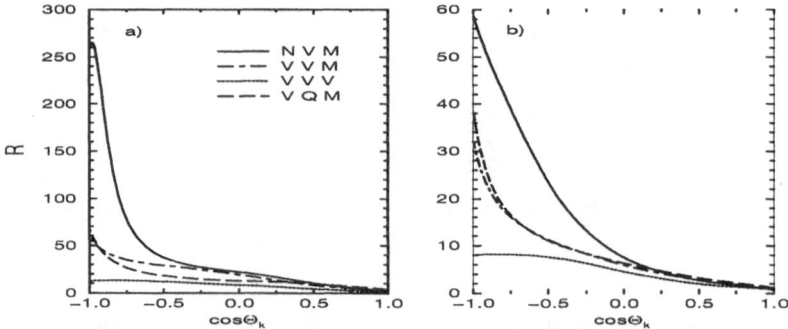

Figure 4. Same as Fig. 2, but for the longitudinal to transverse differential cross sections ratio $R = d\sigma_L/d\sigma_T$, for a) $s = 5$ GeV^2, $Q^2 = 1$ (GeV/c)^2, b) $s = 4$ GeV^2, $Q^2 = 0.5$ (GeV/c)^2.

the increasing virtuality of the photons as one moves to backward hemisphere. The curves for the transverse term reflect the variations in the free parameters values (Table 4.2) obtained by fitting the model with different sets of form factors. Of the two interference terms the longitudinal-transverse component ($d\sigma_I$ in Eq. (6)) offers particularly appealing prospects for measurments.

As in previous investigations [1, 3, 4, 9] another quantity of interest is the ratio of the longitudinal to transverse components $R = d\sigma_L/d\sigma_U$ depicted in Fig. 4. An attractive feature here is the energy dependence of this quantity.

5 Summary and conclusions

In this paper we have investigated the sensitivity of the associated strangeness electroproduction [4, 9] to the strange hadrons electromagnetic form factors [5, 6, 7, 8] by taking advantage of the most comprehensive works on both topics.

Our investigation shows that the most relevant observable is the longitudinal component of the cross section as well as the ratio of the longitudinal to transverse components. Experimental results from JLAB are highly desirable to go further in this investigation.

The model [4, 9] used here is obtained by fitting basically old data going back to the 70s or earlier. The present formalism should be applied to the upcoming data to pin down the reaction mechanism and then study the form factors.

We wish to emphasize that the resonance content of the models has been optimized for the original model [4] which uses the neutron form factor for the neutral hyperons. Moreover, the electroproduction data represent less then 25% of the database used in the fitting procedures.

Given all these considerations, the result of this work is putting forward the relative effects with respect to different ingredients of the models. Nevertheless, the elementary reaction operator discussed here appears to embody the main features of the reaction mechanism: besides reproducing simultaneously all the available data for reactions (3) to (5), it has been proven to produce reasonable results in the investigation of strangeness production on the deuteron [13] and on ^{12}C [14].

A more extensive study requires not only more accurate and copious electroproduction data but implies also photoproduction data. Indeed, one needs to disentangle several ingredients of the approach, namely,

- Resonance content of the reaction mechanism,
- Coupling constants,
- Off-shell parameters,
- Form factors.

Radiative capture and photoproduction data, especially for polarization observables, are expected [1, 4, 9, 15] to be strong enough constraints on the first three points. Then, the electroproduction process can be investigated with the same phenomenological ingredients to study the form factors of strange hadrons.

On the theoretical side, more progress is needed in the delicate quest of introducing hadronic form factors [16] without losing the gauge invariance, and investigating the difficult problem of unitarity [17] which deserves substantial effort.

Acknowledgement. We would like to thank Dr. Saro Ong for enlightening discussions.

References

1. B. Saghai: Nucl. Phys. **A639**, 217c (1998), and references therein.

2. E. B. Dally et al.: Phys. Rev. Lett. **45**, 232 (1980); S. R. Amendolia et al.: Phys. Lett. **B178**, 435 (1986)

3. Ch. Charles, J. C. David, C. Fayard, M. Guidal, G. H. Lamot, and B. Saghai: Few-Body Systems Suppl. **8**, 452 (1995)

4. J. C. David, C. Fayard, G. H. Lamot, and B. Saghai: Phys. Rev. **C53**, 2613 (1996)

5. R. A. Adelseck and L. E. Wright: Phys. Rev. **C38**, 1965 (1988)

6. R. A. Williams, C. R. Ji, and S. R. Cotanch: Phys. Rev. **C46**, 1617 (1992)

7. F. Cardarelli, I.L. Grach, I.M. Narodetskii, E. Pace, G. Salmé, and S. Simula: Phys. Rev. **D53**, 6682 (1996)

8. R.A. Williams and Ch. Puckett-Truman: Phys. Rev. **C53**, 1580 (1996)

9. T. Mizutani, C. Fayard, G.H. Lamot, and B. Saghai: Phys. Rev. **C58**, 75 (1998)

10. M. F. Gari and W. Krümpelmann: Phys. Rev. **D45**, 1817 (1992)

11. R.A. Adelseck and B. Saghai: Phys. Rev. **C42**, 108 (1990)

12. P. Brauel et al.: Z. Phys. **C3**, 101 (1979), and references therein

13. T. S. H. Lee, V. Stoks, B. Saghai, and C. Fayard: Nucl. Phys. **A639**, 247c (1998)

14. T. S. H. Lee, Zhong-Yu Ma, H. Toki, and B. Saghai: Phys. Rev. **C58**, 1551 (1998)

15. B. Saghai and F. Tabakin: Phys. Rev. **C53**, 66 (1996); *ibid* **C55**, 917 (1997)

16. H. Haberzettl, C. Bennhold, T. Mart, and T. Feuster: Phys. Rev. **C58**, 40 (1998)

17. T. Feuster and U. Mosel: nucl-th/9803057

Few-Body Systems Suppl. 11, 241–245 (1999)

Few-
Body
Systems
© by Springer-Verlag 1999

Effective Lagrangian Study of the $\gamma p \to K^+\Lambda$
(Spin 3/2 Resonances and Their Off-Shell Effects)

T. Mizutani

Department of Physics, Virginia Polytechnic Institute and State University,
Blacksburg, VA 24061, USA

Abstract. The purpose of the present discussion is to supplement the talk, by
B. Saghai at this workshop, on the study of the electromagnetic production of
strangeness on the nucleon based upon effective Lagrangian methods. Here we
focus on the proper treatment of the spin 3/2 resonances and their associated
effects due to the spin 1/2 component of the corresponding fields when they
are *off the mass shell.*

1 Introduction

To date, in most works exploiting the effective Lagrangian approaches to investigate the photo- (and electro-) production of the strangeness on the nucleon,
one includes only the tree level contributions consisting of the s-channel nucleon
and its excited resonant state exchanges, the u-channel hyperonic resonance exchanges, and a few t-channel strange meson exchanges. Note that this is also
known as the *isobar* approximation. As the s-channel incident photon energy
becomes higher, the more massive baryonic resonances with spin higher than
1/2 need to be included in this scheme. Until recently, the inclusion of higher
spin resonances had never been exercised except in the work of Renard and
Renard [1] in the early 70's. However, this work never recognised the propagation of the spin 1/2 components of the Rarita-Schwinger spin 3/2 field *off-shell.*
A recent approach [2] took a somewhat different point of view and included
spin 3/2 and 5/2 resonances, but only in the s-channel due to an unwanted
singularity in the u-channel arising from how these higher spin objects were
handled.

A proper spin 3/2 resonance treatment was discussed [3] and exploited by
the Rensselaer Polytechnic (RPI) group for the photo- (electro-) production
of the pion in the $\Delta(1232)$ resonance region [4] In what follows we shall extend their approach to the K-meson production. Our result for the reaction
the $\gamma^{(*)}p \to K^+\Lambda$ have been reported in Ref. [5]. In the present discussion we

shall include one additional aspect in its yet exploratory stage, but not investigated in [5]. Incorporating resonances with spin higher than 3/2 has not been investigated within the context of covariant schemes.

2 Spin 3/2 Resonances and Their Off-shell Effects

First, we refer the reader to appropriate articles on the proper treatment of spin 3/2 fields for more details [3],[6],[7].

Following the convention found in Bjorken and Drell[8], we write the spin 3/2 (isospin 1/2 for our present discussion) Rarita-Schwinger [vector-spinor field] (resonance) as R^μ. Without interaction this field should obey the free Dirac equation (with M_R: the mass of the resonance)
with the subsidiary condition to be satisfied by the *on-shell* spin 3/2 resonance:

$$\gamma_\mu R^\mu = 0 \tag{1}$$

ensuring the correct number of spin components (the often mentioned additional condition :$\partial_\mu R^\mu = 0$, is automatically met by the two equations above). Then the most general form of the corresponding free Lagrangian reads,

$$\mathcal{L}_{free} = \overline{R}^\alpha \Lambda_{\alpha\beta} R^\beta, \tag{2}$$

where

$$\begin{aligned}
\Lambda_{\alpha\beta} &= -[(-i\slashed{\partial} + M_R)g_{\alpha\beta} - iA(\gamma_\alpha \partial_\beta + \gamma_\beta \partial_\alpha) - \frac{i}{2}(3A^2 + 2A + 1)\gamma_\alpha \slashed{\partial}\gamma_\beta \\
&\quad - M_R(3A^2 + 3A + 1)\gamma_\alpha \gamma_\beta].
\end{aligned} \tag{3}$$

Here $A(\neq -1/2)$ is a free parameter. Since this Lagrangian can be shown to be invariant under the following point (or contact) transformation

$$R^\mu \to R^\mu + a\gamma^\mu \gamma^\nu R_\nu; \quad A \to A + (A - 2a)/(1 + 4a), \tag{4}$$

($a \neq -1/4$, but otherwise arbitrary), observables resulting from this Lagrangian are free of parameter A. Note that on the right hand side of the above transformation for R^μ, the part proportional to $\gamma^\nu R_\nu$ may be easily seen to behave as a Dirac spin 1/2 field. Thus *off the mass shell* the Rarita-Schwinger field R^μ always is mixed with spin 1/2 components. This has an important bearing on what we are going to discuss in the following.

First, the propagator for the R^μ field is the inverse of $\Lambda_{\mu\nu}$, and its simplest form may be found by setting $A = -1$,

$$P_{\mu\nu}(q) = \frac{\slashed{q} + M_R}{3(q^2 - M_R^2)} \left[3g_{\mu\nu} - \gamma_\mu \gamma_\nu - \frac{2q_\mu q_\nu}{M_R^2} - \frac{q_\nu \gamma_\mu - q_\mu \gamma_\nu}{M_R} \right], \tag{5}$$

where q is the four momentum of the resonance. It is important to note [3] that this propagator describing the on- and off-shell propagation of the spin 3/2 resonance contains the spin 1/2 contribution. It has been shown that by dropping the spin 1/2 component would lead to the propagator which has no inverse [3].

Next, the interactions associated with the spin 3/2 field R^μ and respects the invariance under the point transformation of the free Lagrangian must be of the form: $\mathcal{L}_I = j^\mu \Theta_{\mu\nu} R^\nu + h.c.$, where the most general form for Θ was found to be [6], [9]

$$\begin{aligned}
\Theta_{\mu\nu}(V) &= (g^\alpha_\mu + V\gamma_\mu\gamma^\alpha)((g^\alpha_\nu + A/2 \cdot \gamma^\alpha\gamma_\nu) \\
&\rightarrow g_{\mu\nu} - (V + 1/2)\gamma_\mu\gamma_\nu \ (A = -1).
\end{aligned} \tag{6}$$

Here V is a free parameter, and while the physical quantities are independent of A as the full Lagrangian is invariant under the point transformation [10], they do depend upon V. From the above expression one sees that V enters to modify only the spin 1/2 part of the Rarita-Schwinger field. Hence, it is clear that the spin 3/2 resonance pole part of the amplitudes should be independent of V. For this reason V is called the *Off-Shell Parameter*. Our objective is then to study the effect of this off-shell freedom on the electromagnetic production of the strangeness on the nucleon. Note that this cannot be fulfilled in combination with an inappropriate resonance propagator which ignores the spin 1/2 contribution. Since our spin 3/2 fields are not elementary, but composites of more fundamental constituents, the corresponding off-shell parameters should not be fixed to be a definite value: a point of view adopted by [3].

Our phenomenological interaction Lagrangian consists of the following pieces:

$$\mathcal{L}_{KYR} = \frac{g_{KYR}}{M_K}\left[\overline{R}^\nu \Theta_{\nu\mu}(Z)Y\partial^\mu K + \overline{Y}(\partial^\mu K^\dagger)\Theta_{\mu\nu}(Z)R^\nu\right], \tag{7}$$

$$\mathcal{L}^{(1)}_{\gamma pR} = \frac{ieg_1}{2M_p}\left[\overline{R}^\nu \Theta_{\mu\lambda}(Y)\gamma_\nu\gamma^5 NF^{\nu\lambda} + \overline{N}\gamma^5\gamma_\nu\Theta_{\lambda\mu}(Y)R^\mu F^{\nu\lambda}\right], \tag{8}$$

$$\mathcal{L}^{(2)}_{\gamma pR} = \frac{-eg_2}{4M_p^2}\left[\overline{R}^\mu \Theta_{\mu\nu}(X)\gamma^5(\partial_\lambda N)F^{\nu\lambda} - (\partial_\lambda\overline{N})\gamma^5\Theta_{\nu\mu}(X)R^\mu F^{\nu\lambda}\right] \tag{9}$$

$$\mathcal{L}^{(3)}_{\gamma pR} = \frac{-eg_3}{4M_p^2}\left[\overline{R}^\mu \Theta_{\mu\nu}(W)\gamma^5 N(\partial_\lambda F^{\nu\lambda}) - (\partial_\lambda F^{\nu\lambda})\overline{N}\gamma^5\Theta_{\nu\mu}(W)R^\mu\right] \tag{10}$$

The above expression may be found in [5] except for the last piece, the necessity of which has been communicated to us by Rick Davidson [11]. This part contributes only to the electroproduction process, which may be obvious from the presence of the divergence of the electromagnetic tensor $F^{\nu\lambda}$ in the interaction: classically this quantity vanishes when there is no source for electric charge/current.

3 Result and Discussion

With an introduction of the spin 3/2 resonances discussed above, we have constructed the tree level amplitude for $\gamma^{(*)}p \rightarrow K^+\Lambda$, and performed a χ^2 fit to the existing data with the coupling strengths of the intermediate resonances (to the initial and final states, in combination) as well as the off-shell parameters: W, X, Y, Z to be varied. The result without $\mathcal{L}^{(3)}_{\gamma p R}$ has been worked out in [5], and the reader is referred to that article for details. In summary, the correct treatment of the spin 3/2 resonances has improved the fit to the existing data, in particular, the undesirable increase in the integrated cross section for $E^{Lab}_\gamma >$ 1.5 GeV in previous models ([2], for example) has been considerably tempered, and the large angle differential cross section at $E^{Lab}_\gamma = 2.0$ GeV has changed drastically, in favor of the yet preliminary data from Bonn [12]. Predictions for several polarisation observables both in photo- and electro-productions show a marked difference between the approaches with and without the off-shell considerations in spin 3/2 resonances. Regarding the off-shell effects arising from $\mathcal{L}^{(3)}_{\gamma p R}$ which only affects the electroproduction, our preliminary results indicate a rather minor effect: we basically need ample data points to firmly constrain the corresponding parameter(s): coupling strength and the off-shell parameter W.

Acknowledgement. The content of this note is a result of a pleasant collaboration with Claude Fayard, Georges Lamot, and Bijan Saghai. We would like to thank Rick Davidson for the possibility of the third electromagnetic coupling of the spin 3/2 particles. A very kind hospitality at ECT* at Trento was more than a pleasure to the present author.

References

1. F. M. Renard and Y. Renard: Nucl. Phys. **B25**, 490 (1971); Y. Renard: *ibid.* **B40**, 499 (1972); Y. Renard: Thèse de Doctorat d'Etat ès-Sciences Physiques, Université des Sciences et Techniques du Languedoc, 1971 (in French)

2. J.C. David, C. Fayard, G.H. Lamot, B. Saghai: Phys. Rev. **C53**, 2613 (1996)

3. M. Benmerrouche, R. M. Davidson, Nimai C. Mukhopadhyay: Phys. Rev. **C39**, 2339 (1989)

4. R.M. Davidson, Nimai C. Mukhopadhyay, R. S. Wittman: Phys. Rev. **D43**, 71 (1991)

5. T. Mizutani, C. Fayard, G.H. Lamot, B. Saghai: Phys. Rev. **C58**, 75 (1998)

6. V. Pascalutsa: hep-ph/9412321

7. T. H. Hemmert, B. R. Holstein: hep-ph/9712496

8. J. D. Bjorken, S. D. Drell, *Relativistic Quantum Mechanics* (McGraw-Hill, New York, 1964)

9. L.M. Nath, B. Etemadi, J.D. Kimel: Phys. Rev. **D3**, 2153 (1971)

10. S. Kamefuchi, L. O'Raifeartaigh, A. Salam: Nucl. Phys. **28**, 529 (1961)

11. R. M. Davidson: private communication

12. SAPHIR Collaboration (J. Barth *et al.*) Contributed paper to the *6th Conference on the Intersections of Particle and Nuclear Physics* (CIPANP 97), Big Sky, MT, 27 May - 2 June 1997 (nucl-th/9707025); D. Menze, private communication

Few-Body Systems Suppl. 11, 246–254 (1999)

Few-
Body
Systems
© by Springer-Verlag 1999

SU(3) Chiral Dynamics with Coupled Channels: Eta and Kaon Production

Norbert Kaiser[1]

[1] Physik Department T39, Technische Universität München, D-85747 Garching, Germany

Abstract. We identify the next-to-leading order s-wave amplitudes of the SU(3) chiral meson-baryon Lagrangian with a coupled-channel potential which is iterated in a separable Lippmann-Schwinger equation. The strangeness $S = -1$ resonance $\Lambda(1405)$ and the $S_{11}(1535)$ nucleon resonance emerge as unstable bound states of anti-kaon/nucleon and kaon/Σ-hyperon. Our approach to meson photoproduction introduces no new parameters. By adjusting a few finite range parameters we are able to simultaneously describe a large amount of low energy data. These include the cross sections of K^-p elastic and inelastic scattering, the cross sections of eta meson and kaon photoproduction from protons as well as those of pion induced production of etas and kaons.

1 Introduction

Over the last few years there has been renewed interest in the photoproduction of η and K mesons from nucleons. At MAMI (Mainz) very precise differential cross sections for $\gamma p \to \eta p$ have been measured from threshold at 707 MeV up to 800 MeV photon lab energy [1]. The nearly isotropic angular distributions show clearly the dominance of the s-wave amplitude E_{0+} in this energy range. Together with the ELSA data [2] taken at higher beam energies they cover the whole energy range of the nucleon resonance $S_{11}(1535)$. The latter has the outstanding feature of a strong ηN decay which is made responsible for the observed large η-production cross sections. At ELSA (Bonn) there is an ongoing program to measure strangeness production with photons from proton targets. Cross sections for the reactions $\gamma p \to K^+\Lambda, K^+\Sigma^0, K^0\Sigma^+$ have been measured with improved accuracy from the respective thresholds up to 1.5 GeV together with angular distributions and hyperon recoil polarizations [3, 4].

Most theoretical models used to describe the abovementioned reactions are based on an effective Lagrangian approach including Born terms and various (meson and baryon) resonance exchanges [5, 6, 7]. Whereas resonance models work well for η-production the situation is more difficult for kaon production where several different kaon-hyperon final states are possible. As shown in [6] resonance models lead to a notorious overprediction of the $\gamma p \to K^0 \Sigma^+$ and $\gamma n \to K^+ \Sigma^-$ cross section. We will use here quite a different approach to η and K photoproduction (and the related pion induced reactions) not introducing any explicit resonance. Our starting point is the SU(3) chiral effective meson-baryon Lagrangian at next-to-leading order. The explicit degrees of freedom are only the baryon and pseudoscalar meson octet with interactions controlled by chiral symmetry and a low energy expansion. As shown in [8] the effective Lagrangian predicts a strong attraction in the s-wave channels $\overline{K}N$ (isospin 0) and $K\Sigma$ (isospin 1/2). If this attraction is iterated to infinite orders in a potential approach one can dynamically generate the $\Lambda(1405)$ and the $S_{11}(1535)$ as unstable meson-baryon bound states with all properties attributed to these resonances. The extension to meson photoproduction does not introduce any further parameter. It is then quite non-trivial to find a good description of so many available photon and pion induced data for this multi-channel problem with just a few free parameters. Of course in s-wave approximation the comparison with data is necessarily restricted to the near threshold region. The systematic inclusion of all p-wave amplitudes into the coupled channel scheme is underway [9]. This will allow to study (more exclusive) observables like angular distributions and recoil polarizations which arise from s- and p-wave interference terms.

2 Coupled Channel Approach

The effective chiral Lagrangian provides a non-linear realization of the chiral symmetry group $SU(3)_L \times SU(3)_R$ on the effective low energy degrees of freedom, the pseudoscalar Goldstone bosons (π, K, η) and the octet baryons (N, Λ, Σ, Ξ). The leading order (relativistic) term reads (for details see ref.[10])

$$\mathcal{L}^{(1)}_{\phi B} = \mathrm{tr}(\overline{B}(i\gamma_\mu D^\mu - M_0)B) + F\,\mathrm{tr}(\overline{B}\gamma_\mu\gamma_5[u^\mu, B]) + D\,\mathrm{tr}(\overline{B}\gamma_\mu\gamma_5\{u^\mu, B\})\,, \quad (1)$$

and at next-to-leading order the terms relevant for s-wave scattering are

$$\begin{aligned}
\mathcal{L}^{(2)}_{\phi B} =\ & b_D\,\mathrm{tr}(\overline{B}\{\chi_+, B\}) + b_F\,\mathrm{tr}(\overline{B}[\chi_+, B]) + b_0\,\mathrm{tr}(\overline{B}B)\,\mathrm{tr}(\chi_+) \\
& + 2d_D\,\mathrm{tr}(\overline{B}\{(v\cdot u)^2, B\}) + 2d_F\,\mathrm{tr}(\overline{B}[(v\cdot u)^2, B]) \\
& + 2d_0\,\mathrm{tr}(\overline{B}B)\,\mathrm{tr}((v\cdot u)^2) + 2d_1\,\mathrm{tr}(\overline{B}v\cdot u)\,\mathrm{tr}(v\cdot uB)\,.
\end{aligned} \quad (2)$$

Whereas the systematic approach to chiral dynamics is chiral perturbation theory, a renormalized perturbative loop-expansion, its range of applicability can be very small in cases where strong resonances lie closely above (or even slightly below) the reaction threshold. Prominent examples for this are the isospin $I = 0$, strangeness $S = -1$ resonance $\Lambda(1405)$ in K^--proton scattering, or the $S_{11}(1535)$ nucleon resonance which has an outstandingly large coupling to the ηN-channel and therefore is an essential ingredient in the description of η-photoproduction. Let us now describe the (non-perturbative) potential approach to meson-baryon scattering and meson photoproduction of ref.[10]).

The indices i and j label the meson-baryon channels involved. They are coupled through a potential in momentum space

$$V_{ij} = \frac{\sqrt{M_i M_j}}{4\pi f^2 \sqrt{s}} C_{ij} , \qquad (3)$$

where the relative coupling strengths C_{ij} (explicit expressions are given [10]) are, up to a factor $-f^{-2}$, (with $f = 92.4$ MeV) the corresponding s-wave amplitudes calculated from the $SU(3)$ chiral meson-baryon Lagrangian eqs.(1,2). The potential V_{ij} is iterated to all orders in a separable Lippmann-Schwinger equation of the form

$$T_{ij} = V_{ij} + \sum_n \frac{2}{\pi} \int_0^\infty dl \frac{l^2}{k_n^2 + i0 - l^2} \left(\frac{\alpha_n^2 + k_n^2}{\alpha_n^2 + l^2} \right)^2 V_{in} T_{nj} , \qquad (4)$$

with T_{ij} the resulting T-matrix connecting the in- and outgoing channels j and i. The propagator used in eq.(4) is proportional to a non-relativistic energy denominator with k_n the on-shell relative momentum. In order to make the dl-integration convergent a "form factor" $[(\alpha_n^2 + k_n^2)/(\alpha_n^2 + l^2)]^2$ parameterizing finite range aspects of the potential has to be introduced. The cut-off parameters α_n will be determined in a fit to many data keeping in mind physically reasonable values $\alpha_n \simeq 1$ GeV. The separable Lippmann-Schwinger equation eq.(4) for the multi-channel T-matrix T_{ij} can be solved in closed form by simple matrix inversion

$$T = (1 - V \cdot G)^{-1} \cdot V , \qquad (5)$$

where G is the diagonal matrix with entries $G_n = (k_n^2 - \alpha_n^2)/(2\alpha_n) - i k_n$ and the total (s-wave) cross section for the reaction $j \to i$ is calculated via

$$\sigma_{\text{tot}} = 4\pi \frac{k_i}{k_j} |T_{ij}|^2 . \qquad (6)$$

We now extend the same formalism to s-wave meson photoproduction. The basic assumption is that the s-wave photoproduction process can be described by a separable Lippmann-Schwinger equation. The s-wave photoproduction potential (named B_{0+}) is derived from the chiral effective Lagrangian without any new parameters. Consequently meson-baryon interactions and meson photoproduction are strongly tied together and the fits of e.g. the finite range parameters α_n are controlled by both sets of data. For the description of the photoproduction reactions $\gamma p \to \eta p$, $K^+\Lambda$, $K^+\Sigma^0$, $K^0\Sigma^+$ we have to know the photoproduction potentials B_{0+} for $\gamma p \to \phi B$, where ϕB refers to the meson-baryon states with total isospin $I = 1/2$ or $I = 3/2$ and isospin projection $I_3 = +1/2$. We label these states by an index which runs from 1 to 6, which refers to $|\pi N\rangle^{(1/2)}$, $|\eta N\rangle^{(1/2)}$, $|K\Lambda\rangle^{(1/2)}$, $|K\Sigma\rangle^{(1/2)}$, $|\pi N\rangle^{(3/2)}$ and $|K\Sigma\rangle^{(3/2)}$, in that order. The resulting expressions involve as parameters only the axial vector coupling constants $F = 0.5$ and $D = 0.75$ and they read

$$B_{0+}^{(1)} = \frac{eM_N}{8\pi f\sqrt{3s}}(D + F)(2X_\pi + Y_\pi), \quad B_{0+}^{(2)} = \frac{eM_N}{8\pi f\sqrt{3s}}(3F - D)Y_\eta ,$$

$$B_{0+}^{(3)} = \frac{e\sqrt{M_N M_\Lambda}}{8\pi f\sqrt{3s}}(-D-3F)X_K, \ B_{0+}^{(4)} = \frac{e\sqrt{M_N M_\Sigma}}{8\pi f\sqrt{3s}}(D-F)(X_K+2Y_K),$$

$$B_{0+}^{(5)} = \frac{e\sqrt{2}M_N}{8\pi f\sqrt{3s}}(D+F)(Y_\pi - X_\pi), \ B_{0+}^{(6)} = \frac{e\sqrt{2M_N M_\Sigma}}{8\pi f\sqrt{3s}}(D-F)(X_K-Y_K), \quad (7)$$

where X_ϕ and Y_ϕ are dimensionless functions depending on the center of mass energy E_ϕ and the mass m_ϕ of the photoproduced meson. X_ϕ takes the form

$$X_\phi = \frac{1}{2} - \frac{1}{4M_0}\left(2E_\phi + \frac{m_\phi^2}{E_\phi}\right) + \left(1+\frac{m_\phi^2}{2M_0 E_\phi}\right)\frac{m_\phi^2}{2E_\phi\sqrt{E_\phi^2 - m_\phi^2}}\ln\frac{E_\phi + \sqrt{E_\phi^2 - m_\phi^2}}{m_\phi},$$

$$(8)$$

and it sums up the contributions of all tree diagrams to the s-wave photoproduction multipole of a positively charged meson. The logarithmic term comes from the meson pole diagram in which the photon couples to the positively charged meson. If the photoproduced meson is neutral the corresponding sum of diagrams leads to a simpler expression,

$$Y_\phi = -\frac{1}{3M_0}\left(2E_\phi + \frac{m_\phi^2}{E_\phi}\right), \qquad (9)$$

for the reduced s-wave multipole. Infinitely many rescatterings of the photoproduced meson-baryon state due to the strong interaction are summed up via the Lippmann-Schwinger equation. The "full" electric dipole amplitude $E_{0+}^{(i)}$ for channel i is then given by

$$E_{0+}^{(i)} = \sum_j [(1-V\cdot G)^{-1}]_{ij}B_{0+}^{(j)}, \qquad (10)$$

where V is the matrix of the strong interaction potential eq.(3) and G the diagonal propagator matrix defined in eq.(5). From $E_{0+}^{(i)}$ one can finally compute the total (s-wave) photoproduction cross section for the meson-baryon final state i,

$$\sigma_{\text{tot}}^{(i)} = 4\pi\frac{k_i}{k_\gamma}|E_{0+}^{(i)}|^2, \qquad (11)$$

with k_γ the photon center of mass energy.

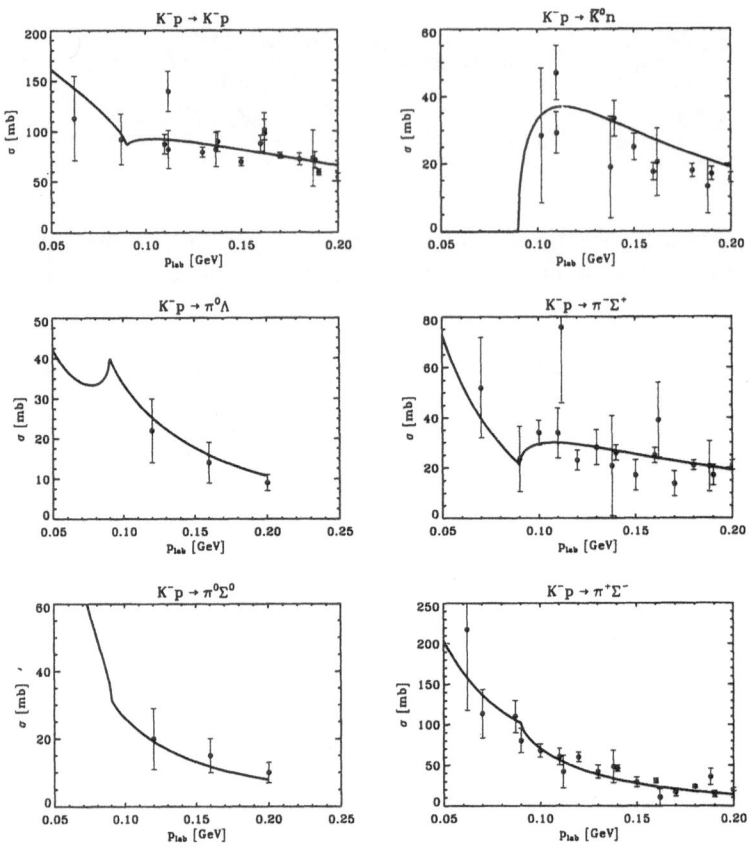

Fig.1: Total cross sections for the six K^-p elastic and inelastic scattering channels.

3 Results

3.1 K^-p Scattering

Fig.1 shows the results for the six K^-p elastic and inelastic channels: $K^-p \rightarrow K^-p$, $\overline{K^0}n$, $\pi^0\Lambda$, $\pi^+\Sigma^-$, $\pi^0\Sigma^0$, $\pi^-\Sigma^+$. One finds good agreement with the available low energy data below 200 MeV kaon lab momentum. For the threshold branching ratios γ, R_c, R_n (defined in eq.(21) of [8]), we find here $\gamma = 2.33\,(2.36 \pm 0.04)$, $R_c = 0.65\,(0.66 \pm 0.01)$, $R_n = 0.23\,(0.19 \pm 0.02)$, where the numbers given in brackets are the empirical values.

3.2 Eta and Kaon Photoproduction

The most precise data available are those for η-photoproduction off protons $(\gamma p \rightarrow \eta p)$ taken at MAMI [1] and they can indeed be perfectly reproduced as seen in Fig.2. Since the measured angular distributions of the $\gamma p \rightarrow \eta p$ differential cross sections [1] are almost isotropic one can safely identify the total cross sections with the s-wave cross sections. For the s-wave multipole at the ηp threshold we find $E_{0+}^{thr}(\gamma p \rightarrow \eta p) = (7.62 + 14.12\,i) \cdot 10^{-3} m_\pi^{-1}$. Interestingly, the initial photoexcitation of the $K^+\Lambda$ state makes after infinitely many rescatterings to ηp the largest contribution to it. The ratio of imaginary part to real part is 1.85, somewhat larger than the typical values around 1 found in resonance model fits [1]. However, since η-photoproduction is totally s-wave dominated in the first 100 MeV above threshold, the real and imaginary parts cannot be disentangled experimentally.

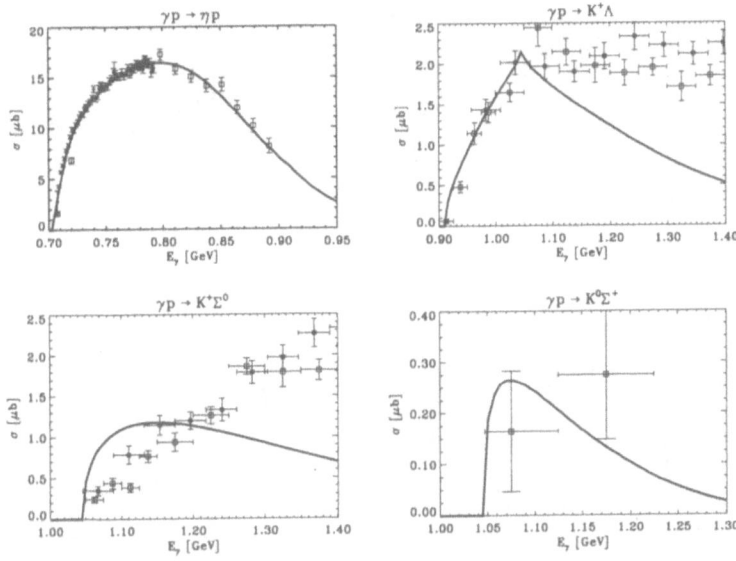

Fig.2: The total cross sections for eta and kaon photoproduction from protons.

In Fig.2 we present furthermore results for kaon photoproduction, $\gamma p \rightarrow K^+\Lambda$, $K^+\Sigma^0$, $K^0\Sigma^+$ together with the data from ELSA [3, 4]. Interestingly, the $\gamma p \rightarrow K^+\Lambda$ s-wave total cross section shows a strong cusp at the opening of the $K\Sigma$-threshold which is a consequence of unitarity and multi-channel dynamics. The present $\gamma p \rightarrow K^+\Lambda$ data do not clearly show such a cusp and it may be covered by the p-waves which become sizeable above $E_\gamma = 1.1$ GeV. If one divides out of the total cross sections the two-body phase space factor, $|E_{0+}(\gamma p \rightarrow K^+\Lambda)|^2$ indeed shows a maximum around $E_\gamma = 1046$ MeV, the $K\Sigma$ threshold. Only more data allowing for a multipole decomposition can clarify whether a strong cusp is present in the $\gamma p \rightarrow K^+\Lambda$ s-wave multipole at the $K\Sigma$-threshold. Note that the analogous pion induced reaction $\pi^-p \rightarrow K^0\Lambda$ (see Fig.3) shows the $K\Sigma$-cusp more clearly. The threshold value of the s-wave multipole is $E_{0+}^{thr}(\gamma p \rightarrow K^+\Lambda) = (-1.78 - 3.50\,i) \cdot 10^{-3} m_\pi^{-1}$ and at the cusp one finds $E_{0+}^{K\Sigma-thr}(\gamma p \rightarrow K^+\Lambda) = (-4.06 - 2.36\,i) \cdot 10^{-3} m_\pi^{-1}$. The first few data points for $\gamma p \rightarrow K^+\Sigma^0$ are somewhat overshot (see Fig.2) and we find $E_{0+}^{thr}(\gamma p \rightarrow K^+\Sigma^0) = (4.15 + 3.11\,i) \cdot 10^{-3} m_\pi^{-1}$ for the s-wave multipole at threshold, which may be too large in magnitude. Finally, we show in Fig.2 the

$\gamma p \rightarrow K^0 \Sigma^+$ channel. This curve is really a prediction with no data included in the fit and indeed one can reproduce the first two recent data points from ELSA [4]. For the threshold value of E_{0+} we find $E_{0+}^{thr}(\gamma p \rightarrow K^0 \Sigma^+) = (1.34 + 3.38\,i) \cdot 10^{-3} m_\pi^{-1}$. In comparison to [6] one finds that our approach does not have the problem of overpredicting the $\gamma p \rightarrow K^0 \Sigma^+$ channel, at least in s-wave approximation.

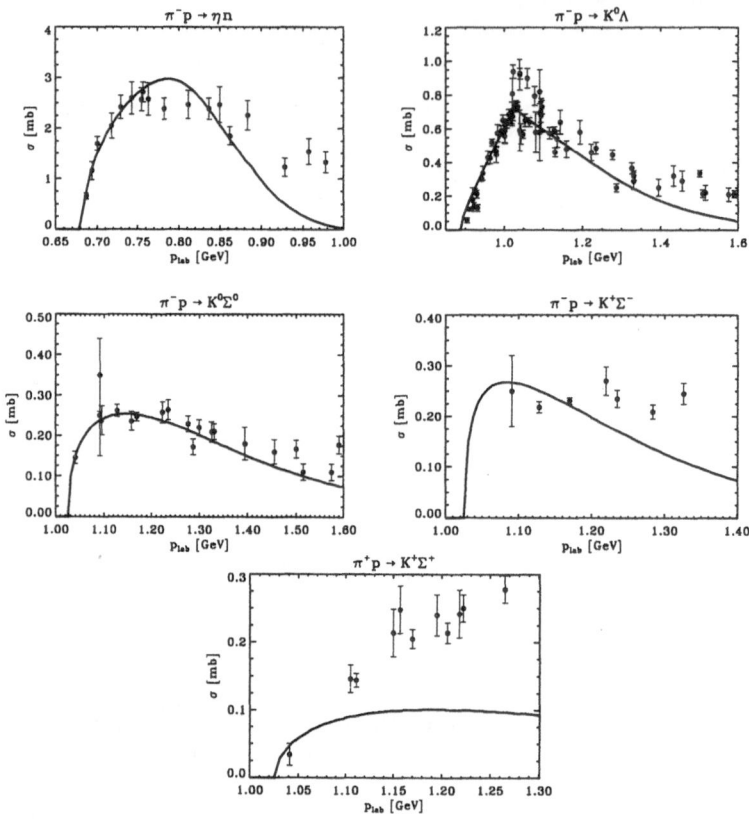

Fig.3: The total cross sections for pion induced eta and kaon production.

It is evident that the p-wave amplitudes E_{1+}, M_{1-}, M_{1+} become important for kaon photon production at energies larger than 200 MeV above threshold. On the other hand p-waves also show up via non-isotropic differential cross sections and non-vanishing hyperon recoil polarizations. The systematic inclusion of all strong and electromagnetic p-wave multipoles is in progress [9].

3.3 Pion Induced Eta and Kaon Production

In Fig.3, we show our results for pion induced eta and kaon production, $\pi^- p \rightarrow \eta n$, $\pi^- p \rightarrow K^0 \Lambda$, $K^0 \Sigma^0$, $K^+ \Sigma^-$ and $\pi^+ p \rightarrow K^+ \Sigma^+$. The agreement of the coupled channel calculation with the data is partly very good even for energies considerably above threshold, except for the (pure isospin 3/2) $\pi^+ p \to K^+ \Sigma^+$ channel. This is a highly non-trivial consistency check of our coupled channel approach which simultaneously describes strong and electromagnetic meson-baryon interactions. Of course, also here p-wave may become important once one considers the corresponding differential cross sections.

3.4 Nature of the $S_{11}(1535)$-resonance

The peaks of the observed total cross sections in $\gamma p \to \eta p$ (see Fig.2) and $\pi^- p \to \eta n$ (see Fig.3) suggest the presence of an isospin 1/2 nucleon s-wave resonance, the $S_{11}(1535)$. Actually the resonance parameters (mass M^* and width Γ) are determined from a fit of the data using a Breit-Wigner parameterization with typical values $M^* = 1.48\ldots 1.55$ GeV and $\Gamma \simeq 200$ MeV [1]. However the closeness of the ηN threshold ($M_N + m_\eta = 1486$ MeV) causes peculiar features. The ηN partial decay width is so strongly energy dependent that the Breit-Wigner curve $|2(M^* - \sqrt{s}) - i\Gamma(\sqrt{s})|^{-2}$ decreases monotonically from the ηN threshold onwards without showing a resonance peak. Furthermore, speed plots (absolute values of derivatives of partial wave amplitudes with respect to $\sqrt{s}$) derived from πN dispersion analysis show no structure in the S_{11} partial wave at $\sqrt{s} \simeq 1535$ MeV, but only the strong ηN cusp and the second $S_{11}(1650)$ resonance.

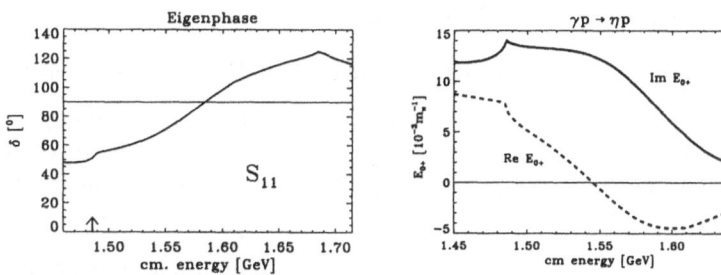

Fig.4: Eigenphase of the four-channel S-matrix and the complex amplitude $E_{0+}(\gamma p \to \eta p)$.

Let us now take a closer look at these issues in the present coupled channel approach. In Fig.4 we show that eigenphase of the four-channel S-matrix which below the ηN-threshold joins the elastic S_{11} πN phase. It passes through 90° at $\sqrt{s} = 1584$ MeV with a slope that can be translated into a full decay width of $\Gamma = 198$ MeV. These numbers are in good agreement with those attributed to the $S_{11}(1535)$. However, as one goes up in energy the phase starts to decrease once it has reached 125° at the $K\Sigma$-threshold. As required for a clean resonance the phase does not change by 180°. Finally, we show in Fig.4 the real and imaginary part of the s-wave multipole $E_{0+}(\gamma p \to \eta p)$. The curves show a resonance-like behavior with the real part of E_{0+} passing through zero at $\sqrt{s} = 1545$ MeV. Obviously, in this restricted energy range a clean resonance and a strong background are indistinguishable from each other. We conclude that within our coupled channel approach, the questionable status of the $S_{11}(1535)$ (resonance or a strong background) is reconfirmed.

Another interesting issue concerns the possible medium modification of these dynamically generated resonances. The $\Lambda(1405)$ as an antikaon-nucleon bound state is strongly affected by the Pauli blocking mechanism occuring in nuclear matter. As a result one finds [11] that the $\Lambda(1405)$ resonance dissolves already at 1/8 nuclear matter density. Along with this goes a strong density dependence of the K^-p scattering length with a sign change of its real part. Such dynamical features indeed allow to explain the observed properties of kaonic atoms. On the other hand Pauli blocking has almost no effect on the dynamical $K\Sigma$ boundstate and thus the $S_{11}(1535)$ in the coupled channels approach is neither shifted nor broadened in nuclear matter. In fact the data of η-photoproduction off nuclei are best described with a $S_{11}(1535)$ resonance which keeps its free mass and decay width. For more details on this subject see [11].

References

1. B. Krusche et al.: Phys. Rev. Lett. **74**, 3736 (1995)

2. B. Schoch, Prog. Part. Nucl. Phys. **34**, 43 (1995)

3. M. Bockhorst et al.: Z. Phys. **C63**, 37 (1994)

4. SAPHIR Collaboration, ELSA, Universität Bonn, work in progress

5. H. Tanabe, M. Kohno and C. Bennhold: Phys. Rev. **C39**, 741 (1989); R.A. Adelseck and B. Saghai: Phys. Rev. **C42**, 108 (1990); **C45** (1992) 2030; L. Tiator, C. Bennhold and S.S. Kamalov: Nucl. Phys. **A580**, 455 (1994)

6. T. Mart, C. Bennhold and C.E. Hyde-Wright: Phys. Rev. **C51**, R1074 (1995)

7. Ch. Sauermann, B.L. Friman and W. Nörenberg: Phys. Lett. **B341**, 261 (1995)

8. N. Kaiser, P.B. Siegel and W. Weise: Nucl. Phys. **A594**, 325 (1995); N. Kaiser, P.B. Siegel and W. Weise: Phys. Lett. **B362**, 23 (1995)

9. J. Caro Ramon, N. Kaiser and W. Weise: work in progress

10. N. Kaiser, T. Waas and W. Weise: Nucl. Phys. **A612**, 297 (1997) and refs. therein

11. T. Waas, N. Kaiser and W. Weise: Phys. Lett. **B365**, 12, (1996); **B379**, 34 (1996); T. Waas and W. Weise: Nucl. Phys. **A625**, 287 (1997)

Few-Body Systems Suppl. 11, 255–262 (1999)

Few-
Body
Systems
© by Springer-Verlag 1999

The Diquark Model for Exclusive Reactions

P. Kroll* †

Fachbereich Physik, Universität Wuppertal, D-42097 Wuppertal, Germany

Abstract. The present status of the diquark model for exclusive reactions at moderately large momentum transfer is reviewed. That model is a variant of the Brodsky-Lepage approach in which diquarks are considered as quasi-elementary constituents of baryons. Recent applications of the diquark model, relevant to high energy physics with electromagnetic probes, are discussed.

1 Introduction

Exclusive processes at large momentum transfer are described in terms of hard scatterings among quarks and gluons [1]. In this so-called hard scattering approach (HSA) a hadronic amplitude is represented by a convolution of process independent distribution amplitudes (DA) with hard scattering amplitudes to be calculated within perturbative QCD. The DAs specify the distribution of the longitudinal momentum fractions the constituents carry. They represent Fock state wave functions integrated over transverse momenta. The convolution manifestly factorizes long (DAs) and short distance physics (hard scattering). It however turned out that most processes are not dominated by the perturbative contribution at experimentally accessible values of momentum transfer. Non-perturbative dynamics still plays a crucial role in that kinematical region and, hence, the HSA although likely the correct asymptotic picture for exclusive reactions, needs modifications.

In a series of papers [2]-[7] such a modification has been proposed in which baryons are viewed as being composed of quarks and diquarks. The latter are treated as quasi-elementary constituents which partly survive medium hard collisions. Diquarks are an effective description of correlations in the wave functions and constitute a particular model for non-perturbative effects. The diquark model may be viewed as a variant of the HSA appropriate for moderately large momentum transfer and it is designed in such a way that it evolves into the standard pure quark HSA asymptotically. In so far the standard HSA

---

*Supported in part by the TMR Network ERB 4061 PL 95 0115
† *E-mail address:* kroll@theorie.physik.uni-wuppertal.de

and the diquark model do not oppose each other, they are not alternatives but rather complements. The existence of diquarks is a hypothesis. However, from experimental and theoretical approaches there have been many indications suggesting the presence of diquarks. For instance, they were introduced in baryon spectroscopy, in nuclear physics, in astrophysics, in jet fragmentation and in weak interactions to explain the famous $\Delta I = 1/2$ rule. Diquarks also provide a natural explanation of the equal slopes of meson and baryon Regge trajectories. For more details and for references, see [3]. It is important to note that QCD provides some attraction between two quarks in a colour $\{\bar{3}\}$ state at short distances as is to be seen from the static reduction of the one-gluon exchange term.

Even more important for our aim, diquarks have also been found to play a role in inclusive hard scattering reactions. The most obvious place to signal their presence is deep inelastic lepton-nucleon scattering. Indeed the higher twist contributions, convincingly observed by the NMC [8], can be modelled as lepton-diquark elastic scattering. Baryon production in inclusive pp collisions also reveals the need for diquarks scattered elastically in the hard interaction [9]. For instance, kinematical dependences or the excess of the proton yield over the antiproton yield find simple explanations in the diquark model. No other explanation of these phenomena is known as yet.

2 The Diquark Model

As in the standard HSA a helicity amplitude for the reaction $AB \to CD$ is expressed as a convolution of DAs and hard scattering amplitudes (s, $-t$, $-u \gg m_i^2$)

$$M(s,t) = \int \mathrm{d}x_C \mathrm{d}x_D \mathrm{d}x_A \mathrm{d}x_B \Phi_C^*(x_C) \Phi_D^*(x_D) T_H(x_i, s, t) \Phi_A(x_A) \Phi_B(x_B) \quad (1)$$

where helicity labels are omitted for convenience. Implicitly it is assumed in (1) that the valence Fock states consist of only two constituents, a quark and a diquark (antiquark) in the case of baryons (mesons). In so far the specification of the quark momentum fraction x_i suffices; the diquark (antiquark) carries the momentum fraction $1 - x_i$. If an external particle is point-like, e.g. a photon, the corresponding DA is to be replaced by $\delta(1 - x_i)$. As in the standard HSA contributions from higher Fock states are neglected. This is justified by the fact that that such contributions are suppressed by powers of α_s/t as compared to that from the valence Fock state (if only S-wave hadrons are involved).

In the diquark model spin 0 (S) and spin 1 (V) colour antitriplet diquarks are considered. Within flavour SU(3) the S diquark forms an antitriplet, the V diquark an sextet. Assuming zero relative orbital angular momentum between quark and diquark and taking advantage of the collinear approximation, the valence Fock state of a ground state octet baryon B with helicity λ and momentum p can be written in a covariant fashion (omitting colour indices)

$$|B; p, \lambda\rangle = f_S\, \Phi_S^B(x)\, B_S\, u(p, \lambda) + f_V\, \Phi_V^B(x)\, B_V (\gamma^\alpha + p^\alpha/m_B)\gamma_5\, u(p, \lambda)/\sqrt{3} \quad (2)$$

where u is the baryon's spinor. The two terms in (2) represent configurations consisting of a quark and either a scalar or a vector diquark, respectively. The couplings of the diquarks with the quarks in a baryon lead to flavour functions which e.g. for the proton read

$$B_S = u\, S_{[u,d]} \qquad\qquad B_V = [u V_{\{u,d\}} - \sqrt{2} d\, V_{\{u,u\}}]/\sqrt{3}\,. \qquad (3)$$

The DAs $\Phi^B_{S(V)}$ are conventionally normalized as $\int dx \Phi = 1$. The constants f_S and f_V play the role of the configuration space wave functions at the origin. The DAs containing the complicated non-perturbative bound state physics, cannot be calculated from QCD at present. It is still necessary to parameterize the DAs and to fit the eventual free parameters to experimental data. Hence, both the models, the standard HSA as well as the diquark model, only get a predictive power when a number of reactions involving the same hadrons is investigated. In the diquark model the following DAs have been proven to work satisfactorily well in many applications [5]-[7]:

$$\Phi^B_S(x) = N^B_S\, x(1-x)^3 \exp\left[-b^2(m_q^2/x + m_S^2/(1-x))\right] \qquad (4)$$
$$\Phi^B_V(x) = N^B_V\, x(1-x)^3 (1 + 5.8\, x - 12.5\, x^2) \exp\left[-b^2(m_q^2/x + m_V^2/(1-x))\right].$$

The constants N^B_S and N^B_V are fixed through the normalization (e.g. for the proton $N^p_S = 25.97$, $N^p_V = 22.92$). The DAs exhibit a mild flavour dependence via the exponential whose other purpose is to guarantee a strong suppression of the end-point regions. The parameters appearing in the exponentials are not considered as free parameters since the final results (form factors, amplitudes) depend on their actual values only mildly. The following values for the parameters are chosen: $b = 0.498\,\mathrm{GeV}^{-1}$, $m_u = m_d = 350\,\mathrm{MeV}$, $m_S = m_V = 580\,\mathrm{MeV}$. It is to be stressed that the quark and diquark masses only appear in the DAs (4); in the hard scattering kinematics they are neglected.

The hard scattering amplitudes T_H, determined by short-distance physics, are calculated from a set of Feyman graphs relevant to a given process. Diquark-gluon and diquark-photon vertices appear in these graphs which, following standard prescriptions, are defined as

$$
\begin{aligned}
\mathrm{S\,g\,S}: &\quad i\, g_s t^a\, (p_1 + p_2)_\mu \\[4pt]
\mathrm{V\,g\,V}: &\quad -i\, g_s t^a \Big\{ g_{\alpha\beta}(p_1 + p_2)_\mu - g_{\beta\mu}\left[(1+\kappa)\, p_2 - \kappa\, p_1\right]_\alpha \\[4pt]
&\qquad\qquad -g_{\mu\alpha}\left[(1+\kappa)\, p_1 - \kappa\, p_2\right]_\beta \Big\}
\end{aligned}
\qquad (5)
$$

where $g_s = \sqrt{4\pi\alpha_s}$ is the QCD coupling constant. κ is the anomalous magnetic moment of the vector diquark and $t^a = \lambda^a/2$ the Gell-Mann colour matrix. For the coupling of photons to diquarks one has to replace $g_s t^a$ by $-\sqrt{4\pi\alpha}e_D$ where α is the fine structure constant and e_D is the electrical charge of the diquark in units of the elementary charge. The couplings $D g D$ are supplemented by appropriate contact terms required by gauge invariance.

The composite nature of the diquarks is taken into account by phenomenological vertex functions. Advice for the parameterization of the 3-point functions (diquark form factors) is obtained from the requirement that asymptotically the diquark model evolves into the standard HSA. Interpolating smoothly between the required asymptotic behaviour and the conventional value of 1 at $Q^2 = 0$, the diquark form factors are actually parametrized as

$$F_S^{(3)}(Q^2) = \frac{Q_S^2}{Q_S^2 + Q^2}, \qquad F_V^{(3)}(Q^2) = \left(\frac{Q_V^2}{Q_V^2 + Q^2}\right)^2. \tag{6}$$

The asymptotic behaviour of the diquark form factors and the connection to the hard scattering model is discussed in more detail in Ref. [3, 4]. In accordance with the required asymptotic behaviour the n-point functions for $n \geq 4$ are parametrized as

$$F_S^{(n)}(Q^2) = a_S F_S^{(3)}(Q^2), \qquad F_V^{(n)}(Q^2) = \left(a_V \frac{Q_V^2}{Q_V^2 + Q^2}\right)^{n-3} F_V^{(3)}(Q^2). \tag{7}$$

The constants $a_{S,V}$ are strength parameters. Indeed, since the diquarks in intermediate states are rather far off-shell one has to consider the possibility of diquark excitation and break-up. Both these possibilities would likely lead to inelastic reactions. Therefore, we have not to consider these possibilities explicitly in our approach but excitation and break-up lead to a certain amount of absorption which is taken into account by the strength parameters. Admittedly, that recipe is a rather crude approximation for $n \geq 4$. Since in most cases the contributions from the n-point functions for $n \geq 4$ only provide small corrections to the final results that recipe is sufficiently accurate.

The diquark hypothesis has striking consequences. It reduces the effective number of constituents inside baryons and, hence, alters the power laws. In elastic baryon-baryon scattering, for instance, the usual power s^{-10} becomes $s^{-6} F(s)$ where F represents the net effect of diquark form factors. Asymptotically F provides the missing four powers of s. In the kinematical region in which the diquark model can be applied ($-t, -u \geq 4\,\text{GeV}^2$), the diquark form factors are already active, i.e. they supply a substantial s dependence and, hence, the effective power of s lies somewhere between 6 and 10. The hadronic helicity is not conserved in the diquark model at finite momentum transfer since vector diquarks can flip their helicities when interacting with gluons. Thus, in contrast to the standard HSA, spin-flip dependent quantities like the Pauli form factor of the nucleon can be calculated.

3 Electromagnetic Nucleon Form Factors

This is the simplest application of the diquark model and the most obvious place to fix the various parameters of the model. The Dirac and Pauli form factors of the nucleon are evaluated from the convolution formula (1) with the DAs (4) and the parameters are determined from a best fit to the data in the space-like region. The following set of parameters

$$\begin{aligned} f_S &= 73.85\,\text{MeV}, \quad Q_S^2 = 3.22\,\text{GeV}^2, \quad a_S = 0.15, \\ f_V &= 127.7\,\text{MeV}, \quad Q_V^2 = 1.50\,\text{GeV}^2, \quad a_V = 0.05, \quad \kappa = 1.39; \end{aligned} \tag{8}$$

provides a good fit of the data [5]. α_s is evaluated with $\Lambda_{QCD} = 200$ MeV and restricted to be smaller than 0.5. The parameters Q_S and Q_V, controlling the size of the diquarks, are in agreement with the higher-twist effects observed in the structure functions of deep inelastic lepton-hadron scattering [8] if these effects are modelled as lepton-diquark elastic scattering. The Dirac and the Pauli form factors of the proton are very well reproduced. The predictions for the two neutron form factors are also in agreement with the data. However, more accurate neutron data are needed in the Q^2 region of interest in order to determine the model parameters better. The nucleon's axial form factor [5] and its electromagnetic form factors in the time-like regions [6] have also been evaluated. Both the results compare well with data. Even electroexcitation of nucleon resonances has been investigated [10, 11]. In the case of the $N\Delta$ form factor the model results agree very well with the data presented in [12] while the model seems to provide to large values for the Coulomb form factor [13].

4   Real Compton Scattering (RCS)

$\gamma p \to \gamma p$ is the next reaction to which the diquark model is applied to. Since the only hadrons involved are again protons RCS can be predicted in the diquark model without any adjustable parameter. The results of the diquark model for RCS are shown in Fig. 1a for three different photon energies [4, 7].

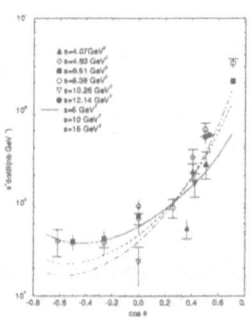
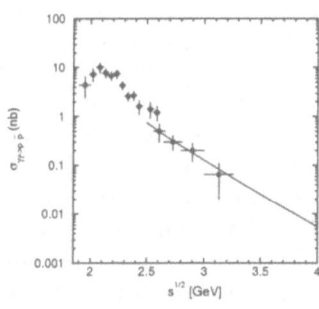

Figure 1. (left) The scaled cross section for RCS off protons vs. $cos\theta$ for three different photon energies. The experimental data are taken from [14]. (right) The integrated $\gamma\gamma \to p\bar{p}$ cross section ($|\cos\theta| \geq 0.6$). The solid line represents the diquark model prediction [6]. Data are taken from CLEO [18].

Note that in the very forward and backward regions the transverse momentum of the outgoing photon is small and, hence, the diquark model which is based on perturbative QCD, is not applicable. Despite the rather small energies at which data [14] are available, the diquark model is seen to work rather well. The predicted cross section does not strictly scale with s^{-6}. The results obtained within the standard HSA are of similar quality [15]. A purely soft, overlap-like contribution can also explain these data [16]. The diquark model also predicts interesting photon asymmetries and spin correlation parameters (see the discussion in [4]). Even a polarization of the proton, of the order of 10%, is obtained [4]. This comes about as a consequence of helicity flips generated by vector diquarks and of perturbative phases produced by propagator poles appearing within the domains of the momentum fraction integrations. The appearance of phases to leading order of α_s is a non-trivial prediction of perturbative QCD [17]; it is characteristic of the HSA and is not a consequence of the diquark hypothesis.

Two-photon annihilation into $p\bar{p}$ pairs is related to RCS by crossing. The only difference is that now the diquark form factors are needed in the time-like region. The continuation of the diquark form factors from the space-like to the time-like region is described in [6]. The diquark model predictions for the integrated $\gamma\gamma \to p\bar{p}$ cross section is compared to the CLEO data [18] in Fig. 1b. At large energies the agreement between predictions and experiment is good. The predictions for the angular distributions are in agreement with the CLEO data too. The diquark model predictions are also in agreement with the recent VENUS data [19].

5 Virtual Compton scattering (VCS)

This process is accessible through $ep \to ep\gamma$. An interesting element in that reaction is that, besides VCS, there is also a contribution from the Bethe-Heitler (BH) process where the final state photon is emitted from the electron. Electroproduction of photons offers many possibilities to test details of the dynamics: One may measure the s, t and Q^2 dependence as well as that on the angle ϕ between the hadronic and leptonic scattering planes. This allows to isolate cross sections for longitudinal and transverse virtual photons. One may also use polarized beams and targets and last but not least one may measure the interference between the BH and the VC contributions. The interference is sensitive to phase differences.

At s, $-t$ and $-u \gg m_p^2$ (or small $|\cos\theta|$ where θ is the scattering angle of the outgoing photon in the photon-proton center of mass frame) the diquark model can also be applied to VCS [7]. Again there is no free parameter in that calculation. The model can safely be applied for $s \geq 10\,\text{GeV}^2$ and $|\cos\theta| \leq 0.6$. For the future CEBAF beam energy of $6\,\text{GeV}$ the model is at its limits of applicability. However, since the diquark model predictions for real Compton scattering agree rather well with the data even at $s \geq 5\,\text{GeV}^2$ (see Fig. 1a) one may expect similarly good agreement for VCS. Predictions for the VCS cross section are given in [7].

Of interest is also the electron asymmetry in $ep \to ep\gamma$:

$$A_L = \frac{\sigma(+) - \sigma(-)}{\sigma(+) + \sigma(+)} \qquad (9)$$

where $\pm$ indicates the helicity of the incoming electron. A_L measures the imaginary part of the longitudinal – transverse interference. According to the model,

A_L is large in the region of strong BH contamination (see Fig. 2). In that region, A_L measures the relative phase between the BH amplitudes and the VCS ones. The magnitude of the effect shown in Fig. 2 is sensitive to details of the model and, therefore, should not be taken literally. Despite of this our results may be taken as an example of what may happen. The measurement of A_L, e.g. at CEBAF, will elucidate the underlying dynamics of VCS strikingly.

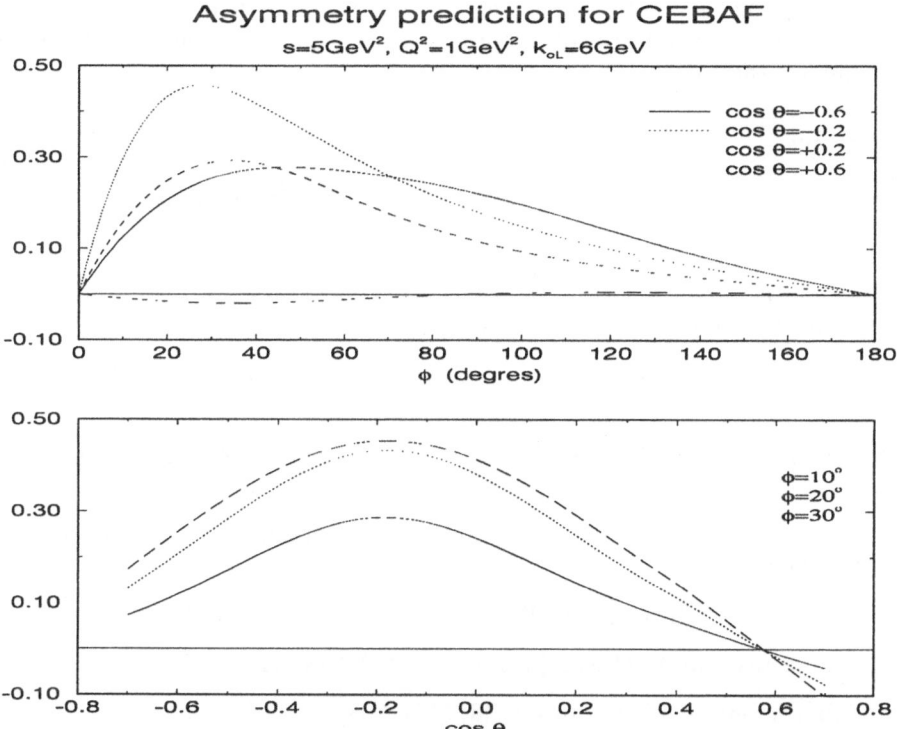

Figure 2. Diquark model predictions for the electron asymmetry in $ep \to ep\gamma$ [7].

6 Summary and outlook

The diquark model which represents a variant of the HSA, combines perturbative QCD with non-perturbative elements. The diquarks represent quark-quark correlations in baryon wave functions which are modelled as quasi-elementary constituents. This model has been applied to many photon induced exclusive processes at moderarely large momentum transfer (typically $\simeq 4\,\text{GeV}^2$). From the analysis of the nucleon form factors the parameters specifying the diquark and the DAs, are fixed. Compton scattering and two-photon annihilations of $p\bar{p}$ can then be predicted. The comparison with existing data reveals that the diquark model works quite well and in fact much better then the pure quark HSA.

Predictions for the VCS cross section and for the $ep \to ep\gamma$ cross section have also been made for kinematical situations accessible at the upgraded

CEBAF and perhaps at future high energy accelerators like ELFE@HERA. According to the diquark model the BH contamination of the photon electroproduction becomes sizeable for small azimuthal angles. The BH contribution also offers the interesting possibility of measuring the relative phases between the VC and the BH amplitudes. The electron asymmetry A_L is particularly sensitive to relative phases. In contrast to the standard HSA the diquark model allows to calculate helicity flip amplitudes, the helicity sum rule does not hold at finite Q^2. One example of an observable controlled by helicity flip contributions is the Pauli form factor of the proton. Also in this case the diquark model accounts for the data.

References

1. G.P. Lepage and S.J. Brodsky: Phys. Rev. **D22**, 2157 (1980)

2. M. Anselmino, P. Kroll and B. Pire: Z. Phys. **C36**, 89(1987)

3. P. Kroll: *Proceedings of the Adriatico Research Conference on Spin and Polarization Dynamics in Nuclear and Particle Physics*, Trieste, 1988

4. P. Kroll, W. Schweiger and M. Schürmann: Int. Jour. of Mod. Phys. **A6**, 4107(1991)

5. R. Jakob, P. Kroll, M. Schürmann and W. Schweiger: Z. Phys. **A347**, 109 (1993)

6. P. Kroll, Th. Pilsner, M. Schürmann and W. Schweiger: Phys. Lett. **B316**, 546 (1993)

7. P. Kroll, M. Schürmann and P. Guichon: Nucl. Phys. **A598**, 435 (1996)

8. M. Virchaux and A. Milsztajn: Phys. Lett. **B274**, 221 (1992)

9. A. Breakstone et al.: Z. Phys. **C28**, 335(1985)

10. P. Kroll, M. Schürmann and W. Schweiger: Z. Phys. **A342**, 429 (1992)

11. J. Bolz, P. Kroll and J.G. Körner: Z. Phys. **A350**, 145 (1994)

12. L.M. Stuart et al.: Phys. Rev. **D58**, 032003 (1998)

13. V. Burkert and L. Elouadrhiri: Phys. Rev. Lett. **75**, 3614 (1995)

14. M.A. Shupe et al.: Phys. Rev. **D19**, 1921(1979)

15. A.S. Kronfeld and B. Nižić: Phys. Rev. **D44**, 3445 (1991)

16. A.V. Radyushkin: Phys. Rev. **D58**, 114008 (1998)

17. G.R. Farrar et al.: Phys. Rev. Lett. **62**, 2229 (1989)

18. M. Artuso et al. (CLEO collaboration): Phys. Rev. **D50**, 5484 (1994).

19. H. Hamasaki et al. (VENUS collaboration): Phys. Lett. **B407**, 185(1997)

Few-Body Systems Suppl. 11, 263–266 (1999)

Few-
Body
Systems
© by Springer-Verlag 1999

Single Meson Photoproduction via Higher Twist Mechanism and IR Renormalons

S. S. Agaev

High Energy Physics Lab., Baku State University, Z.Khalilov st. 23, 370148 Baku, Azerbaijan

Abstract. Single pseudoscalar and vector mesons hard semi-inclusive photo-production $\gamma h \to MX$ via higher twist (HT) mechanism is calculated using the QCD running coupling constant method. The structure of infrared renormalon singularities of the HT subprocess cross section and the Borel sum for it are found. The problem of normalization of HT process cross section in terms of the meson elm form factor is discussed.

One of the fundamental achievements of QCD is the prediction of asymptotic scaling laws for large-angle exclusive processes and their calculation in the framework of pQCD [1-3]. In the context of the factorized QCD an expression for an amplitude of an exclusive process can be written as integral over $\mathbf{x}, \mathbf{y}$ of hadron wave functions (w.f.)[1] $\Phi_i(\mathbf{x}, \hat{Q}^2)$ (an initial hadron), $\Phi_f^*(\mathbf{y}, \hat{Q}^2)$ (a final hadron) and amplitude $T_H(\mathbf{x}, \mathbf{y}; \alpha_S(\hat{Q}^2), Q^2)$ of the hard-scattering subprocess [2]. This approach can be also applied for calculation of HT corrections to inclusive processes. The HT corrections to a single meson photoproduction was studied in [4], where for computation of integrals over $\mathbf{x}, \mathbf{y}$, the frozen coupling approximation was used. In our work we consider the hard semi-inclusive photoproduction of single pseudoscalar and vector mesons $\gamma h \to MX$ using the running coupling constant method.

The two HT subprocesses, namely $\gamma q_1 \to Mq_2$ and $\gamma \bar{q}_2 \to M\bar{q}_1$ contribute to the photoproduction of the single meson M in the reaction $\gamma h \to MX$. The amplitude for the subprocess $\gamma q_1 \to Mq_2$ can be found by means of the Brodsky-Lepage method [2],

$$M = \int_0^1 \int_0^1 dx_1 dx_2 \delta(1 - x_1 - x_2) T_H(x_1, x_2; \alpha_S(\hat{Q}^2), \hat{s}, \hat{u}, \hat{t}) \Phi_M(x_1, x_2; Q^2)$$

$$(1)$$

---

[1] Strictly speaking, $\Phi_M(\mathbf{x}, \hat{Q}^2)$ is a hadron distribution amplitude and it differs from a hadron wave function. But in this paper we use these two terms on the same footing.

In Eq.(1) Φ_M is the meson M w.f. In this work we use the following w.f.; for the pion and ρ-meson

$$\Phi_M(x, \mu_0^2) = \Phi_{asy}^M(x) \left[a + b(2x - 1)^2\right]. \tag{2}$$

where, $a = 0, b = 5$ (pion), $a = 0.7, b = 1.5$ (longitudinally and transversely polarized ρ-meson) and for the kaon

$$\Phi_K(x, \mu_0^2) = \Phi_{asy}^K(x) \left[a + b(2x - 1)^2 + c(2x - 1)^3\right], \tag{3}$$
$$a = 0.4, \quad b = 3, \quad c = 1.25.$$

with $\Phi_{asy}^M(x) = \sqrt{3} f_M x(1 - x)$ being the meson M asymptotic w.f. The meson decay constants f_M take the values $f_\pi = 0.093$ GeV, $f_K = 0.112$ GeV, $f_\rho^L = 0.2$ GeV, $f_\rho^T = 0.16$ GeV.

The details of calculation of the HT subprocess cross section are described in our work [5], where expressions for $d\hat{\sigma}^{HT}/d\hat{t}$ can be found. The subprocess cross section depends on quantities $I_{1,2}$, $K_{1,2}$,

$$I_1(K_1) = \int_0^1 \int_0^1 \frac{dx_1 dx_2 \delta(1 - x_1 - x_2) \alpha_S(\hat{Q}_1^2(\hat{Q}_2^2)) \Phi_M(x_1, x_2)}{x_2(x_1)}, \tag{4}$$

$$I_2(K_2) = \int_0^1 \int_0^1 \frac{dx_1 dx_2 \delta(1 - x_1 - x_2) \alpha_S(\hat{Q}_1^2(\hat{Q}_2^2)) \Phi_M(x_1, x_2)}{x_1 x_2}, \tag{5}$$

where for I_1, I_2 the renormalization and factorization scale is $\hat{Q}_1^2 = x_2 \hat{s}$, while for K_1, K_2 it is given by $\hat{Q}_2^2 = -x_1 \hat{u}$.

In the frozen coupling approximation one puts $\hat{Q}_{1,2}^2$ equal to their mean values $\hat{s}/2, -\hat{u}/2$ and removes $\alpha_S(\hat{Q}_{1,2}^2)$ as the constant factor in Eqs. (4-5). After such manipulation integrals (4-5) are trivial and can be easily computed. In this approach the single meson photoproduction cross section can be normalized in terms of the meson elm form factor only if the meson w.f. is symmetric under replacement $2x - 1 \leftrightarrows 1 - 2x$ (pion, ρ-meson).

In the framework of the running coupling method, for example, I_1 takes the form

$$I_1(\hat{s}) = \int_0^1 \frac{\alpha_S((1 - x)\hat{s}) \Phi_M(x, \mu_0^2) dx}{1 - x}. \tag{6}$$

The quantity $\alpha_S((1 - x)\hat{s})$ has the infrared singularity at $x \to 1$ and as a result the integral (6) diverges. This divergence is induced by IR renormalons. Indeed, the Borel transform $B[I_1](u)$ of $I_1(\hat{s})$ has IR renormalon poles at $u = 1, 2, 3, 4$ (for w.f. (2)) [5]. The integral (6) can be regularized by means of the principal value prescription. The Borel sum (resummed expression) of I_1 is

$$[I_1(\hat{s})]^{res} = \frac{4\sqrt{3}\pi f_M}{\beta_0} \left[(a + b) \frac{Li(\lambda)}{\lambda} - (a + 5b) \frac{Li(\lambda^2)}{\lambda^2}\right.$$
$$\left. + 8b \frac{Li(\lambda^3)}{\lambda^3} - 4b \frac{Li(\lambda^4)}{\lambda^4}\right], \tag{7}$$

where $Li(\lambda)$ is the logarithmic integral, for $\lambda > 1$ defined in its principal value

$$Li(\lambda) = P.V. \int_0^\lambda \frac{dx}{\ln x}, \quad \lambda = \hat{s}/\Lambda^2. \tag{8}$$

Similar expressions can be found for I_2 and $K_{1,2}$.

Important question here is the normalization of the meson photoproduction cross section in terms of the meson elm form factor. The pion and kaon form factors have been calculated by means of the running coupling method in our papers [6-7]. Using the expressions obtained in this works and our recent results, it is not difficult to conclude that, in the running coupling approach the HT subprocess cross section cannot be normalized in terms of the meson form factor neither for mesons with symmetric w.f. nor for non-symmetric ones.

Some of our numerical results for the photon-proton process are plotted in Fig.1. Here, for calculation of ratios r_M, R_M the $\Sigma_{M+(-)} = d\sigma(\gamma p \to M^{+(-)}X)$ inclusive cross sections and the difference $\Delta_M = \Sigma_{M+} - \Sigma_{M-}$ are used. In $\Sigma_{M+(-)}$, Δ_M the dominant leading twist LT ($\gamma q \to gq$ with $q \to M$) and HT ($\gamma q \to Mq$) contributions to the photoproduction have been taken into account.

As seen from Fig.1(a), effect of IR renormalons is considerable for K^-, whereas for π^- we have $r_M \simeq 1$. In Fig.1(b) the ratio $R_M = |\Delta_K^{HT}/\Delta_K^{LT}|$ is shown. For all particles the LT cross section difference is positive $\Delta_M^{LT} > 0$, since $\Sigma_{M+}^{LT} \sim u_p(x, -\hat{t})e_u^2$, while $\Sigma_{M-}^{LT} \sim d_p(x, -\hat{t})e_d^2$. The smaller quark charge e_d and the smaller distribution function d_p both suppress Σ_{M-}^{LT}. The HT cross section difference may change sign at small p_T and become negative $\Delta_M^{HT} < 0$. Therefore, we plot the absolute value of R_M. The similar picture has been also found for other mesons. The rapidity dependence of R_M at $\sqrt{s} = 25$ GeV, $p_T = 3$ GeV/c plotted in Fig.1(c) illustrates not only the tendency of the HT contributions to be enhanced in the region of negative rapidity, but also reveals an interesting feature of the HT terms; as it is seen from Fig.1(c) the ratio R_M is an oscillating function of the rapidity. This property of the HT

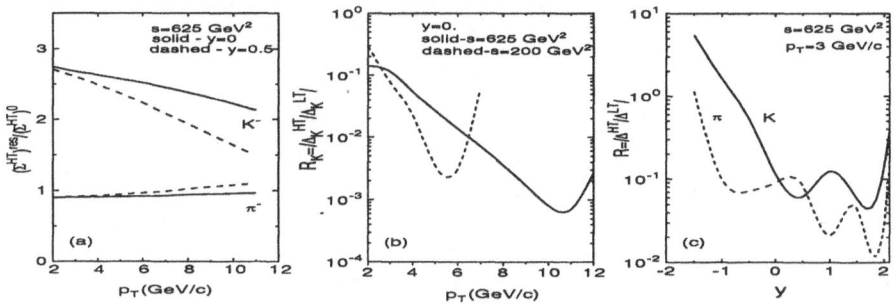

Figure 1. a) Ratio $r_M = (\Sigma_M^{HT})^{res}/(\Sigma_M^{HT})^0$, where $\Sigma^{res,(0)}$ are HT contributions to the photoproduction cross section calculated using the running and frozen coupling approximations, respectively. b) R_M for the kaon, as a function of p_T; c) R for π, K as a function of y.

terms has important phenomenological consequences in the case of ρ-meson photoproduction; comprehensive analysis of these effects can be found in ref.[5].

Summing up we can state that:

i) for mesons with non-symmetric w.f. in the framework of the frozen coupling approximation the higher twist subprocess cross section cannot be normalized in terms of a meson electromagnetic form factor;

ii) in the context of the running coupling constant method the HT subprocess cross section cannot be normalized in terms of meson's elm form factor neither for mesons with symmetric w.f. nor for non-symmetric ones;

iii) the resummed HT cross section differs, in some cases, considerably from that found using the frozen coupling approximation;

iv) HT contributions to the single meson photoproduction cross section have important phenomenological consequences, specially in the case of ρ-meson photoproduction. In this process the HT contributions wash the LT results off, qualitatively changing the LT predictions.

References

1. A.V. Efremov and A.V. Radyushkin: Theor. Math. Phys. **42**, 97 (1980); Phys. Lett. **B94**, 245 (1980)

2. G.P. Lepage and S.J. Brodsky: Phys. Rev. **D22**, 2157 (1980)

3. V.L. Chernyak and A.R. Zhitnitsky: Phys. Rep. **112**, 173 (1980)

4. J.A. Bagger and J.F. Gunion: Phys. Rev. **D25**, 2287 (1982)

5. S.S. Agaev: Eur. Phys. J. **C1**, 321 (1998)

6. S.S. Agaev: Phys. Lett. **B360**, 117 (1995); E. Phys. Lett. **B369**, 379 (1996); Mod. Phys. Lett. **A10**, 2009 (1995)

7. S.S. Agaev: Mod. Phys. Lett. **A11**, 957 (1996); ICTP preprint IC/95/291, September 1995, hep-ph/9611215

Few-Body Systems Suppl. 11, 267–274 (1999)

Few-
Body
Systems
© by Springer-Verlag 1999

Double Pion Photoproduction Experiments

B. Krusche *

II Physikalisches Institut, University of Giessen Heinrich-Buff-Ring 16, D-35392 Giessen, Germany

Abstract. The recent advances in the experimental investigation of double pion photoproduction reactions in particular at the MAMI accelerator in Mainz are reviewed. Special emphasis is given to the energy range covering the so called second resonance region with the $P_{11}(1440)$, the $D_{13}(1520)$ and the $S_{11}(1535)$ resonances. In this range all possible double pion isospin channels from the nucleon have now been measured which has largely enhanced our understanding of this reaction. In addition to the results from the nucleon first preliminary results for $2\pi^{\circ}$-photoproduction from heavy nuclei are reported.

1 Introduction

Photoproduction of pions is a very valuable tool for the study of nucleon resonances. Apart from the decay to the nucleon groundstate via single pion emission, sequential decays involving intermediate nucleon states or decays via correlated pion pairs are possible, giving rise to multiple pion photoproduction. This process becomes more important at higher excitation energies and above ≈ 2 GeV multiple pion production almost exhausts the total photoabsorption cross section. Double pion photoproduction is very important already in the so called second resonance region at photon energies between 500 MeV and 1 GeV. The cross sections for single meson production (pions and η-meson) and double pion production are almost equal at incident photon energies between 600 and 800 MeV (see figure 1). Moreover most of the rise of the total cross section from the dip above the Δ-resonance to the peak of the second resonance bump is due to double pion production.

A detailed interpretation of this structure, which includes contributions from the $P_{11}(1440)$, $D_{13}(1520)$ and $S_{11}(1535)$ resonances, clearly requires the understanding of double pion production. This is not only important for the investigation of resonances of the free nucleon. Measurements of the total photoabsorption cross section from nuclei showed a strong depletion of the res-

---

*E-mail address: bernd.krusche@exp2.physik.uni-giessen.de

onance structure above the Δ-region [2, 3]. Different explanations have been invoked to account for this experimental finding, among them a damping of resonance excitation in the nuclear medium [4]. However, it was not even clear if the P_{11}, D_{13}, and S_{11}-resonances play an important role for double pion production, which makes such a large contribution to the 'bump'.

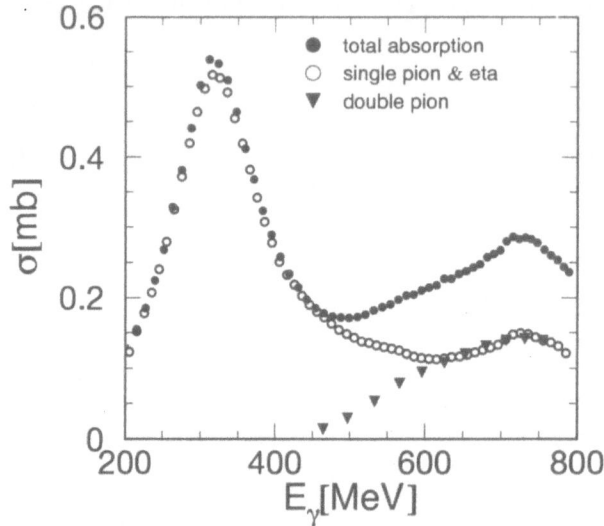

Figure 1. *Single and double meson photoproduction cross sections from the proton in the first and second resonance region. Full circles represent the total photoabsorption cross section, open circles the sum of single pion and η-photoproduction and triangles the sum of all double pion photoproduction reactions.*

A great number of background terms complicate the picture and require models for the interpretation of the data. An overview of the theory of double pion photoproduction is given in this workshop by E. Oset [1].

Until very recently only final states with at least two charged particles were studied experimentally mostly by bubble chamber experiments [5]. A lot of new interest into double pion production came up in the last few years when due to the recent progress in accelerator and detector technology the isospin channels with neutral particles in the final state became measurable. In particular a very active experimental program centered around the DAPHNE-[6] and TAPS-detectors [7] was carried out at the Mainz MAMI-accelerator at photon energies below 800 MeV. This series of experiments [8, 9, 10, 11] covers all isospin channels from the nucleon except $\gamma n \rightarrow n \pi^+ \pi^-$ and produced some first results for quasifree $2\pi^o$-production from heavy nuclei [11]. At higher photon energies preliminary results were reported from the GRAAL-facility in Grenoble ($\gamma p \rightarrow p \pi^o \pi^o$) [12] and from the SAPHIR-detector at ELSA ($\gamma p \rightarrow p \pi^+ \pi^-$) [13].

2 Double pion photoproduction from the nucleon

2.1 Final state with two charged pions

Among the possible double pion production reactions $\gamma p \to p\pi^+\pi^-$ is the only channel that was previously measured with any reasonable precision. Total cross sections and invariant mass distributions of the $\pi^+\pi^-$-, $p\pi^+$- and $p\pi^-$-pairs are available in the literature [5]. They were analysed in an early attempt to extract the dominant production mechanisms by Lüke and Söding [17].

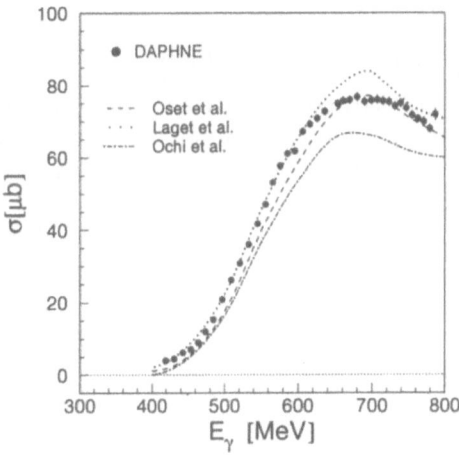

Figure 2. *Total cross section for the reaction $\gamma p \to p\pi^+\pi^-$. The data points are from ref. [8]. The dashed, dotted and dash-dotted curves correspond to the calculations from [14, 15, 16].*

The total cross section for all isospin channels is very small between threshold at ≈ 310 MeV and ≈ 400 MeV. It rises sharply for $\gamma p \to p\pi^+\pi^-$ from ≈ 400 MeV to a maximum at ≈ 650 MeV (see fig. 2). This rise reflects the $\pi\Delta(1232)$ threshold smeared by the width of the Δ-resonance. It is accompanied by a strong peak at the mass of the Δ in the invariant mass distribution of the $p\pi^+$-pair while the $p\pi^-$-invariant mass shows no such signal. A large contribution to the cross section is therefore assigned to the $\gamma p \to \Delta^{++}\pi^-$-channel while the $\gamma p \to \Delta^0\pi^+$-channel is almost negligible. The $\Delta\pi$ intermediate state could be populated by the decay of a resonance e.g. the $P_{11}(1440)$. However, a more detailed analysis [17] showed that the excitation of resonances from the second resonance region contributes only very little and instead the reaction is dominated by the Δ-Kroll-Ruderman term and the pion pole term.

More recent analyses of this reaction [14, 15, 16], taking into account the new precise data from the DAPHNE-detector [8] have solidified this picture. Although the level of agreement between the predictions and the data is somewhat different (see fig. 2), all models find the reaction dominated by the Δ-Kroll-

Ruderman term. However, even though the direct contribution from higher resonances is negligible, it was pointed out by Oset and coworkers [14] that the peak like structure between 600 and 800 MeV is due to an interference of the Δ-Kroll-Ruderman term with the sequential decay of the $D_{13}(1520)$-resonance via $\gamma p \to D_{13} \to \Delta\pi$. This allows the extraction of the coupling constant of the D_{13}-decay into $\Delta\pi$.

At higher photon energies (up to 2 GeV) total cross sections and invariant mass distributions were obtained with the SAPHIR-detector at ELSA [13]. The preliminary invariant mass distributions show clear signals for $\Delta \to N\pi$ and $\rho \to \pi^+\pi^-$ contributions.

2.2 Final state with charged and neutal pion

It came as a complete surprise when the first ever measurement of the $\gamma p \to n\pi^+\pi^o$ reaction [8] came up with a total cross section that is strongly underestimated by the model predictions from ref. [14, 15] (see figure 3).

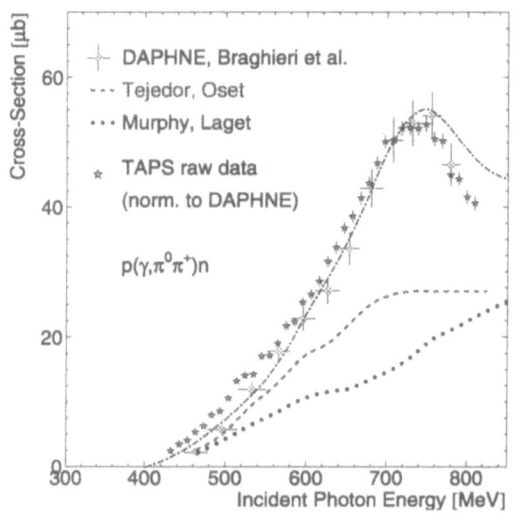

Figure 3. *Total cross section for the reaction $\gamma p \to n\pi^+\pi^o$. The data points are from experiments with the DAPHNE [8] and TAPS-detectors [18]. The TAPS-data are preliminary, the absolute scale is normalized to the DAPHNE-data (final normalisation under analysis). The curves correspond to the same calculations as in fig. 2.*

In the meantime also the reaction $\gamma n \to p\pi^-\pi^o$ was studied in quasifree kinematics from a deuterium target [9]. Corrections due to nuclear effects are expected to be small since it was shown that they are negligible for the $\gamma p \to p\pi^+\pi^-$ reaction which was studied from the free and the bound proton [9].

The cross sections for $\gamma p \to n\pi^+\pi^o$ and $\gamma n \to p\pi^-\pi^o$ are almost identical and also the latter is significantly underestimated by the model predictions.

Obviously an important contribution is missing in the models. Recently Ochi et al. [16] in their model treated the ρ-meson propagator in a dynamical way and introduced a $\rho\pi\pi$ form factor since the ρ-meson contributing to double pion decays is not on-shell in the energy range of interest. They found a large contribution to the $n\pi^+\pi^o$ and $p\pi^-\pi^o$ final states from the ρ-Kroll-Ruderman term which brought the $\gamma p \rightarrow n\pi^+\pi^o$ cross section up to the measured values. This term cannot contribute to the other final states since the photon does not couple to the neutral ρ-meson. However, their model describes the other reaction channels, in particular double π^o-production, less well than others.

In order to clarify the situation invariant mass distributions of the meson- and meson-nucleon-pairs are needed. The extraction of such distributions from the TAPS-data for the $n\pi^+\pi^o$-channel [19] and from the DAPHNE-data for the $p\pi^-\pi^o$-channel [20] are under way.

2.3 Final state with two neutral pions

The final states with two neutral pions are particularly well suited for the study of resonance decays. Since the photon does not couple to the neutral pion and the ρ-meson does not decay into a pair of neutral pions, almost all background terms are excluded. Surprisingly, the two models from refs. [14, 15] made very

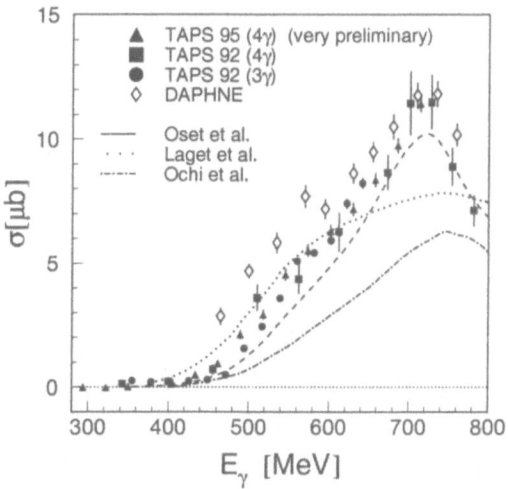

Figure 4. *Total cross section for the reaction $\gamma p \rightarrow p\pi^o\pi^o$. The data points are from experiments with the DAPHNE- [8] and TAPS-detectors [10, 21] The 1995 TAPS-data are preliminary. The curves correspond to the same calculations as in fig 2.*

different predictions. One of them [14] predicted the sequential decay of the $D_{13}(1520)$ resonance, the other [15] the decay of the $P_{11}(1440)$ resonance via a correlated pair of pions in a relative s-wave as dominating process. Already

272

the total cross section agrees somewhat better with the result from ref. [14] (see figure 4). The question was solved by the invariant mass distributions

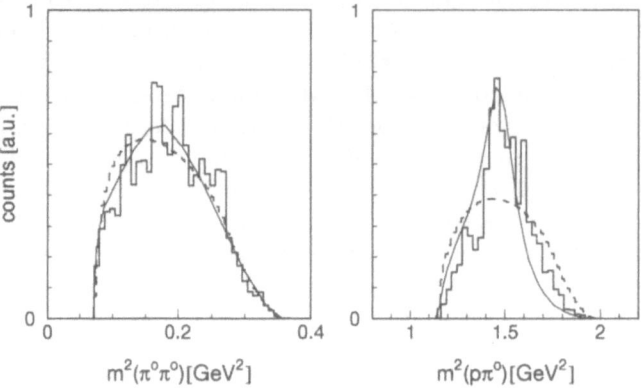

Figure 5. *Distribution of the invariant mass squared for the $\pi^\circ\pi^\circ$- and $p\pi^\circ$-pairs from the $\gamma p \to p\pi^\circ\pi^\circ$ reaction. The histogram shows the data measured with the TAPS-detector [11], the dashed line the expectation for pure phase-space behavior and the solid line the result from ref. [14].*

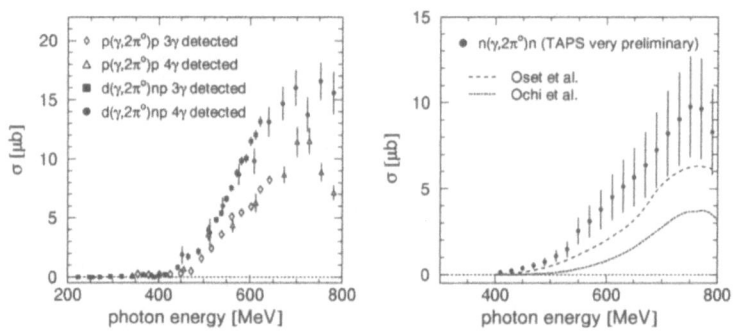

Figure 6. *Double π°-photoproduction from the neutron. On the left hand site the cross sections from proton [10] and deuteron target [11] are compared. The right hand side shows the extracted preliminary neutron cross section compared to the model predictions from [14] (dashed) and [16] (dash-dotted).*

measured with TAPS (see fig. 5). While the $\pi^\circ\pi^\circ$-mass distribution is explained by phase space, the $p\pi^\circ$-distribution shows a peak at the mass of the $\Delta(1232)$. This favors the sequential resonance decay over the correlated decay. In the meantime similar results have been obtained at the GRAAL-facility in Grenoble [12]. Those data will provide invariant mass distributions up to higher energies.

The final state with two neutral pions from the deuteron was measured with the TAPS-detector in quasifree kinematics . Since again nuclear effects are expected to be small the cross section for $\gamma n \rightarrow n\pi^o\pi^o$ can be extracted in a simple participant-spectator approach by a comparison of proton and deuteron cross section. The neutron cross section is similar to the proton cross section and somewhat underestimated by the models (see fig. 6).

3   Quasifree double π^o-photoproduction from heavy nuclei

The suppression of the second resonance bump observed in the photoproduction from nuclei has motivated exclusive experiments in this energy region. In a study of η-photoproduction [23], which is strongly dominated by the $S_{11}(1535)$ resonance, no unexplained depletion of the in-medium strength was observed. The data are in agreement with models taking into account trivial nuclear

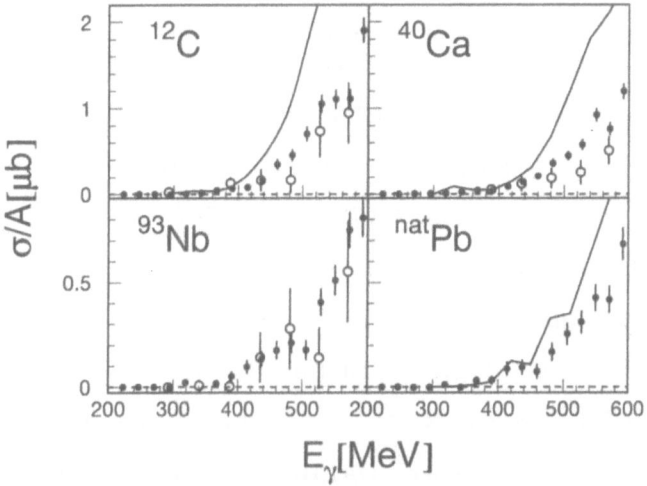

Figure 7. *Total cross section of quasifree double π^o-photoproduction from nuclei normalized to the mass number [11]. The open symbols are obtained from events with two identified π^o-mesons, the full points from events with one pion and at least one more photon. The latter, which are practically background free below the η-production threshold, provide better statistics. The curves show the prediction from ref. [22].*

effects like Fermi smearing and final state interaction effects of the mesons. Double π^o-production is of particular interest since the experiments on the nucleon have shown that it is dominated by the excitation of the $D_{13}(1520)$. Preliminary cross sections measured with TAPS are significantly overestimated (see figure 7) by the same models which nicely reproduced η-photoproduction. New experiments yielding much better statistics which will allow the extraction of angular and energy distributions are planned with TAPS at MAMI for 1999.

274

References

1. E. Oset: *in these Proceedings*

2. Th. Frommhold et al.: Phys. Lett. **B295**, 28 (1992); Z. Phys. **A350**, 249 (1994)

3. N. Bianchi et al.: Phys. Lett. **B299**, 219 (1993)

4. M.M. Giannini and E. Santopinto Phys. Rev. **C49**, R1258 (1994)

5. ABBHHM-collaboration: Phys. Rev. **175**, 892 (1969); G. Gialanella et al.: Nuovo Cim. **LXIII A**, 892 (1969); A. Piazza et al.: Nuovo Cim. **III**, 403 (1970); Cambridge Bubble Chamber group: Phys. Rev. **155**, 1477 (1967), Phys. Rev. **163**, 1510 (1967); H.R. Crouch et al.: Phys. Rev. Lett. **13**, 636 (1964); H.G. Hilpert et al.: Phys. Rev. Lett. **23**, 707 (1966)

6. G. Audit et al.: Nucl. Inst. Meth. **A301**, 473 (1991)

7. R. Novotny et al.: IEEE Trans. Nucl. Sci. **38**, 379 (1991); H. Ströher: Nucl. Phys. News **6**, 7 (1996)

8. A. Braghieri et al.: Phys. Lett. **B363**, 46 (1995)

9. A. Zabroddin et al.: Phys. Rev. **C55**, R1617 (1997)

10. F. Härter et al.: Phys. Lett. **B401**, 229 (1997)

11. B. Krusche et al.: *in preparation*

12. E. Guinault: PhD thesis, Paris 1997 (unpublished)

13. F.J. Klein: PhD thesis, Bonn 1996 (unpublished)

14. J.A. Gomez Tejedor and E. Oset: Nucl. Phys. **A600**, 413 (1996)

15. L.Y. Murphy and J.M. Laget: DAPHNIA/SPhN **96-10** (1996)

16. K. Ochi et al.: *nucl-th/9703058*; M. Hirata et al.: *nucl-th/9711031*

17. D. Lüke and P.Söding: *Springer Tracts in Modern Physics* **59**, 39 (1971)

18. M. Kotulla: diplome thesis, Giessen 1997 (unpublished)

19. W. Langgärtner: diploma thesis, Giessen, in preparation

20. P. Pedroni: *private communication*

21. M. Wolf: PhD thesis, Giessen, in preparation

22. M. Effenberger et al.: Nucl. Phys. **A614**, 501 (1997), and *private communication*

23. M. Röbig-Landau et al.: Phys. Lett. **B373**, 45 (1996)

Few-Body Systems Suppl. 11, 275–283 (1999)

Few-
Body
Systems
© by Springer-Verlag 1999

Double Pion Production Reactions

E. Oset[1], J. A. Gómez Tejedor[2], F. Cano[3], J.C. Nacher[1], S. Kamalov[4], L. Alvarez-Ruso[1], E. Hernández[5]

[1] Departamento de Física Teórica and IFIC, Centro Mixto Universidad de Valencia - CSIC, 46100 Burjassot (Valencia), Spain

[2] Departamento de Física Aplicada, Universidad Politécnica de .Valencia, Valencia, Spain

[3] Dipartimento di Física, University of Trento, Trento, Italy

[4] Institut für Theoretische Physik, Universität Mainz, Germany

[5] Departamento de Física, Universidad de Salamanca, Spain

Abstract. We report on reactions producing two pions induced by real and virtual photons or nucleons. The role of different resonances in these reactions is emphasized. Novel results on coherent 2π photoproduction in nuclei are also reported.

1 Double pion photoproduction on the nucleon.

The $\gamma N \to \pi^+\pi^- N$ reaction is attracting attentions of both theoretical and experimental groups and is bound to play a significant role in photonuclear reactions much as $\gamma N \to \pi N$ played in the past.

Apart from the work of the Valencia group which I will report here, there is work by other groups. A simplified model containing many of the important features of the reaction was worked out in [1] and improved in [2], [3]. The model of [3] contains more mechanisms than the one of [2], presently under revision [4]. On the other hand, the model of [2] incorporates an approximate unitarization prescription which allows one to go to higher energies with the model. In ref. [5] a simplified model is also used incorporating, however, some ρ decay channels. This work has been revised in [6] in view that some mass distribution was in disagreement with the data, and a new parametrization is offered, which relies on a range parameter for the ρ of the order of $200\,MeV$, which would not accommodate easily other known facts of phenomenology such as the isovector πN s-wave amplitude.

The model of [3] contains parameters determined solely from γ and π couplings to nucleons and resonances plus known properties of resonance decay with some undetermined sign borrowed from quark models.

The $(\gamma, \pi\pi)$ reaction has also been studied at threshold with the aim of testing chiral perturbation theory [7, 8], particularly the $\gamma p \to \pi^0\pi^0 p$ reaction where chiral loops are very important.

The $\gamma p \to \pi^+\pi^- p$ reaction was studied in ref. [3] using effective Lagrangians, which incorporate the couplings of the photon and pion to the nucleon and resonances. The N and the $\Delta(1232)$, $N^*(1440)$ and $N^*(1520)$ (or N'^*) resonances were taken into account. Furthermore, the ρ as an intermediate state coupling to two pions was also considered. The model reproduces fairly well the experimental cross section [9]. The model is further improved [10] to account for $s - d$ waves in the $N'^* \to \Delta\pi$ decay, while at the same time reduces from 67 to 20 the number of Feynman diagrams needed to study the reaction in the range of Mainz energies $E_\gamma \leq 800 \ MeV$. In ref. [10] this simplified model is used to evaluate cross sections for all other charge channels: $\gamma p \to \pi^+\pi^0 n$, $\gamma p \to \pi^0\pi^0 p$, $\gamma n \to \pi^+\pi^- n$, $\gamma n \to \pi^-\pi^0 p$, $\gamma n \to \pi^0\pi^0 n$. The agreement with the data is overall good but some discrepancies remain in the peak of the $\gamma p \to \pi^+\pi^0 n$ reaction and its charge conjugate one, the $\gamma n \to \pi^-\pi^0 p$ reaction, recently measured [11].

The relevance of this reaction for the resonance field is the novel information that it provides on the $N^*(1520)$ resonance, which I try to explain here. In Fig. 1a, I show the dominant diagram in the $\gamma p \to \pi^+\pi^- p$ reaction. It is the $\Delta N\pi\gamma$ Kroll Ruderman or gauge term. On the other hand in Fig. 1b, I show a diagram where the $N^*(1520, J^\pi = \frac{3}{2}^-)$ is photoexcited from the nucleon and then it decays into $\Delta\pi$, the Δ decaying later into $N\pi$.

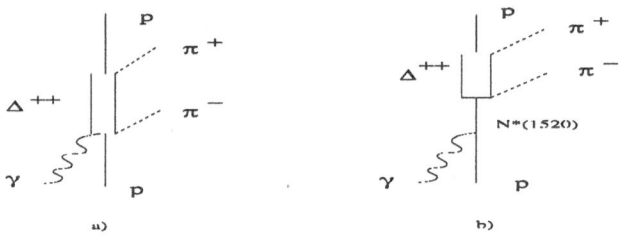

Figure 1. Feynman diagrams

From the 1/2 and 3/2 experimental $N^*(1520)$ helicity amplitudes we can construct an effective Lagrangian from where we obtain a transition operator given by

$$-i\delta H = ig_\gamma S^\dagger \epsilon + g_\sigma(S^\dagger \times \sigma)\epsilon \, , \qquad (1)$$

where $S^\dagger$ is a spin transition operator from spin 1/2 to spin 3/2. Furthermore, we write the $N'^* \to \Delta\pi$ transition operator as

$$-i\delta H = -[\tilde{f} + \frac{\tilde{g}}{\mu^2}(S^\dagger q)(Sq)]T^{\dagger\lambda} \ + h.c. \, , \qquad (2)$$

where $T^\dagger$ is the isospin $1/2$ to $3/2$ transition operator and μ the pion mass. The choice of eq. (2) is not arbitrary. It allows $N'^* \to \Delta\pi$ decay in s and d waves and provides a q dependence of the amplitudes (q is the CM pion momentum) which provides the best agreement with experiment. By means of eq. (2) we can write the s and d wave decay amplitudes in $N'^* \to \Delta\pi$. We find

$$A_s = -\sqrt{4\pi}(\tilde{f} + \tfrac{1}{3}\tilde{g}\tfrac{q^2}{\mu^2}) ,$$

$$A_d = \frac{\sqrt{4\pi}}{3} \tilde{g} \frac{q^2}{\mu^2} , \tag{3}$$

and the width is given by

$$\Gamma_{N'^* \to \Delta\pi} = \frac{1}{4\pi^2} \frac{m_\Delta}{m_{N'^*}} q(|A_s|^2 + |A_d|^2). \tag{4}$$

From the analysis of the $\pi N \to \pi\pi N$ reaction of ref. [12] one has information on Γ_s, Γ_d plus also another ingredient, the relative sign of A_s to A_d which is positive. With this information we obtain A_s and A_d up to a global sign (a sign relative to the $\gamma N \to N^*$ amplitudes). This sign is the first novel thing that the $\gamma p \to \pi^+\pi^- p$ reaction provides. Indeed we can see in Fig. 2 the results with the two signs and we observe that while one of the signs is in good agreement with the experiment (a), the other choice leads to unacceptable results (b).

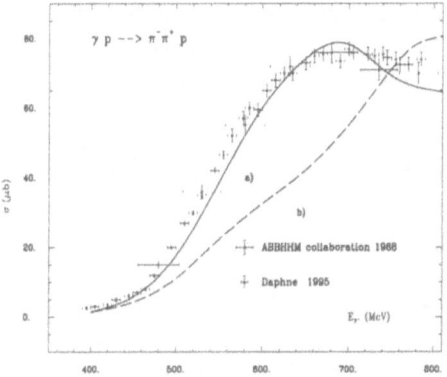

Figure 2. Total cross section for $\gamma p \to \pi^+\pi^- p$ reaction for different global sign.

The reason for the so different results with the two signs is that the $N^*(1520)$ mechanism of Fig. 1b interferes with the dominant one of Fig. 1a. The two amplitudes (by taking the s-wave part of the $N'^* \to \Delta\pi$ decay) have the same momentum and spin structure, and the $N^*(1520)$ piece can be accounted for by making a simple substitution in the $N\Delta\pi\gamma$ Kroll Ruderman piece:

$$e\frac{f^*}{\mu} \to -(g_\gamma - g_\sigma)(\tilde{f} + \frac{1}{3}\tilde{g}\frac{q^2}{\mu^2})D_{N'^*}(s) , \tag{5}$$

where $D_{N'^*}$ is the $N^*(1520)$ propagator. We can see that with the value $g_\gamma - g_\sigma = 0.157 > 0$ and $\tilde{f} + \frac{1}{3}\tilde{g}\frac{q^2}{\mu^2} > 0$ one gets a constructive interference before the N'^* pole and a destructive one after it. This is what can be observed in Fig. 2.

I shall not discuss here the other channels. Some results and comments can be found in the talk of Krusche in this Workshop [13]. The $\gamma p \to \pi^0 \pi^0 p$ is well reproduced and here the $N^*(1520)$ term does not show up through the interference but as the main term by itself. On the other hand there are some discrepancies in the $\gamma p \to \pi^+ \pi^0 n$ channel which we can not explain so far.

2 Repercussion of the $N^*(1520)$ findings on quark models.

With the values of $\tilde{f} = 0.911$ and $\tilde{g} = -0.552$ obtained from a fit to the s and d wave partial decay widths of $N^*(1520) \to \Delta\pi$ and the global sign given by the $\gamma p \to \pi^+ \pi^- p$ experiment, the amplitudes A_s, A_d of eqs. (3) provide a definite q dependence of these amplitudes.

As mentioned, this q dependence is the one providing an optimal fit to the experiment. We have checked that any other q˙dependence of the s-wave amplitude, consistent with the value for the on shell decay width, provides a worse agreement.

At this point it is worth mentioning the repercussion of these results in the quark models. This has been shown recently [15]. In this work a nonrelativistic constituent quark model using the input of Badhuri's model [16], adapted by Silvestre-Brac to the baryonic sector [17], is employed, and the decay amplitudes $B \to B'\pi$ are evaluated. For this purpose one starts with a coupling of pions to quarks

$$H_{qq\pi} = \frac{f_{qq\pi}}{\mu} \bar{\psi}_q \gamma^\mu \gamma_5 \boldsymbol{\tau} \psi_q \partial_\mu \phi, \tag{6}$$

and makes a nonrelativistic expansion keeping recoil terms

$$H_{qq\pi} = f_{qq\pi} [\boldsymbol{\sigma} \boldsymbol{q} e^{-i\mathbf{qr}} - \frac{\omega_\pi}{2m_q} \boldsymbol{\sigma} (\boldsymbol{p} e^{-i\mathbf{qr}} + e^{-i\mathbf{qr}} \boldsymbol{p})]. \tag{7}$$

Now, when evaluating the N^*, Δ transition matrix elements, since one has a radial excitation in the N^* state, one needs to expand the exponential in the first term of eq. (7) (direct term) up to order q. On the other hand, the second term of eq. (7) (recoil term) already gives a contribution keeping the unity in the expansion of the exponential. Hence, we find

$$\text{DIR} \propto q^2 \quad ; \quad \text{REC} \propto 1. \tag{8}$$

A direct evaluation of the s and d-wave amplitudes for the $N'^* \to \Delta\pi$ decay gives

$$\frac{A_d}{A_s} = \frac{\text{DIR}}{2\text{REC} - \text{DIR}}, \tag{9}$$

which implies

$$A_d \propto \quad \text{DIR} \propto \boldsymbol{q}^2$$

$$A_s + A_d \propto \quad \text{REC} \propto 1.$$

(10)

Hence, the non relativistic constituent quark model keeping recoil terms makes very clear predictions on the q dependence of the amplitudes. Now, by looking at the q dependence demanded by the $\gamma p \to \pi^+\pi^- p$ reaction, expressed in eq. 3, we obtain

$$A_d = \frac{\sqrt{4\pi}}{3} \tilde{g} \frac{\boldsymbol{q}^2}{\mu^2} \; ; \; A_s + A_d = -\sqrt{4\pi}\, \tilde{f} \;, \tag{11}$$

which is the exactly the $\boldsymbol{q}$ dependence provided by the quark model with recoil terms, eq. (10).

The global sign of these amplitudes prefered by the $\gamma p \to \pi^+\pi^- p$ experimental data is also the one provided by the quark model. This is another accomplishment of these quark models, but one should recall that not all variants of nonrelativistic, or relativized quark models will satisfy these new constraints. This is important to note since problems still remain when one comes to absolute values of these amplitudes[15]. Extra work is needed to explain these discrepancies, but it is important that these improvements are done respecting the new constraints found thanks to the $\gamma p \to \pi^+\pi^- p$ reaction. Actually, a treatment similar to the present one but making an expansion in terms of ω_π/E_q instead of ω_π/m_q seems to lead to very much improved results, while keeping the consistency with the findings discussed in this section [21].

3 Meson exchange current and coherent 2π photoproduction

Assume the $\gamma N \to \pi\pi N$ reaction occurs inside a nucleus and one of the pions, say the π^-, is produced off shell and absorbed by a second nucleon. One obtains then meson exchange current mechanisms which contribute to the (γ, π^+) reaction in nuclei and which would be represented by diagrams like those in Fig. 1 with the π line attached to a nucleon line. This mechanism has already been explored in [22] where it was found to contribute significantly to the $\gamma\,^3He \to t\,\pi^+$ reaction at large momentum transfer.

In addition, the coherent 2π photoproduction process in nuclei has been studied in [23] and has shown very interesting features tied to the isospin structure of the amplitudes. A photon coupling to a nucleon can have an isoscalar and isovector component. Assume we have the coherent reaction occurring in isospin $I = 0$ nuclei

$$\gamma + A \to \quad \begin{array}{l} \pi^+\pi^- \quad + A_{g.s.} \\ \pi^0\pi^0 \quad + A_{g.s.} \end{array} \tag{12}$$

and let us take the isoscalar part of the amplitude. This will force the $\pi^+\pi^-$ $(\pi^0\pi^0)$ system to have $I = 0$ and, because of symmetry, even angular momentum

$L = 0, 2 \ldots$ The isovector part will force the $\pi^+\pi^- (\pi^0\pi^0)$ system into $I = 1$. This is forbidden for the $\pi^0\pi^0$ system, so only the $\pi^+\pi^-$ can go with $I = 1$, which forces $L = 1, 3 \ldots$ The dynamics of the elementary reaction is such that the $\gamma N \to \pi^+\pi^- N$ reaction is dominated by the diagram of Fig. 1a, where the photon behaves as an isovector, while this mechanism is forbidden for $\pi^0\pi^0$ production. Obvious consequences of that are that the $\pi^+\pi^-$ system is largely suppressed when the pions travel together ($L = 0$ and hence $I = 0$).

Similarly the $\pi^0\pi^0$ system is only produced in $I = 0$ and hence the pions prefer to travel together. On the other hand the strength of the isoscalar part of the $\gamma N \to \pi\pi N$ amplitude is much smaller than the isovector part in the model of [10] and the consequence of it is that the maximum of the $\pi^0\pi^0$ cross section is about three orders of magnitude smaller than the maximum of the $\pi^+\pi^-$ one. These are very strong tests of the model which should encourage the experimentalists to perform such reactions.

4 Two pion electroproduction.

The model of ref. [9] can be extended to virtual photons coming from the (e, e') vertex. These reactions are presently under experimental investigation at TJNAF [24, 25]. We have studied the 2π production processes where there is a $\Delta\pi$ in the final state, ie. $ep \to e'\Delta^{++}\pi^-$ and $ep \to e'\Delta^0\pi^+$ (with $\Delta^0 \to \pi^- p$). Only 8 diagrams of the model of [10] contain a Δ in the final state and we depict these diagrams below in Fig. 3.

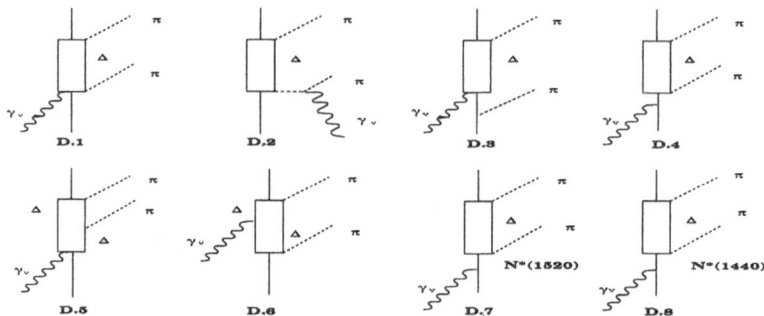

Figure 3. Feynman diagrams used in the model for $\gamma_v p \to \pi\Delta$.

The evaluation of the amplitudes for these reactions requires the extension of the model of [10] to account for the zeroth component of the electromagnetic current and the implementation of adequate form factors. This task has been undertaken in [26]. In Fig. 4 we show the results obtained for the cross section of the virtual photons, defined in the standard way

$$\frac{d\sigma}{d\Omega'_e dE'_e} = \Gamma(\sigma^T_{\gamma_v} + \epsilon\sigma^L_{\gamma_v}) = \Gamma\sigma_{\gamma_v} , \qquad (13)$$

with Γ and ϵ the flux factor and the polarization of the virtual photon [27]. The results are shown for the $ep \rightarrow e'\Delta^{++}\pi^-$ reaction as a function of $Q^2 = -q^2$ and averaged over the range $0.3 < Q^2 < 1.4\,GeV^2$ in order to compare with the data. As one can see, the agreement is fair but more precise data are expected to come soon which will impose stronger constraints on the theory.

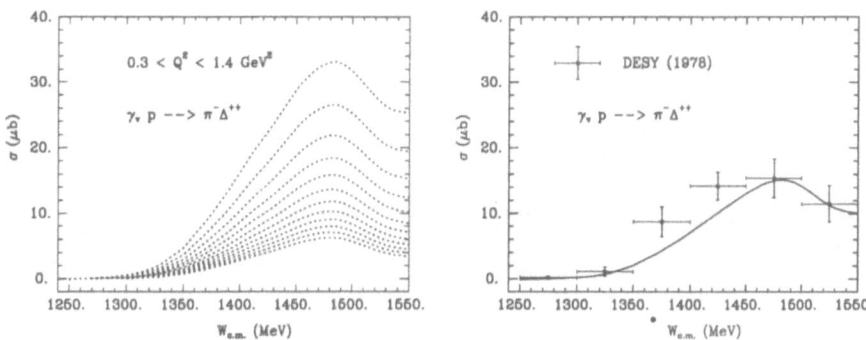

Figure 4. Cross sections for $\gamma_v p \rightarrow \Delta^{++}\pi^-$ as a function of the $\gamma_v p$ center of mass energy.

5 Application of isoscalar N^* excitation in the $NN \rightarrow NN\pi\pi$ reaction.

We have developed recently a model for the $NN \rightarrow NN\pi\pi$ reaction which contains terms coming from chiral Lagrangians, Δ excitation and Roper excitation [28]. The model is depicted in Fig. 5. The excitation on the second nucleon and antisymmetry are incorporated in addition. Summarizing the results we find that the N^* excitation terms (4 - 7), where the N^* decays into $N(\pi\pi)^{I=0}_{s-wave}$ are largely dominant close to threshold in the channels where the two pions can be in $I = 0$ in the final state, like the $pp \rightarrow pp\pi^+\pi^-$ reaction. On the other hand in the $pp \rightarrow pn\pi^+\pi^0$ reaction the $\Delta\Delta$ excitation terms are the most important. The comparison of these two channels allows us to appreciate the role played by N^* excitation in some of the isospin channels which, as we can see, is essential to understand the experiment at low energies. In Fig. 6, we show the cross sections for the $pp \rightarrow pp\pi^+\pi^-$ and $pp \rightarrow pn\pi^+\pi^0$ reactions. The calculations are done with plane waves, but the results are increased at lower energies when final state interaction is considered, and the agreement with experiment is improved. In the figures, the total cross sections are given by the solid lines, corresponding to two different options for the $N(\pi\pi)^{I=0}_{s-wave}$ decay[28].

One of the important ingredients in this reaction is that the largest strength for N^* excitation comes from isoscalar exchange. The strength of this transition was obtained from a theoretical analysis [29] of the (α, α') reaction on proton targets exciting the Roper resonance [30].

282

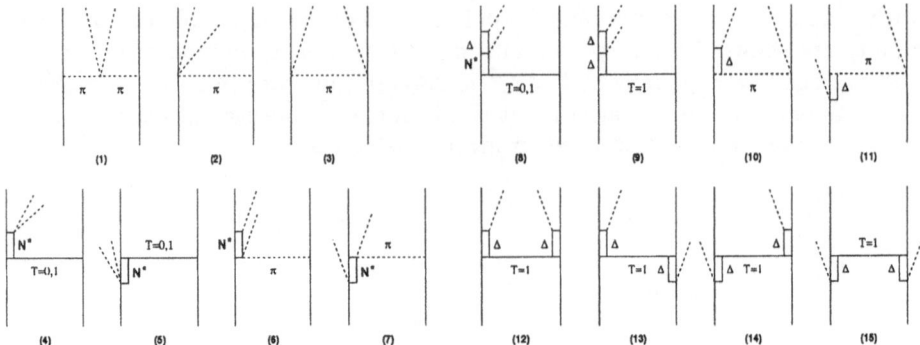

Figure 5. Complete set of Feynman diagrams of our model.

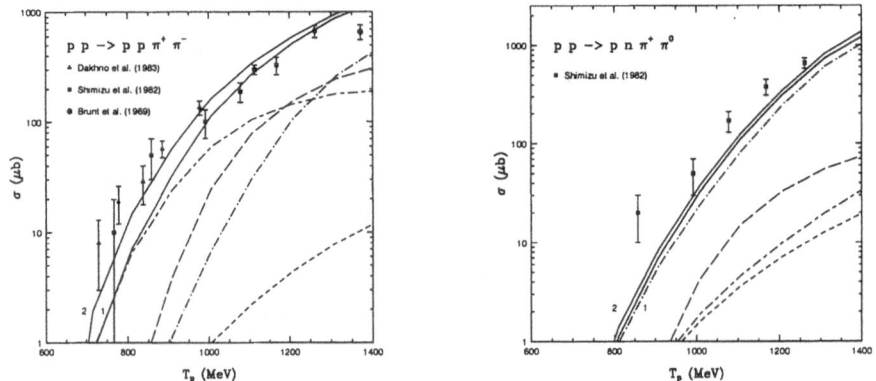

Figure 6. Total cross sections for two of the channels, as a function of the incoming proton kinetic energy in lab. frame.

In conclusion we have seen several reactions involving two pions in the final state. In all of them the N^* resonances play an important role and we have clarified the links between some resonance properties and observables in 2π production reactions. Further investigations both theoretical and experimental, extending the work at higher energies, look also like a fertile land to extend our knowledge about N^* resonance properties.

References

1. L. Lüke and P. Söding: *Springer Tracts in Modern Physics*, vol. 59 (Springer, Berlin, 1971)

2. L. Y. Murphy and J. M. Laget: *DAPHNIA, SPhN 95-42 preprint*

3. J. A. Gómez-Tejedor and E. Oset: Nucl. Phys. C **571**, 667 (1994)

4. J. M. Laget: private communication

5. K. Ochi, M. Hirata and T. Takaki: Phys. Rev. C **56**, 1472 (1997)

6. M. Hirata, K. Ochi and T. Takaki: *nucl-th/ 9711017*.

7. M. Benmerouche and E. Tomusiak: Phys. Rev. Lett. **73**, 667 (1994)

8. V. Bernard, N. Kaiser, U. G. Meissner and A. Schmidt: Nucl. Phys. A **580**,475 (1994)

9. A. Braghieri et al.: Phys. Lett. B **363**, 46 (1995)

10. J. A. Gómez-Tejedor and E. Oset: Nucl. Phys. A **600**, 413 (1996)

11. A. Zabrodin et al.: Phys. Rev. C **55**, R1617 (1997)

12. D. M. Manley and E. K. Saleski: Phys. Rev. D **45**, 4002 (1992)

13. B. Krusche: *in these Proceedings*

14. F. Härter et al.: Phys. Lett. B **401**, 229 (1997)

15. J. A. Gómez-Tejedor, F. Cano and E. Oset: Phys. Lett. B **379**, 39 (1996)

16. R. K. Bhaduri, L.E. Cohler and J. Nogami: Nouvo Cimento A **65**, 376 (1981)

17. B. Silvestre-Brac and C. Gignoux: Phys. Rev. D **32**, 743 (1985)

18. R. Koniuk and N. Isgur: Phys. Rev. D **21**, 1868 (1980)

19. S. Capstick and W. Roberts: Phys. Rev. D **49**, 4570 (1994)

20. F. Iachello: Proceedings of the *Forth CEBAF/INT Workshop on N* Physics*, Editors: T.-S. H. Lee and W. Roberts, (World Scientific, 1997)

21. F. Cano and P. González: private communication.

22. J. A. Gómez-Tejedor, S. Kamalov and E. Oset: Phys. Rev. C **54**, 3160 (1996)

23. S. S. Kamalov and E. Oset: Nucl. Phys. A **625**, 873 (1997)

24. V. Burkert: *in these Proceedings*

25. L. Elouadrhiri: *in these Proceedings*

26. J. C. Nacher and E. Oset: *nucl-th/9804006*

27. E. Amaldi, S. Fubini and G. Furlan: *Pion Electroproduction, Springer Tracts in Modern Physics*, Vol. 83 (Springer, Berlin, 1989)

28. L. Alvarez-Ruso, E. Oset and E. Hernández: Nucl. Phys. A **633**, 519 (1998)

29. S. Hirenzaki, P. Fernández de Córdoba and E. Oset: Phys. Rev. C **53**, 277 (1996)

30. H. P. Morsch et al.: Phys. Rev. Lett. **69**, 249 (1992)

Few-Body Systems Suppl. 11, 284–291 (1999)

Few-
Body
Systems
© by Springer-Verlag 1999

Baryon Resonances in the Double Pion Channel at Jefferson Lab (CEBAF): Experimental and Physical Analysis Status and Perspectives

M. Ripani*

Istituto Nazionale di Fisica Nucleare,
Via Dodecaneso 33, I-16146 Genova, Italy

Abstract. The excited baryons made from light quarks are known to decay in single meson as well as in multimeson final states. In particular, the double pion production is sensitive to many excited states of proton and neutron. Quark models predict such decays and also that some resonances could decouple from single meson channels and appear predominantly in multipion production reactions via electromagnetic excitation: the so called "missing resonances". These issues are part of the CLAS collaboration scientific program at Jefferson Laboratory, where the reaction $eN \to e'N\pi\pi$ is being used in the mass region between threshold and 2.2 GeV to investigate baryon resonances and test quark models. In this contribution I will present a framework for the physical interpretation of the data, especially focusing on the approach developed by the Genova-Moscow collaboration. Some very preliminary raw mass distributions collected with CLAS are then shown.

1 Introduction

As established in several years of experimental and theoretical investigation[1, 2], mesons and baryons appear to gather in mass multiplets that can be interpreted as the manifestation of ground state and excitation spectrum of a system with internal structure. The multiplet structure is seen as the reflection of symmetry properties of the Hamiltonian describing the system. Looking at the ground state, one can see that baryons are organized in an octet of spin 1/2 particles containing proton and neutron, while the well-known $\Delta(1232)$ excitation of the nucleon appears to be member of a spin 3/2 decuplet. Octet and decuplet can be in turn put together in a 56-plet, where the 56 comes from spin states counting. Octet and decuplet naturally arise assuming a Hamiltonian symmetric under the "flavour" group $SU(3)$ describing the basic u, d, s

---

*E-mail address : ripani@ge.infn.it

lightest quarks. The spin $SU(2)$ symmetry does the remaining job, leading to the 56-plet of $SU(3) \otimes SU(2)$. Addition of internal quark motion leads to a sequence of orbital bands, like those obtained using an harmonic oscillator confining potential. Flavour symmetry breaking, basically due to the mass difference between u, d and s quarks, leads to the splitting between baryon states with different strangeness. Moreover, to explain the nucleon-Δ 300 MeV mass difference, spin-spin interactions are introduced: they break SU(2) symmetry, producing a configuration mixing and shifting the Δ mass from that of the nucleon, as required. Finally, the color degree of freedom is assumed to be frozen in singlet states, such that the resulting hadrons are white, or colorless.

The use of any quark model incorporating the basic features of approximate SU(6) symmetry with explicit flavour-breaking terms and spin-spin interaction, with a spatial wavefunction obtained from some confining potential, is able to account quite reasonably for some general properties of baryon states observed experimentally. In particular, the ground state and the first excited states are usually well accounted for as far as their main static properties are concerned. However, besides the well-known discrepancies between electromagnetic properties like calculated and measured form factors, there is also a major issue regarding the number of states: the symmetric quark model predicts a number of states in the second orbital band which is higher than seen in experiments. This is referred to as the problem of "missing states" and stimulated different formulations: in quark models[3] with hyperfine mixing and explicit meson couplings, it turns out that some states could have a very weak pion coupling, while decaying predominatly in multipion channels, as observed on the other hand in many high-lying measured states; as the sources of experimental information are mainly reactions with the pion as projectile or the single pion as final channel, photoproduced off the nucleon, it would not be surprising to find that baryon states with very small pion coupling were absent from those data sets. Other models[4, 5, 6, 7] based on various meson creation assumptions found similar results. An alternative explanation given for instance by the Quark Cluster Model [8] is on the contrary based on a reduction of the spatial degrees of freedom. From this introduction, it is quite clear that to test different model pictures it is necessary to increase the experimental information on the multipion production, but using an electromagnetic probe, to avoid the weak pion coupling situation that could affect hadron facilities, without forgetting that the experimental investigation is made difficult by the often large non-resonant background, as discussed in the following sections. Needless to say, Jefferson Lab with the CLAS detector[9, 10], with its high luminosity, acceptance and good momentum resolution, is the ideal place for performing such kind of studies: experiments[11] are currently conducted at Jefferson Lab with namely this goal.

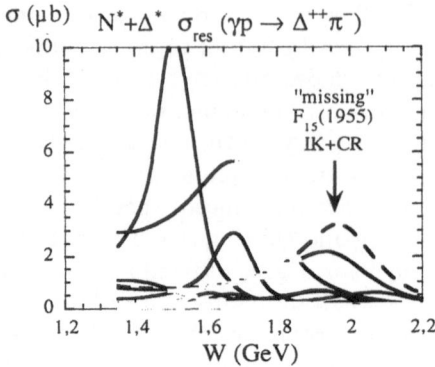

Figure 1. Breit-Wigner cross section in $\gamma p \to \Delta^{++}\pi^-$ reaction for known resonances[17] and for a particular missing state[3,6].

2 Phenomenology

Main contributions to the double pion production are isobar channels like $\Delta(1236)\pi$ and ρN[12]: $eN \to e'\Delta\pi \to e'N'\pi\pi$, $eN \to e'\rho N \to e'N'\pi\pi$. All isobar production channels can proceed through continuum processes, or through the excitation of baryon resonances with a cascade like $eN \to e'N^* \to e'\Delta\pi \to e'N'\pi\pi$. The double pion production data come mainly from bubble chamber experiments with real photons[12, 13], where data about various charge channels were collected. Another experiment at DESY[14] measured the electroproduction of $p\pi^+\pi^-$ off the proton with very poor statistics and large binning. Recent photoproduction measurements up to slightly above the $D_{13}(1520)$ have been performed at Mainz[15, 16], using the DAPHNE large angle detector, while data in a wider energy range have been collected in Bonn using the SAPHIR[16, 17] detector. In fig. 1, data[18] about known resonance excitations (full curves) together with predictions[3, 6] for missing states photo-excitation and subsequent decay (dashed curve) are used to give an estimate of the effect of a missing state: we can expect that the cross section should manifest some sensitivity.

3 Data analysis and interpretation

At new detector like CLAS, high luminosity, large geometrical acceptance, good efficiency for both charged and neutral particles are opening a new era of unprecedented accuracy in the measurement of exclusive reactions, allowing a more sophisticated data analysis with respect to the past. The main feature evident from all the two pion production data collected in the past experiments[12, 14, 15, 16] is the presence of the isobar "quasi-two-body" states $\Delta\pi$ and ρN. A typical approach for separating such different isobar contributions is to simply fit their bumps in the invariant masses, obtaining approximate cross sections. This was the data analysis adopted in most of the past experiments with electromagnetic probes[12, 14], being interested essentially in the gross features and being the data affected by high statistical uncertainty. However, the correct description of a three-body collision is based on five independent kinematical variables in the most general case[19] and moreover the isobar quasi-two-body production and subsequent decay involves all of them[20]. Investigations of double pion production from pion beams have been in fact conducted using isobar model approaches containing the partial wave expansion for each quasi-two-body process and fitting the data in the full kinematical space[21]. Any resonance analysis with the goal of extracting the baryon resonance decay branches in a quasi-two-body channel or the product of the e.m. transition matrix elements with the strong decay one (the "electrostrong properties" [22]), in a way as model independent as possible, needs such an isobar partial wave separation from the data as an input, similar to what done in previous analysis[23, 24]. Therefore in a preliminary simple study done on the $\Delta\pi$ channel, pseudo-events were generated using only the geometrical partial wave expansion[25], with no explicit dynamics, then refitted to retrieve the partial wave coefficients. The outcome was that even in this simple case the fitting code was not able to retrieve the large number of independent helicity amplitudes that arise with increasing angular momentum. Different solutions could be in principle pursued: one way is to add polarisation observables, in order to have a more constrained fit; a second possibility is to use orbital waves constrained by threshold behavior; a further possibility is to use simple model assumptions for the continuum and the resonances. In fact, it is important to consider that the N^* study in two-pion production is affected by strong non-resonant processes and therefore model-independent methods of analysis may be not effective. For all these reasons the choice in the Genova-Moscow collaboration[26] was to give up the requirement of minimal model dependence and use some partial wave content suggested from a model as input to the analysis, as described in the next section.

4   Our approach for the quasi-two-body channels

After the old work that followed the bubble chamber first experiments[12, 27], recent approaches to describe double pion photoproduction have been presented in a few papers[28, 29] based on a variety of tree-level diagrams and a few baryon resonances. The restricted number of resonances included however makes them strictly applicable only for W lower than 1.6 - 1.7 GeV; moreover non-resonant terms have been evaluated only at the photon point and not always corrected for unitarity absorption effects.

The Genova-Moscow approach to calculate cross sections is described in more detail in [30, 26]. I report here the general features. Following the data, we also use a coherent superposition of $\gamma_{r,v} p \to \pi^- \Delta^{++1}$ and $\gamma_{r,v} p \to \rho p^1$ quasi-two-body subchannels. All remaining processes are described in phase

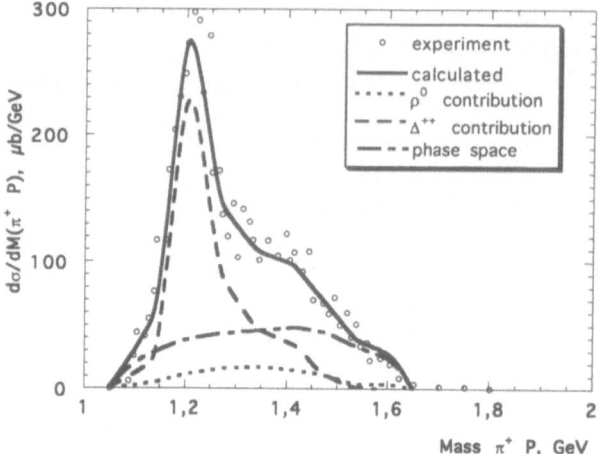

Figure 2. Invariant mass distribution for $p\pi^+$ pair from SAPHIR data[36] and the Genova-Moscow fit: the meaning of the different contributions is reported in the picture legend .

space approximation. The $\gamma_{r,v}p \rightarrow \pi^-\Delta^{++}$ reaction is described by a superposition of N^*, Δ^* excitation in s-channel and a minimal set of non-resonant processes obeying gauge invariance conditions, similar to what done in the previous literature[12, 27]. Non-resonant amplitudes are derived from an effective Lagrangian[27], as done for other meson production channels[31]. New features of this approach are: (1) the treatment of particle's off-shell behaviour through introduction of vertex functions that result from a combination of electromagnetic form factors and strong form factors specified via a cut-off parameter[30, 26]; data[32] have been used to determine part of them, while the remaining terms in the calculation were derived imposing gauge invariance[30, 26]; (2) the initial and final state absorption due to competitive channels follows [33], but the elastic hadronic amplitudes are reconstructed using resonant contributions taken from [34], plus a smooth background parametrised in the same fashion of [35]. It is important to stress here that "missing" resonances with strong two pion coupling should be introduced consistently in both the e.m. amplitudes as well as in the absorption: this evaluation is currently under way in our Genova-Moscow collaboration. Results from this calculation for $\gamma_{r,v}p \rightarrow \pi^-\Delta^{++}$ reaction are extensively reported in the contribution presented by V. Mokeev at this Workshop. Basically, main findings are that leaving the absorption as a free parameter it is possible to get a very good fit of the data, but resonance extraction becomes more uncertain; using the above parametrisation of initial and final state interaction, data are not completetely reproduced[26], but this discrepancy opens room to

---

[1]Indexes r,v stand for real and virtual photons respectively.

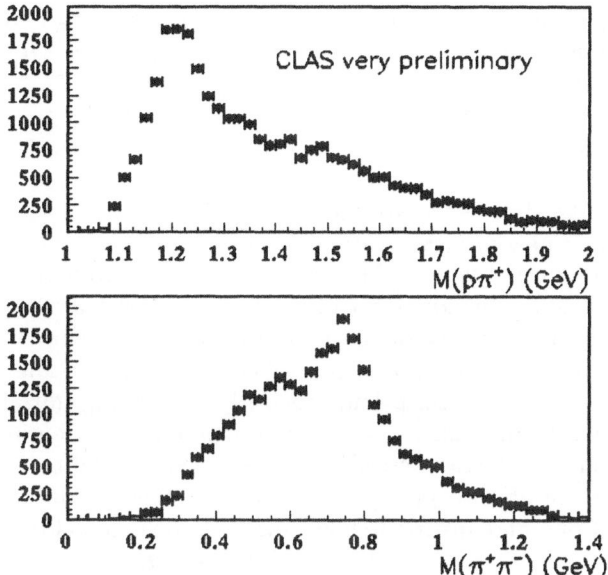

Figure 3. Invariant mass distributions for the $p\pi^+$ (top) and the $\pi^+\pi^-$ (bottom) pairs from CLAS for W>1.7 GeV. These are raw data without any energy or momentum transfer binning; acceptance correction were also not applied.

interesting effects like possible missing states contributions.

As the experiment does not measure isobar production directly of course, but only the two pion final state, in order to have a complete tool for the analysis next step was to merge together the $\Delta\pi$ and ρN production channels plus a phase space in a full three-body calculation. A new feature of the Genova-Moscow approach in this respect is the introduction of decay strong form factors for the Δ and ρ decay[36]. In fig. 2, I present an example of the results obtained fitting invariant mass distributions from recent photoproduction data[17]. The fit is pretty good, therefore providing a promising tool for a quite reliable extraction of the different isobar components in the reaction.

5 A quick look at the first CLAS data

In this talk I showed also some very preliminary data from CEBAF-CLAS experiment E-93-006. In fig. 3 a snapshot from a sample of CLAS data is reported, showing the invariant mass distributions for the $p\pi^+$ and the $\pi^+\pi^-$ pairs for W>1.7 GeV. The data were neither binned in W nor in Q^2 and they were not corrected for the detector acceptance; therefore they represent only the raw output from CLAS with the intent of giving essentially an idea of its capabilities. However, the contributions of the Δ^{++} and of the ρ^0 meson, respectively, are recognizable. The data already collected contain about an order of magnitude more events and nearly the same amount will be accumulated in other planned running periods, thereby allowing a quite large binning and investigation of details such as decay angular distributions with much higher accuracy than the past.

6 Summary and conclusions

New experiments like those currently conducted at Jefferson Laboratory are providing a wealth of new accurate data about exclusive electromagnetic reactions. Two pion production is one of the main subjects of investigation, being related to baryon resonances coupled to this channel. A specific approach for the isobar channels that appear in the two pion production has been developed in the framework of the Genova-Moscow collaboration, taking particular care about the $\Delta\pi$, channel description, especially concerning initial and final state absorption, gauge invariance, vertex functions and multiple resonances. ρ meson production was instead described through a simple diffraction ansatz. These calculations are able to give good account of existing data about differential cross sections and invariant mass distributions, therefore promising to allow a quite complete data analysis and a first evaluation of resonance contributions. In the first data from CLAS it is already possible to recognize the isobar formation with good statistics, opening the route to more detailed studies of the involved dynamics.

References

1. F.E. Close: *An Introduction to Quarks and Partons*, Academic Press, 1979

2. M.M. Giannini: Rep. Prog. Phys. **54**, 453 (1990)

3. R. Koniuk and N. Isgur: Phys. Rev. Lett.**44**, 845 (1980); Phys. Rev. **D21**, 1868 (1980)

4. R. Koniuk: Nucl. Phys. **B195** , 452 (1982)

5. F. Stancu and P. Stassart: Phys. Rev. **D47**, 2140 (1993)

6. S. Capstick, W. Roberts: Phys. Rev. **D49**, 4570 (1994)

7. S. Capstick: In these Proceedings

8. K.F. Liu and C.W. Wong: Phys. Rev. **D28**, 170 (1983)

9. J.J. Domingo: Proc. 5th Workshop on *Perspectives in Nuclear Physics at Intermediate Energies*, Trieste, Italy, May 6-10, 1991, eds. S. Boffi, C. Ciofi degli Atti and M. Giannini, (World Scientific, 1992) p. 260

10. V.D. Burkert: Nucl. Phys. **A623** (1997) 59c

11. J. Napolitano et al.: CEBAF Experiment E-93-033; M. Ripani, V. Burkert et al.: CEBAF Experiment E-93-006

12. Cambridge Bubble Chamber Group: Phys. Rev. **155**, 1477(1967); ABB-HHM Collaboration: Phys. Rev. **175**, 1669 (1968); D. Lüke and P. Söding: Springer Tracts in Mod. Phys. **59** (1971)

13. A. Piazza et al.: Lett. al Nuovo Cim., **12**, 403 (1970)

14. V. Eckart et al.: Nucl. Phys. **B55**, 45 (1973); P. Joos et al.: Phys. Lett. **B52**, 481 (1974); K. Wacker et al.: Nucl. Phys. **B144**, 269 (1978)

15. A. Braghieri et al.: Phys. Lett. **B363**, 46 (1995)

16. B. Krusche: In these Proceedings

17. F.J. Klein: Bonn University thesis BONN-IR-96-08

18. Particle Data Group: Phys. Rev. **D54**, 1 (1996)

19. Byckling and Kajantie, *Particle kinematics*, Wiley and Sons (1973)

20. H.M. Pilkuhn: *Relativistic Particle Physics*, Springer Verlag (1979)

21. D.M. Manley et al.: Phys. Rev. **D30**, 904 (1984)

22. N. Mukhopadhyay: In these Proceedings

23. F.A. Berens, A. Donnachie: Nucl. Phys. **B84**, 342 (1975)

24. R.A. Arndt, I.I. Strakovsky, R.L. Workman: Phys. Rev. **C53**, 430(1996)

25. M. Jacob and G.C. Wick: Ann. of Phys. **7**, 404 (1959)

26. V. Mokeev: In these Proceedings

27. A. Bartl, W. Majerotto, D. Schildknecht: Nuovo Cim. **12A**, 703 (1972)

28. J.A. Gomez-Tejedor and E. Oset: Nucl. Phys.**A571**, 667 (1994); J.A. Gomez-Tejedor and E. Oset: Nucl. Phys. **A600**, 413 (1996); E. Oset: In these Proceedings ;

29. L.Y. Murphy, J.M. Laget: DAPNIA-SPHN-96-10, 1996

30. M. Anghinolfi et al.: Preprint INP MSU 98-16/517, Moscow State University

31. M. Benmerrouche et al.: Phys. Rev. **D51**, 3237 (1995)

32. C.J. Bebek et al.: Phys. Rev. **D17**, 1693 (1978); R. Machleidt: Advances in Nucl. Phys. **19**, 189 (1979)

33. K. Gottfried, J.D. Jackson: Nuovo Cim. **34**, 736 (1964)

34. D.M. Manley, E.M. Salesky: Phys. Rev. **D45**, 4002 (1992)

35. S.A. Dytman, T.P. Vrana, T-S.H. Lee: nucl-th/9702033.

36. D.G. Cassel et at.: Phys. Rev. **D24**, 2787 (1981)

Few-Body Systems Suppl. 11, 292–297 (1999)

Few-
Body
Systems
© by Springer-Verlag 1999

Description of Two-Pion Production on Proton by Real and Virtual Photons in N^* Excitation Region.

V. Mokeev[1], M. Anghinolfi[2], M. Battaglieri[2], G. Fedotov[1], E. Golovach[1],
B. Ishkhanov[1], M. Osipenko[1], G. Ricco[2], M. Ripani[2], D. Rodionov[1],
V. Sapunenko[1,2], M. Taiuti[2], V. Zamiralov[1]

[1] Skobeltsyn Nuclear Physics Institute of Moscow State University
[2] Istituto Nazionale di Fisica Nucleare (sez. di Genova) e Universitá di Genova

Abstract. An approach has been developed to describe two-pion production on proton by real and virtual photons in N^* excitation region (W < 2 GeV). The mechanisms of two-pion production are considered as a superposition of tree diagrams containing s-channel N^* excitations and non-resonant mechanisms. Absorptive correction effects in the initial γp and final $\pi\Delta$ states are evaluated using the results of πN inelastic amplitudes analysis.

1 Introduction.

Investigation of exclusive reactions on nucleon with real and virtual photons is an important issue of research programs at modern electron facilities as JLAB [1]. From the data [2, 3, 4, 1, 5, 6] one can see that the cross section for this process strongly increases with photon energy and reaches its maximum at W > 1.8 GeV. An important role in two-pion production is played by nucleon resonances N^* having an appreciable branching ratio in this final state. Moreover quark model calculations based on the three constituent quark picture [7, 8, 9] predict more states than observed in experimental search [10]. These "missing" states could be visible in measurements of two-pion production by photons due to strong coupling with the $\Delta^{++}\pi^-$ and ρp decay channels. The N^* study in two-pion production faces serious obstacles due to the strong nonresonant processes and model-independent methods of analysis may be not effective. Therefore a model approach relating the measured two-pion cross section with N^* electromagnetic form factors appears to be the only way to extract information about N^* structure in two-pion production processes. Here, we report our approach aimed to describe $\gamma_{r,v} p \to \pi^+\pi^- p$ exclusive channel at N^* excitation energies (i.e., W < 2 GeV). Contributions of all N^* with masses below 2 GeV and well established decay amplitudes are taken into account as well as contributions of relevant non-resonant processes.

2   The approach.

A few approaches for two-pion production by real photons at $W < 1.6$ GeV were developed during recent years[11, 12]. Our calculation has some features in common with them, while it is different for some other aspects. According to observed Dalitz-distributions of invariant πp versus $\pi\pi$ masses we consider the mechanisms of $\gamma_{r,v} p \to \pi^+\pi^- p^1$ reaction as a coherent superposition of $\gamma_{r,v} p \to \pi^- \Delta^{++1}$ and $\gamma_{r,v} p \to \rho p^1$ quasi-two-body subchannels. All remaining processes are described in phase space approximation.

The $\gamma_{r,v} p \to \pi^- \Delta^{++}$ reaction is described by superposition of N^*, Δ^* excitation in s-channel and a minimal set of non-resonant processes obeying gauge invariance conditions and depicted in Fig. 1[3].

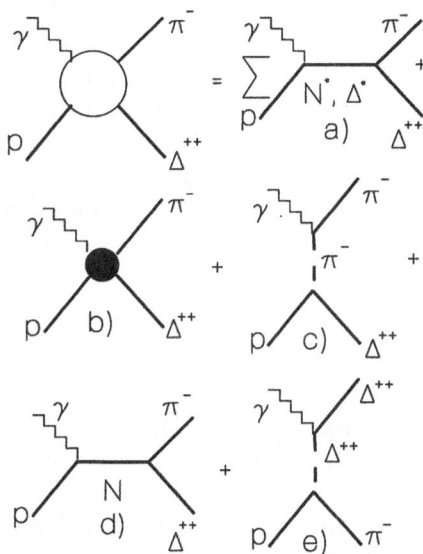

Figure 1. Tree-level diagrams for $\gamma_{r,v} p \to \pi^- \Delta^{++}$ subchannel.

The non-resonant amplitudes are evaluated in the effective Lagrangian formalism[13], which has been recently revived in calculations of other meson production channels[14]. A new feature of our model is the treatment of particle's off-shell behaviour in diagrams shown on Fig. 1b-e[15]. For vertices in diagram

---

[1]Indexes r,v stand for real and virtual photons respectively.

Fig. 1c we implement pion electromagnetic and $\pi N \Delta$ strong formfactor from experimental data analysis[16], while the vertex functions for other mechanisms (Fig. 1d-e) were evaluated from gauge invariance requirements[15]. Absorptive correction effects in the initial and final states are taken into account according to prescription[17]. However this prescription contains free parameters: the coupling constants with open inelastic channels in the initial and final states C_{in} and C_{out} and the interference phase ϕ between resonant and non-resonant amplitudes. We first determined these parameters from the fit of the measured $\gamma p \rightarrow \pi^- \Delta^{++}$ cross section. However, the fit does not provide a physical insight in the nature of absorption and could in principle distort the information about N^* electromagnetic form factors. Therefore we tried to relate absorptive correction coefficients in the initial and final states with $pp \rightarrow pp$ and $\pi^- \Delta^{++} \rightarrow \pi^- \Delta^{++}$ elastic scattering amplitudes. These amplitudes, in turn, were evaluated in a simple isobaric model as a superposition of all relevant N^* excitations and smooth background. Resonance amplitudes were evaluated under Breit-Wigner approximation, using results of analysis[18] for the decay amplitudes of N^*, Δ^* into pp and $\pi\Delta$ systems, while for the background we used [19]. After the evaluation of absorptive correction factors from $pp \rightarrow pp$ and $\pi^- \Delta^{++} \rightarrow \pi^- \Delta^{++}$ elastic scattering amplitudes our model does not have any free parameters to be determined from the fit of $\gamma_{r,v} p \rightarrow \pi^+ \pi^- p$ cross section. N^* excitations in the s-channel of $\gamma_{r,v} p \rightarrow \pi^- \Delta^{++}$ reaction (Fig. 1a)are described in Breit-Wigner approximation[15]. We relate N^* electromagnetic production amplitude with helicity couplings $A_{1/2}$, $A_{3/2}$, $C_{1/2}$, while strong decay amplitudes are evaluated using results of $\pi N \rightarrow \pi\pi N$ reaction analysis[18]. Non-resonant amplitudes of $\gamma_{r,v} p \rightarrow pp$ subchannel are obtained in the frame of diffractive production model[20] and their Q^2-dependences were calculated according to the approach[21].

3   Results and discussion.

The comparison between calculated pion angular distributions in $\gamma_r p \rightarrow \pi^- \Delta^{++}$ reaction and data[2] is presented in Fig. 2. As one can see neglecting absorptive corrections we reproduced the data at W < 1.6 GeV, while for W > 1.6 GeV calculated cross sections drastically exceed the data at high angles. The data fit accounting for the absorptive corrections with C_{in}, C_{out} and ϕ as free parameters allows to reproduce the data in W region from 1.5 to 2 GeV. Extracted parameters C_{in}, C_{out} appear to be maximum for W > 1.7 GeV (C_{in} = C_{out} = 1), while for W < 1.6 GeV the data can be reproduced without corrections at all ($C_{in} = C_{out} = 0$). Therefore we have a strong variation of C_{in}, C_{out} parameters in the small W region between 1.5 and 1.7 GeV, so, despite of the good fit obtained, this procedure appears questionable for the purpose of extracting N^* form factors. Results of calculations with absorptive corrections obtained from $\pi^- \Delta^{++} \rightarrow \pi^- \Delta^{++}$ and $pp \rightarrow pp$ elastic scattering are close to the data. However, for W < 1.6 GeV calculated values are a bit lower than experimental, while for W > 1.8 GeV and for pion emission angles above

80^0 calculated cross sections are of order of 1 - 2 mcbn, but the experimental data are compatible with zero. Possible reasons for the discrepancy at W < 1.6

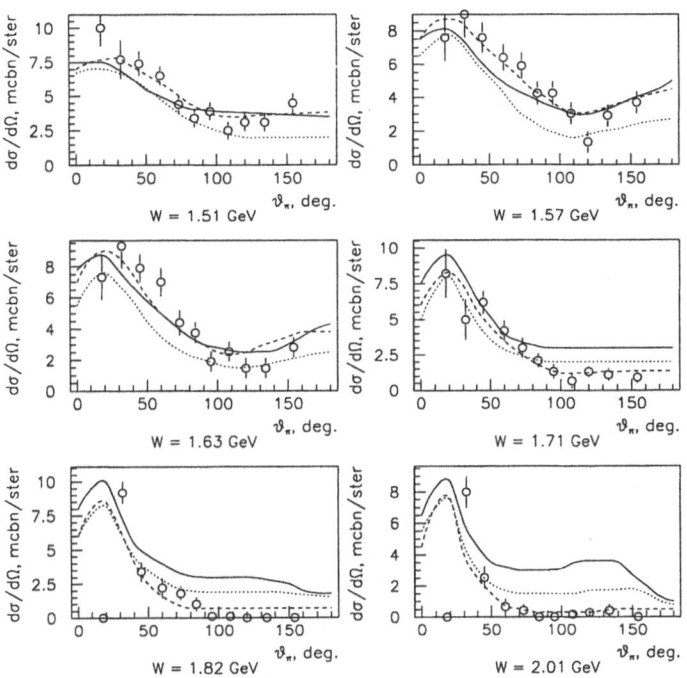

Figure 2. The comparison between calculated and measured [1] pion angular distributions in $\gamma_{r,v} p \to \pi^- \Delta^{++}$ reaction: solid lines are the calculation without absorptive corrections; dashed lines are the calculation with absorptive correction parameters C_{in}, C_{out} and interference phase determined from the data fit; dotted lines are the calculations with absorptive correction factors determined from πN amplitude analysis [18,19].

GeV could be approximations used for the u-channel process description[15], uncertainties in the $F_{\pi N\Delta}$ formfactor parameters as well as the uncertainties in strong couplings of the N^* excitations[10, 18]. On the other hand, quark models[8] predict two low-lying "missing" resonances with masses between 1.8 - 1.9 GeV. Their contribution could act on the angular distribution at W between 1.7 and 2 GeV and could be a reason for the discrepancy of calculated and measured cross sections in Fig. 2. New data of E-93-006 experiment which is now in progress in Hall B at TJNAF[22] are expected to clarify this aspect of "missing" N^* physics. The comparison between calculated and measured[6, 21] Q^2 dependencies of $\gamma_v p \to \pi^- \Delta^{++}$ and $\gamma_v p \to \rho p$ cross sections is presented in[15]. The model is satisfactory in reproducing existing data.

Using quasi-two-body amplitudes of $\gamma_{r,v}p \to \pi^-\Delta^{++}$ and $\gamma_{r,v}p \to \rho p$ reactions as an input we evaluated three-body amplitudes of $\gamma_{r,v}p \to \pi^+\pi^-p$ process and the cross section in complete kinematics. Then we compared evaluated invariant mass distributions $d\sigma/dM_{\pi^+\pi^-}$ and $d\sigma/dM_{\pi^+p}$ with data[4]. An example of this comparison at W = 1.78 GeV is shown in Fig. 3. The model

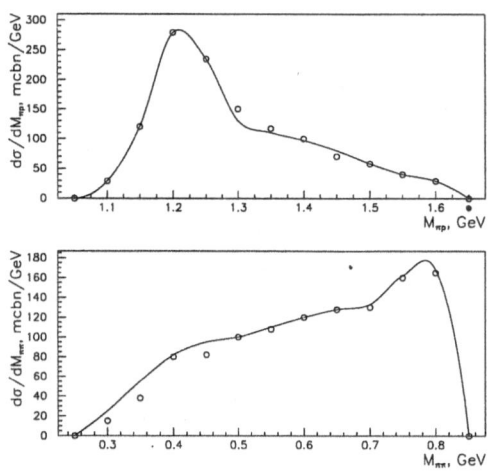

Figure 3. The comparison between calculated invariant mass distributions $d\sigma/dM_{\pi p}$, $d\sigma/dM_{\pi\pi}$ in reaction $\gamma p \to \pi^+\pi^-p$ and data[3]. W = 1.78 GeV.

developed reproduces pretty well $d\sigma/dM_{\pi^+\pi^-}$ and $d\sigma/dM_{\pi^+p}$ cross sections.

4 Summary.

We developed an approach to describe two-pion production by real and virtual photons at N^* excitation energies (1.3 < W < 2 GeV). All well established N^* with masses lower than 2 GeV as well as relevant background processes at tree level diagram level are included. Comparison of our model calculations with existing experimental data about cross sections of two main subchannels, $\gamma_{r,v}p \to \pi^-\Delta^{++}$ and $\gamma_{r,v}p \to \rho p$, is in reasonable agreement with data. The invariant mass distribution in $\gamma_{r,v}p \to \pi^+pi^-p$ reaction is also in good agreement with the experimental data. Using a simple parametrisation for initial and finals tate absorption, all model parameters are determined from other experiments, therefore the approach can be used for extraction of N^* electromagnetic form factors from measured $\gamma p \to \pi^+\pi^-p$ cross sections.

References

1. V.D. Burkert: Nucl. Phys. **A623**, 59c (1997)

2. Cambridge Bubble Chamber Group: Phys. Rev. **155**, 1477 (1967); ABBHM Collaboration: Phys. Rev. **175**, 1669 (1968)

3. D. Lüke, P. Söding: Springer Tracts in Mod. Phys. **59** (1971)

4. F. J. Klein, Bonn University thesis BONN-IR-96-08

5. I. Damman et al.: Nucl. Phys. **B54**, 355, (1973)

6. K. Wacker et al.: Nucl. Phys. **B144**, 269 (1978)

7. R. Koniuk, N. Isgur, Phys. Rev. Lett. **44**, 845 (1980); R. Koniuk, N. Isgur, Phys. Rev. **D21**, 1868 (1980); R. Koniuk, Nucl. Phys. **B195**, 452 (1982)

8. S. Capstick, W. Roberts, Phys. Rev. **D49**, 4570 (1994)

9. R. Bijker, F. Iachello, A. Levitan: Phys. Rev. **C54**, 1935 (1996)

10. Particle Data Group: Phys. Rev. **D54**, 1 (1996)

11. L. Y. Murphy, J. M. Laget: DAPNIA-SPHN-96-10

12. J. A. Gomez-Tejedor, E. Oset: Nucl. Phys. **A571**, 667 (1994); J. A. Gomez-Tejedor, E. Oset: Nucl. Phys. **A600**, 413 (1996)

13. A. Bartl, W. Majerotto, D. Schildknecht: Nuovo Cimento **12A**, 703 (1972)

14. M. Benmerrouche et al.: Phys. Rev. **D51**, 3237 (1995)

15. M. Anghinolfi et al.: Preprint INP MSU 98-16/517, Moscow State University

16. C. J. Bebek et al.: Phys. Rev. **D17**, 1693 (1978); R. Machleidt in Advances in Nuclear Physics, **v. 19** (1979)

17. K. Gottfried, J. D. Jackson: Nuovo Cimento **34**, 736 (1964)

18. D. M. Manley, E. M. Salesky: Phys. Rev. **D45**, 4002 (1992)

19. S. A. Dytman, T. P. Vrana, T-S. H. Lee, nucl-th/9702033

20. P. Söding: Phys. Lett. **19**, 702 (1966)

21. D. G. Cassel et al.: Phys. Rev. **D24**, 2787 (1981)

22. M. Ripani, V. Burkert et al.: CEBAF experiment E-93-006; M. Ripani: Nucl. Phys. **A623**, 110c (1997)

Few-Body Systems Suppl. 11, 298–307 (1999)

Few-
Body
Systems
© by Springer-Verlag 1999

Vector Meson Production

A. Donnachie

Department of Physics and Astronomy, University of Manchester, Manchester
M13 9PL, England

Abstract. The current state of analysis of e^+e^- annihilation below 2.0 GeV
and of the vector component of τ-decay is reviewed, and the evidence for and
against the presence of hybrid vectors discussed. The role of photoproduction as
a complementary approach is outlined, both at high and intermediate energies.
It is stressed that photoproduction does not simply mirror e^+e^- annihilation,
and provides a much richer source of meson spectroscopy.

1 Hunting the Vector Hybrids

Analysis [1, 2] of e^+e^- annihilation below 2.0 GeV and of the vector component
of τ decay, provides definite evidence for at least two isovector states in addition
to the ρ. The mass of the lower state (ρ'_1) is found to be in the range 1.42 to
1.50 GeV, and that of the upper state (ρ'_2) to be in the range 1.68 to 1.72
GeV. The non-strange isoscalar e^+e^- annihilation data are compatible with
the existence of corresponding ω' partners, with ω'_1 in the mass range 1.35 to
1.55 GeV and ω'_2 in the mass range 1.59 to 1.62 GeV [3, 4]. Additionally the
data on $e^+e^- \to K\bar{K}, K^*\bar{K}, \bar{K}^*K$ can be interpreted as evidence for a ϕ'_1 with
mass ~ 1.7 GeV [4].

A natural explanation is that these are the first radial, 2^3S_1, and first orbital, 1^3D_1, excitations of the ρ and ω and the first radial excitation of the ϕ
as the masses are close to those predicted by the quark model [5].

Despite the reasonable agreement of the observed masses with the quark
model predictions, it is now widely accepted that the data on the 4π channels
in e^+e^- annihilation and in τ decay do not appear to be compatible with those
expected for the radial and orbital excitations of the $q\bar{q}$ system. This statement
is of course model dependent as it assumes that we can predict the hadronic
decays of the vector $q\bar{q}$ excitations. The $^3P_0$ model [6, 7, 8, 9, 10] does appear
to allow this with some accuracy. A systematic study of known light $q\bar{q}$ decays
shows that a $^3P_0$-type amplitude dominates, and widths which are predicted to
be large or small are found respectively to be so. More quantitatively calculated
widths agree with data to within $25 - 40\%$.

The success of the $^3P_0$ model for well-known decays justifies its use in predicting other decays, and in particular those of the radial and orbital excitations of the ρ. In its simplest form the $^3P_0$ model contains only two parameters: an inverse length scale β which controls the meson form factors, and the pair creation strength γ. These are not known precisely, but are reasonably well constrained with $\beta \sim 0.4$ GeV, $\gamma = 0.39$ GeV. Assuming that their masses are respectively 1.45 and 1.80 GeV, the partial widths for ρ_S and ρ_D are:

Channel	$\pi\pi$	$\pi\omega$	$\rho\eta$	πh_1	πa_1	$\rho\rho$	$\rho\sigma$	other	total
ρ_S	68	115	18	1	3	10	1	80	295
ρ_D	27	23	13	104	105	6	0	137	415

In the above Table "other" includes $K\bar{K}$, $K^*\bar{K}$ + c.c. and 6π channels, and the σ is the broad S-wave $\pi\pi$ enhancement. Altogether 16 channels have been incorporated in the calculation.

It is not necessary to go through a detailed analysis to show that these $^3P_0$ model results exclude interpreting the e^+e^- and τ decay data in terms of the ρ_S and ρ_D if the model is strictly applied. The key is in the 4π decays. From the Table one can see that the 4π decays of the ρ_S are negligible, and so the ρ_S effectively makes no contribution to the 4π channel. In contrast the 4π decays of the ρ_D are large, and the two dominant ones, $h_1\pi$ and $a_1\pi$, are comparable. Now $h_1\pi$ contributes only to the $\pi^+\pi^-\pi^0\pi^0$ channel in e^+e^- annihilation, but $a_1\pi$ contributes to both this and to $\pi^+\pi^-\pi^+\pi^-$. An immediate consequence is that we would expect $\sigma(e^+e^- \rightarrow \pi^+\pi^-\pi^0\pi^0) > \sigma(e^+e^- \rightarrow \pi^+\pi^-\pi^+\pi^-)$, after subtraction of the $\omega\pi$ cross section from the total $\pi^+\pi^-\pi^0\pi^0$. This contradicts observation. Despite considerable uncertainty in the $\pi^+\pi^-\pi^0\pi^0$ cross section, it is undeniably appreciably smaller than the $\pi^+\pi^-\pi^+\pi^-$ cross section over most of the relevant energy range.

One explanation of this has been to suggest that the $q\bar{q}$ vector states are mixed with a hybrid vector [11, 12] as this decays predominantly to $a_1\pi$ i.e. that e^+e^- annihilation and the corresponding τ decays should be explained in terms of some suitable combination of ρ, ρ_S, ρ_D and hybrid ρ_H. Evidence for the excitation of gluonic degrees of freedom has emerged in other processes. There are two independent indications of an isovector $J^{PC} = 1^{-+}$ exotic resonance $\hat{\rho}(1600)$ in $\pi^- N \rightarrow \pi^+\pi^-\pi^- N$, specifically in the $\rho^0\pi^-$ channel. The E852 collaboration [13] quote a mass of 1593 ± 8 MeV and width of 168 ± 20 MeV, which are consistent with the preliminary claim of the VES collaboration [14] of a resonance at 1620 ± 20 MeV with a width of 240 ± 50 MeV. There is also evidence for this state in the $\eta'\pi$ channel [14, 15]. It has been argued that the $\rho\pi$, $\eta'\pi$ and $\eta\pi$ couplings of this state support the hypothesis that it is a hybrid meson, although other interpretations cannot be eliminated entirely [16]. There is also a possible hybrid state with $J^{PC} = 0^{-+}$ in the 1.8 GeV mass region [17], as both its mass and unusual decay pattern match this interpretation [7, 10, 18]. The interpretation of a peak in the $\eta\pi$ mass spectrum at 1.4 GeV with $J^{PC} = 1^{-+}$ as a resonance [19] has been challenged [20]. However there is additional evidence for the same resonance in $p\bar{p}$ annihilation from the

Crystal Barrel collaboration [21], although it is not seen as a peak in the $\eta\pi$ mass distribution but is deduced from interference in the Dalitz plot. It is possible that the doorway mechanism used in [20] could generate sufficient phase variation without a resonance, but that conclusion is very model dependent.

The relevance of these results for the $J^{PC} = 1^{--}$ channel is that potential hybrid states are occuring in precisely the relevant mass region. Unmixed hybrids are expected around 1.7 to 1.9 GeV [10, 18], but spin-dependent forces may lower the mass of the hybrid ρ and ω, which are spin $S = 0$ in contrast to the conventional $q\bar{q}$ components which are $S = 1$, and cause mixing between hybrid and conventional quarkonia. For example,

$$|V\rangle = cos\phi\{cos\theta|2^3S_1\rangle + sin\theta|1^3D_1\rangle\} + sin\phi|V_H\rangle$$

To explain the predominance of the $\pi^+\pi^-\pi^+\pi^-$ channel in this model it is necessary to take $\phi \sim 0$, so that ρ'_1 is given by

$$|\rho'_1\rangle \sim cos\theta|\rho_S\rangle + sin\theta|\rho_H\rangle$$

with mixing close to maximal i.e. $\theta \sim \pi/4$

The absence of a significant 1^3D_1 component is also apparent in the isoscalar states, as neither ω'_1 nor ω'_2 has a large $b_1\pi$ $(\omega\pi\pi)$ component. In the $^3P_0$ model the $b_1\pi$ decay of the 1^3D_1 is large, about 300 MeV. Experimentally, it appears that ω'_1 has $\Gamma_{b_1\pi} \sim 0$ MeV and ω'_2 has $\Gamma_{b_1\pi} \sim 30$ MeV. As for the ρ_H the ω_H has only one significant decay mode, in this case $\rho\pi$.

The e^+e^- widths of the ω'_1 and ω'_2 are almost the same, which is very indicative of strong mixing. Thus it is again consistent to take

$$|\omega'_1\rangle \sim cos\theta|\omega_S\rangle + sin\theta|\omega_D\rangle$$

with $\theta \sim \pi/4$. The $\rho\pi$ decays are also consistent with $\omega_S - \omega_H$ mixing.

The above scenario is very tempting, but a possible alternative is to invoke an inherent uncertainty in the $^3P_0$ model when applied to the decays of radial excitations to the ground state plus an $s-$wave $\pi\pi$ pair. In the $^3P_0$ model the decay of the ρ_S to $\rho + (\pi\pi)_S$ is strongly suppressed by a cancellation between two terms, one of which is strongly dependent on the model parameters. If this decay of the ρ_S could be sufficiently enhanced within the structure of the model then the 4π problem could possibly be resolved. (Note that the $\rho+(\pi\pi)_S$ decay of the ρ_D is strictly forbidden in the $^3P_0$ model.)

Many radial excitations are known to decay preferentially to the ground state, or a lower radial excitation, plus $(\pi\pi)_S$. The most obvious ones occur in higher quarkonia. The branching fractions of these decays are: $\psi(2S) \to \psi(1S)$, $50.8 \pm 3.7\%$; $\Upsilon(2S) \to \Upsilon(1S)$, $27.3 \pm 1.4\%$; $\Upsilon(3S) \to \Upsilon(1S)$, $6.5 \pm 0.4\%$, $\Upsilon(3S) \to \Upsilon(2S)$, $4.8 \pm 0.7\%$. The same phenomenon occurs in light quarkonia: $\eta'(985) \to \eta(\pi\pi)_S$; $\eta'(1295) \to \eta(\pi\pi)_S$; $\pi(1300) \to \pi(\pi\pi)_S$. Conceptually these decays occur by "vacuum excitation" i.e. the energy released in the decay can excite an isoscalar $S-$wave pion pair from the vacuum. It can be visualized as occurring via colour-singlet multiple-gluon exchange, although quantitatively the details in the heavy and light quarkonia sectors may be very different.

In the $^3P_0$ model, with standard wave-function parameters, the decay radial to ground state plus $(\pi\pi)_S$ is small, a few MeV at most. This is true for ρ_S, ω_S, $\eta'(985)$, $\eta'(1295)$, $\pi(1300)$. It is caused by a node in the $(\pi\pi)_S$ wave function, with the consequence that the decay is *very* sensitive to the parameter β. It is easy to generate widths of 50 to 100 MeV for the decays $\rho_S \to \rho(\pi\pi)_S$ and $\omega_S \to \omega(\pi\pi)_S$.

Thus there are two possible mechanisms for generating a large 4π width for the ρ_S. Further, for the decay $\rho_S \to \rho(\pi\pi)_S$, $\sigma(\pi^+\pi^-\pi^+\pi^-) = 2\sigma(\pi^+\pi^-\pi^0\pi^0)$, thus solving the 4π problem without recourse to an exotic state. However it should be recalled that the decay $\omega_S \to \omega(\pi\pi)_S$ does not appear to be strong, and if this is confirmed it will put limits on the corresponding isovector decays.

There are some additional problems requiring clarification. Firstly, the LASS data [22] on $K^-p \to (\pi^+\pi^-)\Lambda$ imply a $J^{PC} = 1^{--}$ $\pi\pi$ resonance at 1.266 ± 0.014 GeV with a width of 0.166 ± 0.035 GeV. Secondly there are DESY data [23] on $\gamma p \to (e^+e^-)p$ which show interference effects between the hadronic and QED amplitudes implying two $J^{PC} = 1^{--}$ resonances at 1097^{+16}_{-19} MeV and 1266 ± 5 MeV, with widths of 31^{+24}_{-20} and 110 ± 35 MeV respectively. Finally there are CERN Omega data [24] on $\gamma p \to (\pi^+\pi^-\pi^0)p$ which show evidence for structure in $J^{PC} = 1^{--}$ just above the ϕ, at ~ 1100 MeV. The coincidence of masses in these very disparate experiments is remarkable.

2 Diffractive Photoproduction

Naively, diffractive photoproduction simply mirrors e^+e^- annihilation vià Vector Meson Dominance:

$$|\gamma\rangle = \frac{e}{\gamma_\rho}|\rho\rangle + \frac{e}{\gamma_\omega}|\omega\rangle + \frac{e}{\gamma_\phi}|\phi\rangle + \dots..$$

where the e.m. couplings γ_V are given by the e.m. width of the vector V:

$$\Gamma_{e^+e^-} = \frac{\alpha^2}{3}\frac{4\pi}{\gamma_V^2}m_V$$

For example, ρ^0 photoproduction at $t = 0$ is given by

$$\frac{d\sigma}{dt}(t=0) = \alpha\frac{4\pi}{\gamma_\rho^2}\frac{d\sigma}{dt}(\rho^0 p \to \rho^0 p; t=0)$$

Quark additivity allows us to write

$$A(\rho^0 p \to \rho^0 p) \sim \frac{1}{2}\{A(\pi^+ p \to \pi^+ p) + A(\pi^- p \to \pi^- p)\}$$

which allows a parameter-free prediction of ρ photoproduction at all t. The energy-dependence and t-dependence are both correct, but the normalisation is too high, requiring a multiplicative factor of 0.84 [25]. However there are intrinsic uncertainties in VMD due to off-shell or wave-function effects, and the additive quark model is only accurate to about 5%. Thus a factor of 0.84 between the naive prediction and the data is well within acceptable limits.

There is no visible difference between ρ^0 photoproduction and ω^0 photoproduction at high energies, apart from the magnitude of the cross section: $\sigma(\gamma p \to \omega p) = \frac{1}{9}\sigma(\gamma p \to \rho p)$, and VMD works equally well. There is a difference at low energies, where ω photoproduction is dominated by pion exchange.

However the picture falls apart for higher-mass vector mesons. For ϕ photoproduction the VMD prediction is too high by a factor of 2, although the energy dependence is correct. For J/ψ photoproduction VMD is completely wrong.

The ϕ and J/ψ are simple examples of the failure of VMD. There are many other examples where diffractive photoproduction does *not* mirror e^+e^- annihilation.

2.1 The ρ line-shape

In diffractive photoproduction the ρ line shape varies with energy, momentum transfer, and target. This is understood in terms of the Drell-Söding mechanism [26] in which the initial photon dissociates into a pion pair, one of wh. ᵇ scatters diffractively on the target. This provides an additional term which contributes on its own and through interference with the direct production, and completely explains the observed variation of line shape.

2.2 $\rho' \to \pi\pi$

In diffractive photoproduction the relative ρ, ρ' phases are different from those in e^+e^- annihilation. In the latter the interference is destructive, producing a sharp dip in the cross section. In the former it is constructive, producing a peak in the $\pi\pi$ mass distribution. This phase change is seen over a large energy range: from SLAC, through CERN to FNAL. This phase change can be explained [27] by the colour-dipole character of the photoproduction cross section which favours the large r part of the vector meson wave function rather than the short range part which is relevant for e^+e^- annihilation. The radial wave function has a node, the inner part being negative relative to the ground state and the outer part positive.

As Q^2 is increased the photon wave function shrinks and the inner negative part of the radial vector meson wave function becomes increasingly important. This has two effects. Firstly the longitudinal cross section for the ρ' has a zero at small Q^2. This lies between 2 and 3 GeV^2 in the explicit model [27]. The precise value is obviously wave-function dependent, but the effect itself is of greater generality. Secondly as Q^2 increases beyond this point the ratio $\sigma_{\rho'}/\sigma_\rho$ will increase, in qualitative agreement with preliminary HERA data.

An intriguing feature of the model is that it is necessary to assume that the physical ρ_1' and ρ_2' states are mixed states of an active $2S$ excitation and some inert state which cannot be excited diffractively in lowest-order QCD. The latter can be inferred as the hybrid hypothesised from e^+e^- annihilation.

2.3 $\rho' \to \pi^+\pi^-\pi^+\pi^-$ line shape

Comparing the mass distributions $M^2\sigma(e^+e^-)$, $M^{-1}d\sigma(\gamma p)/dM$, which are both Breit-Wigner like, shows that the photoproduction data have a significant enhancement at lower 4π mass. This is not necessarily $J^P = 1^-$ so interpretations include the following:

(i) a coherent non-resonant contribution, as in $\pi\pi$, with $a_1\pi$ as the participating system in this case.

(ii) an incoherent contribution from a state with $J^P \neq 1^-$. e.g. the $\pi^+\pi^-\pi^+\pi^-$ decay of $b_1(1235)$.

(iii) something more exotic e.g. a 4π decay of the LASS state.

2.4 Diffractive production of states with $J^P \neq 1^-$

The most obvious example of this in photoproduction is provided by the diffractive photoproduction of $b_1(1235)$, with $J^P = 1^+$. The cross section is large, $\sim 1\mu b$, i.e. comparable to the diffractive photoproduction of the ρ' states. However the best-known example is provided by $\pi p \to (\rho\pi)p$. The $\rho\pi$ system is produced by the Deck mechanism, which is analogous to the Drell-Söding mechanism. The initial π dissociates into $\rho\pi$, either of which can scatter diffractively. This produces a resonance-like enhancement at ~ 1.2 GeV, and at one time caused great confusion over the interpretation of the $a_1(1285)$ as a resonance.

At sufficiently high energy, where the πp (ρp) scattering is dominated by pomeron exchange, the Deck mechanism leads to the Gribov-Morrison rule [28]: the final $\rho\pi$ system is in the J^P sequence $0^-, 1^+, 2^-, 3^+, \ldots\ldots$ At lower energies, where other exchanges are possible, other J^P final states are allowed [29]. A similar effect will occur at any strongly coupled threshold.

The J^P sequence expected in photoproduction will include 1^+ ($b_1(1235)$: seen at SLAC [30], CERN [31]); 2^+ ($f_2(1270)$:seen at FNAL [32]); 3^- ($\rho_3(1690)$: seen at CERN [33]).

2.5 Major anomalies between photoproduction and e^+e^- annihilation

Significant anomalies in photoproduction are seen in the high-energy photoproduction of $\rho\eta$; $\pi^+\pi^-\pi^+\pi^-$; $\rho\pi$; $\omega\eta$; $K^*\bar{K}$ + c.c.; $\rho\rho\pi$; $\omega\pi\pi$ [34]. In each case the photoproduction cross section is much larger than or has a different shape from that expected from e^+e^- annihilation. The most notable is the $J^P = 1^-$ $\rho\pi$ which is a full order-of-magnitude larger than expected, with $\langle d\sigma/dM \rangle \sim 1\mu b$ GeV^{-1} i.e. an isoscalar cross section of a magnitude more generally expected for an isovector cross section.

3 Medium Energy Photoproduction

As the energy decreases, contributions from π, ρ, ω, f_2, a_2,.....exchange become increasingly important and the allowed J^{PC} for the final mesonic system becomes increasingly diverse. The disadvantage is increased complexity. The advantages include:

(i) the possibility of obtaining phase information through interference with known states.

(ii) the possibility of direct production of orbital excitations, unlike e^+e^- annihilation.

(iii) the possibility of direct production of exotic mesons.

(iv) the possibility to produce charged as well as neutral states.

3.1 Example: the vector hybrids

The intrinsic coupling of the isovector $J^{PC} = 1^{--}$ hybrid, V_H, to ρf_2 is comparable to its coupling to πa_1, the dominant decay channel. The decay to ρf_2 is not seen simply because the hybrid has insufficient mass for this channel. However there is no relative suppression of one or other of the ρ, f_2 is off-shell, which is the case for direct photoproduction of the vector hybrid by f_2 exchange. VMD works well for the ρ so the photon coupling is strong. The $\rho - \rho_H - f_2$ coupling is strong, as is the $f_2 - N$ coupling. This particular mechanism is not so strong for the ρ_S, so the ρ_H, ρ_S mix in photoproduction should be different from that in e^+e^- annihilation. Similarly, the isoscalar $J^{PC} = 1^{--}$ hybrid couples strongly to ωf_2. More interestingly it couples even more strongly to ρa_2 and to $\rho\pi$, its dominant decay mode. So there is the expectation that the isoscalar hybrid will be produced by an isovector photon, giving a greatly enhanced cross section.

3.2 Example: the exotic $J^{PC} = 1^{-+}$

There are two independent indications of an isovector $J^{PC} = 1^{-+}$ exotic $\hat{\rho}(1600)$ in $\pi^- n \to \pi^+\pi^-\pi^- N$, observed in the decay mode $\rho^0\pi^-$: the E852 experiment [13] which quotes a mass of 1593 ± 8 MeV and width of 168 ± 20 MeV; and the VES experiment [14] which quotes a mass of 1620 ± 20 MeV and width of 240 ± 50 MeV. The $\rho\pi$ decay makes it an obvious candidate for photoproduction by pion exchange. The cross section has been estimated [35] and shown to be well within the reach of CEBAF.

3.3 Example: the $\pi(1600)$ and $\pi(1800)$

The $\pi(1600)$ and $\pi(1800)$ are reported by the VES experiment [17], seen in the $\rho\pi$ channel. They are candidates for mixed hybrid-$q\bar{q}$ states: hybrid because otherwise there would be an excess of 0^{-+} $q\bar{q}$ states; $q\bar{q}$ because $\rho\pi$ decay is suppressed for a 0^{-+} hybrid. Once again the $\rho\pi$ decay implies medium energy photoproduction as an appropriate mechanism for further investigation of their properties.

3.4 $\gamma N \to (\rho\pi)N$: an interesting study

The $\rho\pi$ channel is readily identifiable, and provides a diversity of physics topics in a mass range readily accessible to CEBAF.

(i) The 1^{--} at 1100 MeV: is it real?

(ii) Can the 1^{--} ω_S, ω_D and ω_H be cleanly identified?

(iii) Is the $\eta\pi$ 1^{-+} peak seen by the E852 experiment an exotic resonance or an artefact of the production dynamics?

(iv) Are the $\pi(1600)$ and $\pi(1800)$ really hybrid-$q\bar{q}$ mixed states?

3.5 The Vector States

Despite all the effort that has gone into analysing e^+e^- annihilation, τ-decay, diffractive photoproduction, $p\bar{p}$ annihilation etc., there is still considerable uncertainty about the details of the vector states.

3.5.1 The isovector states

Although a considerable quantity of data exists on these, there remain many questions to answer.

(i) The $\rho(1450)$ is well established in $\pi\pi$, $\omega\pi$ and 4π. Crystal Barrel report a strong $\rho\epsilon$ mode for the latter, photoproduction data appear to favour an $a_1\pi$ component and partial-wave analysis in e^+e^- annihilation gives approximately equal weight to both.

(ii) The data require a second state (or states) at $\sim$ 1700 MeV. Its/their nature and decay modes are unclear.

(ii) There are unexplained phenomena near the $p\bar{p}$ threshold, seen in the timelike nucleon form-factors, $e^+e^- \to 6\pi$ and $\gamma p \to 6\pi$.

(iv) The 6π spectra from e^+e^- annihilation and diffractive photoproduction can be interpreted as providing evidence for a $\rho'(2150)$.

3.5.2 The isoscalar states

Analysis of these states suffers from poor quality data, particularly in $\omega\pi\pi$.

(i) The $\omega'(1420)$ is established only in its dominant $\rho\pi$ decay mode. Other decay modes are uncertain.

(ii) The data at higher masses are compatible with the presence of another state, but the overall data situation is so poor that "anything goes".

(iii) The comparatively weak $\omega\pi\pi$ ($b_1\pi$) channel poses problems for $q\bar{q}$ states and encourages the hybrid story. However because of the data situation, this may be misleading.

3.5.3 The $s\bar{s}$ channels

Definitive analysis is difficult because of limited data and because of contamination of the open strangeness channels from ρ' and ω' decays.

(i) The $\phi'(1680)$ is established in its dominant $K^*\bar{K}$ + c.c. decay and claimed in other modes. However the amount of mixing with isoscalar $n\bar{n}$ states is unknown.

(ii) The $K\bar{K}$, $K^*\bar{K}$ + c.c., $K^*\bar{K}^*$ decays of the isovector and isoscalar $n\bar{n}$ states are not specified, other than some evidence for $\rho'(1700) \rightarrow K\bar{K}$.

3.5.4 Other

There are several possible states which come under this heading.

(i) The LASS $\rho'(1270)$ needs confirmation.

(ii) The states reported in [23] at ~ 1100 and ~ 1300 MeV need confirmation.

(iii) A $J^{PC} = 1^{--}$ $\phi\pi$ state, the $C(1480)$, has been claimed [36]. However it has been shown that this cannot be a decay of the $\rho'(1450)$ [37], it may even be a threshold effect, and it is not confirmed in either e^+e^- annihilation or $p\bar{p}$ experiments. So does it exist, and if so then what is it?

References

1. A. Donnachie and A.B. Clegg: Phys. Rev. **D51**, 4979 (1995)

2. N.N. Achasov and A.A. Kozhevnikov: Phys. Rev. **D55**, 2663 (1997)

3. A. Donnachie and A.B. Clegg: Z. Phys. **C42**, 663 (1989)

4. N.N. Achasov and A.A. Kozhevnikov: Phys. Rev. **D57**, 4334 (1998)

5. S. Godfrey and N. Isgur: Phys. Rev. **D32**, 189 (1985)

6. G. Busetto and L. Oliver: Z. Phys. **C20**, 247 (1983); P. Geiger and E.S. Swanson: Phys. Rev. **D50**, 6855 (1994); H.G. Blundell and S. Godfrey: Phys. Rev. **D53**, 3700 (1996)

7. R. Kokoski and N. Isgur: Phys. Rev. **D35**, 907 (1987)

8. A. LeYaouanc, L. Oliver, O. Pene and J. Raynal: Phys. Rev. **D8**, 2223 (1973)

9. E.S. Ackleh, T. Barnes and E.S. Swanson: Phys. Rev. **D54**, 6811 (1996)

10. T. Barnes, F.E. Close, P.R. Page, and E.S. Swanson: Phys. Rev. **D55**, 4157 (1997)

11. A. Donnachie and Yu.S. Kalashnikova: Z. Phys. **C59**, 621 (1993)

12. F.E. Close and P.R. Page: Phys. Rev. **D56**, 1584 (1997)

13. N.M. Cason (E852 Collaboration): Proc. HADRON'97 (Brookhaven, August 1997) in press

14. Yu.P. Gouz (VES Collaboration): Proc. XXVI ICHEP (DALLAS, 1992),ed. J.R. Sanford, p.572

15. D. Ryabchikov (E852 Collaboration): Proc. HADRON'97 (Brookhaven, August 1997) in press

16. P.R. Page: Phys. Lett. **B415**, 205 (1997)

17. A.M. Zaitsev (VES Collaboration): Proc. XXVII ICHEP (Glasgow, 1994), ed. P. Bussey and I. Knowles, p.1409

18. N. Isgur and J.E. Paton: Phys. Rev. **D31**, 2910 (1985)

19. D.R. Thompson et al. (E852 Collaboration): Phys. Rev.Lett. **79**, 1630 (1997)

20. A. Donnachie and P.R. Page: preprint JLAB-THY-98-20

21. A. Abele et al. (Crystal barrel Collaboration): Phys. Lett. in press

22. D. Aston et al. (LASS Collaboration): SLAC-PUB-5657(1991)

23. S. Bartalucci et al.: Nuovo Cimento **49A**, 207 (1979)

24. A. Donnachie, Yu.S. Kalashnikova and A.B. Clegg: Z. Phys. **C59**, 621 (1993)

25. A. Donnachie and P.V. Landshoff: Phys. Lett. **B348**, 213 (1995)

26. S.D. Drell: Rev. Mod. Phys. **33**, 458 (1961); P. Söding: Phys. Lett. **19**, 702 (1966)

27. G. Kulzinger, H.G. Dosch and H.J. Pirner: hep-ph/9806352

28. V. Gribov: Sov.J.Nucl. Phys. **59**, 138 (1967); M.R.O. Morrison: Phys. Rev. **165**, 1699 (1968)

29. G. Ascoli et al.: Phys. Rev. **D8**, 3894 (1973); *ibid* **D9**, 1963 (1974)

30. J. Brau et al.: Phys. Rev. **D37**, 2379 (1988)

31. M. Atkinson et al.: Nucl. Phys. **B243**, 1 (1984)

32. P. Lebrun (E687 Collaboration): Proc. HADRON'97 (Brookhaven, August 1997) in press

33. D. Aston et al.: Phys. Lett. **92B**, 215 (1980); M. Atkinson et al.: Z. Phys. **C30**, 531 (1986)

34. A. Donnachie and A.B.Clegg: Z. Phys. **C48**, 341 (1990)

35. A. Afanasev and P.R. Page: Phys. Rev. **D57**, 6771 (1998)

36. S.I. Bityukov et al.: Phys. Lett. **188B**, 383 (1987)

37. A.B. Clegg and A. Donnachie: Z. Phys. **C40**, 313 (1988)

Few-Body Systems Suppl. 11, 308–315 (1999)

Few-
Body
Systems
© by Springer-Verlag 1999

Spin Information from Vector-Meson Decay in Photoproduction

W.M. Kloet[1]*, Wen-Tai Chiang[2]† Frank Tabakin[2] ** ††

[1] Department of Physics & Astronomy, Rutgers University, Piscataway, New Jersey 08855-0849
[2] Department of Physics & Astronomy, University of Pittsburgh, Pittsburgh, Pennsylvania 15260

Abstract. For the photoproduction of vector mesons, physically meaningful single and double spin observables are defined in the γN center-of-mass system, which we prefer over extracting density matrix elements in the vector meson rest frame, since angle and energy evolution of spin observables can yield insights into special dynamics. Assuming negligible background amplitudes, we show that: (1) measuring the two pseudoscalar meson decay of a photoproduced ρ or ϕ does *not* determine the vector meson's vector polarization, but only its tensor polarization and (2) vector meson decay into lepton pairs is also insensitive to the vector meson's vector polarization, unless one measures the spin of one of the leptons. Similar conclusions hold for all double spin observables which involve observation of two-meson vector meson decay. To access the vector meson's vector polarization, one therefore needs to either measure the spin of the decay leptons, make an analysis of background interference effects or relate the vector meson's vector polarization to other accessible spin observables.

1 Introduction

Vector meson decay products in vector meson electromagnetic production can be measured using new high-flux, continuous beams of polarized electrons and photons, polarized targets, and recoil polarimeters. Most experimental proposals follow the earlier analysis in which the vector meson's decay is described in its rest frame. This talk is based on our recent paper [1] on the photoproduction of vector mesons $\gamma + N \rightarrow V + N'$ (V denotes either a ρ or a ϕ meson).

---

*Email address: kloet@physics.rutgers.edu
†Email address: chiang@tabakin.phyast.pitt.edu
**Email address: frankt@tabakin.phyast.pitt.edu
††Talk presented by Frank Tabakin

The ρ decays predominantly to two pions, the ϕ to two kaons. The amplitude for this process yields a density matrix which describes the spin state of the produced vector meson. The rest frame of the vector meson differs from the production frame and since standard spin observables are defined most naturally in the γN CM frame, it is necessary to treat the decay in that same CM frame. In much of the literature, the decay and density matrix of the vector meson are described in the rest frame of the vector meson and standard spin observables are not invoked. This difference in approach is partly a question of differing motivation. Extraction of the vector meson's density matrix in its rest frame, using either the Gottfried-Jackson, Adair or Helicity axes [2], can provide a framework for determining the character of t-channel exchange mechanisms. On the other hand, to understand the angular and energy evolution of standard spin observables, which are driven by specific resonances and other dynamical mechanisms [3, 4] , requires that spin observables be defined in a consistent manner, especially for cases involving measuring the vector meson decay along with having polarized photons and/or nucleons (e.g. for double or triple spin observables). Our goal is to define single and double spin observables in a manner consistent with the treatment of the decay process. Therefore, we describe all spin observables and decays in the same overall γN CM frame.

In this talk, the following points are stressed:

- For the photoproduction of vector mesons, all single and double spin observables involving vector meson two-body decays should be defined consistently in the γN center of mass frame.

- These definitions yield a procedure for extracting physically meaningful single and double spin observables that are subject to rules concerning their angle and energy evolution.

- As part of this analysis, we assert [1] that measuring the two-meson decay of a photoproduced ρ or ϕ does not determine the vector meson's vector polarization, but only its tensor polarization.

- The vector meson decay into lepton pairs is also insensitive to the vector meson's vector polarization, unless one measures the spin of one of the leptons.

- Similar results are found for all double spin observables which involve observation of vector meson decay. Spin correlations involving vector polarization of vector mesons cannot be extracted directly from the angular distribution of the decay products.

To access the vector meson's vector polarization, one needs to either measure the spin of any decay lepton or make an analysis of background interference effects or relate the vector meson's vector polarization to other accessible spin observables [5].

2 Single spin observables

The reaction $\gamma + N \rightarrow \rho + N'$ is described by twelve independent, complex helicity amplitudes; thus one needs essentially [1] twenty-three independent observables at each energy and angle to ascertain the amplitude $\mathcal{T}$. The eight independent single spin observables are the cross-section $\sigma(\theta)$, the photon beam asymmetry Σ_x^γ, the target asymmetry T_y^N, the recoil polarization $P_y^{N'}$, the vector meson's vector polarization P_y^V, and the vector meson's tensor polarization T_{20}^V, T_{21}^V, and T_{22}^V. Parity makes the following observables vanish $\Sigma_y^\gamma, \Sigma_z^\gamma, T_x^N, T_z^N, P_{x'}^{N'}, P_{z'}^{N'}, P_{x'}^V, P_{z'}^V$ and also requires that $T_{2-2}^V = T_{22}^V$ and $T_{2-1}^V = -T_{21}^V$.

Thus, the 3×3 spin density matrix for the photoproduced vector meson is described by either the four single-spin vector meson observables: $P_y^V, T_{20}^V, T_{21}^V, T_{22}^V$ or by the four independent real quantities, ρ_{00}^V, ρ_{1-1}^V, $\text{Re}(\rho_{10}^V)$, and $\text{Im}(\rho_{10}^V)$. The four spin observables and the four independent density matrix elements, all of which are defined in the γN CM frame, are related by

$$P_y^V = -2\sqrt{2}\, \text{Im}\, \rho_{10}^V \qquad T_{20}^V = \frac{1}{\sqrt{2}}\left(1 - 3\rho_{00}^V\right)$$

$$T_{21}^V = -\sqrt{6}\, \text{Re}\, \rho_{10}^V \qquad T_{22}^V = \sqrt{3}\, \rho_{1-1}^V\, . \tag{1}$$

The vector meson's density matrix, ρ^V,

$$\rho^V = \frac{1}{3}[I + \frac{3}{2}\boldsymbol{S} \cdot \boldsymbol{P}^V + \tau \cdot T^V], \tag{2}$$

is determined by the amplitude product $\mathcal{T}\mathcal{T}^\dagger$ for the case of an unpolarized initial beam and unpolarized target experiment.

One can obtain the single spin observables from the following trace relations

$$\boldsymbol{P}^V = \frac{\text{Tr}[\mathcal{T}\mathcal{T}^\dagger\, \boldsymbol{S}]}{\text{Tr}\,[\mathcal{T}\mathcal{T}^\dagger]} \qquad T_{2\mu}^V = \frac{\text{Tr}[\mathcal{T}\mathcal{T}^\dagger\, \tau_{2\mu}]}{\text{Tr}\,[\mathcal{T}\mathcal{T}^\dagger]}. \tag{3}$$

Here, $\boldsymbol{S}$ is the spin-1 operator and $\tau_{2\mu} \equiv \sqrt{3}\,[S_1 \times S_1]_{2\mu}$. We assume above that no measurement is made of the recoil polarization.

All these single spin observables are defined only in the γN CM frame, as is true for double, triple, quadruple spin observables.

3 Double spin observables

It is possible to have two polarized particles in the reaction. The possible double spin observables are classified as: the beam-vector meson $C^{\gamma V}$, beam-target $C^{\gamma N}$, beam-recoil $C^{\gamma N'}$, target-recoil $C^{NN'}$, target-vector meson C^{NV}, and recoil-vector meson $C^{N'V}$ experiments. The only double spin observable we

---

[1] Aside from quadrant ambiguities [5, 6]

shall discuss is $C^{\gamma V}$, which requires that we describe the polarized photon beam using the photon density matrix:

$$\rho^\gamma = \frac{1}{2}[I + \boldsymbol{P}^S \cdot \boldsymbol{\sigma}^\gamma], \tag{4}$$

where $\boldsymbol{P}^S$ is the Stokes "vector." The final state density matrix $\rho_f = \mathcal{T}\rho_\gamma\mathcal{T}^\dagger$ is expanded as

$$\rho_f = \rho^V + P_x^S \rho^x + P_y^S \rho^y + P_z^S \rho^z,$$

following ref. [7].

In the γN CM frame, using standard [4] beam-vector meson double spin observables, the above density matrix can be expressed in terms of the single-spin photon asymmetry observable $\boldsymbol{\Sigma}$ and the double spin $C^{\gamma V}$ observable

$$\rho^x = \frac{1}{3}[\Sigma_x^\gamma + \frac{3}{2}\sum_j S_j\, C_{xj}^{\gamma V} + \sum_\mu (-1)^\mu \tau_{2-\mu}\, C_{x2\mu}^{\gamma V}]. \tag{5}$$

For Stokes vector: $P_y^S = P_z^S = 0, P_x^S = \pm 1$, the photon beam is linearly polarized either perpendicular $(\hat{y})$ for $P_x^S = +1$, or parallel $(\hat{x})$ for $P_x^S = -1$ to the scattering plane. Other Stokes vector choices correspond to various photon polarizations, such as right or left circular polarized beams.

For the case of a polarized photon beam with no measurement of the initial or final baryon spin states, five independent real quantities, $\rho_{11}^x, \rho_{00}^x, \rho_{1-1}^x$, $\text{Re}(\rho_{10}^x)$, and $\text{Im}(\rho_{10}^x)$ appear in the density matrix. In the γN CM frame, these density matrix elements for a beam-vector meson experiment are related to the single photon spin observable Σ_x^γ and the four vector meson double spin observables by:

$$\Sigma_x^\gamma = \rho_{00}^x + 2\rho_{11}^x$$

$$C_{xy}^V = -2\sqrt{2}\,\text{Im}(\rho_{10}^V) \qquad C_{x20}^V = \frac{1}{\sqrt{2}}(\Sigma_x^\gamma - 3\rho_{00}^x)$$

$$C_{x21}^V = -\sqrt{6}\,\text{Re}(\rho_{10}^x) \qquad C_{x22}^V = \sqrt{3}\,\rho_{1-1}^x. \tag{6}$$

For Stokes vector $P_x^S = P_z^S = 0, P_y^S = \pm 1$, which corresponds to a photon linearly polarized at an angle of $\mp 45°$ with respect to the $\hat{x}$ axis, the final state density matrix ρ^y is:

$$\rho^y = \frac{1}{3}[\Sigma_y^\gamma + \frac{3}{2}\sum_j S_j\, C_{yj}^{\gamma V} + \sum_\mu (-1)^\mu \tau_{2-\mu} C_{y2\mu}^{\gamma V}]. \tag{7}$$

Using parity $\Sigma_y^\gamma = 0$. Four independent real quantities now appear, ρ_{11}^y, $\text{Im}(\rho_{1-1}^y)$, $\text{Re}(\rho_{10}^y)$, and $\text{Im}(\rho_{10}^y)$, which are related in the γN CM frame to the four vector meson double spin observables $C_{yx'}^{\gamma V}, C_{yz'}^{\gamma V}, C_{y21}^{\gamma V}$, and $C_{y22}^{\gamma V}$ by:

$$C_{yx'}^{\gamma V} = 2\sqrt{2}\,\text{Re}\,(\rho_{10}^y) \qquad C_{yz'}^{\gamma V} = 2\,\rho_{11}^y$$

$$C_{y21}^{\gamma V} = \sqrt{6}\,i\,\text{Im}\,(\rho_{10}^y) \qquad C_{y22}^{\gamma V} = -\sqrt{3}\,\rho_{1-1}^y. \tag{8}$$

For Stokes vector $P_x^S = P_y^S = 0, P_z^S = \pm 1$, which corresponds to a circularly polarized photon, with helicity ± 1, e.g. right or left circular polarization, the final state density matrix ρ^z is:

$$\rho^z = \frac{1}{3}[\Sigma_z^\gamma + \frac{3}{2}\sum_j S_j C_{zj}^{\gamma V} + \sum_\mu (-1)^\mu \tau_{2-\mu} C_{z2\mu}^{\gamma V}] . \tag{9}$$

Using parity $\Sigma_z^\gamma = 0$. Another four independent real quantities appear for this case, ρ_{11}^z, $\mathrm{Im}(\rho_{1-1}^z)$, $\mathrm{Re}(\rho_{10}^z)$, and $\mathrm{Im}(\rho_{10}^z)$, which are related in the γN CM frame to the four vector meson double spin observables $C_{zx'}^{\gamma V}$, $C_{zz'}^{\gamma V}$, $C_{z21}^{\gamma V}$, and $C_{z22}^{\gamma V}$ by:

$$C_{zx'}^{\gamma V} = 2\sqrt{2}\,\mathrm{Re}\,(\rho_{10}^z) \qquad C_{zz'}^{\gamma V} = 2\,\rho_{11}^z$$
$$C_{z21}^{\gamma V} = \sqrt{6}\,i\,\mathrm{Im}\,(\rho_{10}^z) \qquad C_{z22}^{\gamma V} = -\sqrt{3}\,\rho_{1-1}^z. \tag{10}$$

Thus we have related the observed density matrices to the usual double spin observables, provided all quantities are given in the γN CM frame.

4 Vector Meson Decay into Two mesons:

Which of these single and double spin observables can be measured using the decay pions or leptons? For $\rho \to \pi\pi$ decay the angular distribution of the two pions in the overall γN (or $\rho N'$) CM system can be measured. The same comment holds for the case of $\phi \to K\bar{K}$. It is nontrivial to separate the direct meson decay mode from other mechanisms for producing the two final pions. To define spin observables that involve the ρ meson, such a separation is required. Assuming that we can isolate the two-meson decay modes, we need to describe that decay in the $\rho N'$ CM system. The two-meson decay, which involves the angles of each meson, can be described[1] in terms of the angles $\bar{\theta}, \bar{\phi}$ between the vector meson's momentum and the relative velocity vector of the two decay pions as described in Fig. 1. In the vector meson rest frame $\bar{\theta}, \bar{\phi}$ reduce to the angles of the decay pion π_1, e.g., $\bar{\theta}, \bar{\phi} \to \theta_1, \phi_1$. The relation between the production and decay planes and the definition of these angles is illustrated in Fig. 1.

4.1 The decay angular distribution

Using the above angles, the two-pion angular distribution from vector meson decay is given by

$$dN^V = W^V(\bar{\theta}, \bar{\phi})\,d\cos\bar{\theta}\,d\bar{\phi}, \tag{11}$$

where dN^V denotes the number of pion pairs in the solid angle $d\cos\bar{\theta}\,d\bar{\phi}$. Boosting along the ρ direction from the $\rho N'$ overall CM frame to the ρ rest

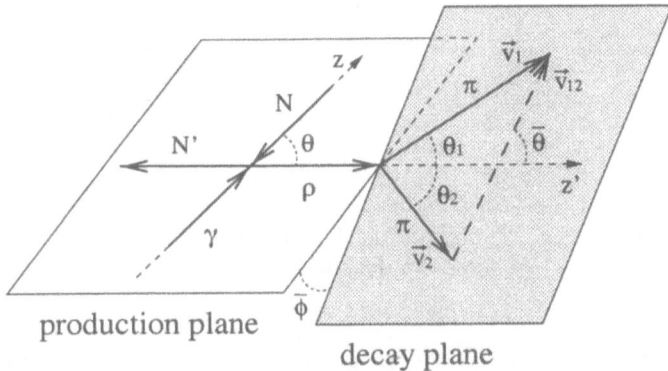

production plane

decay plane

Figure 1. A vector meson is produced in the $\hat{z}'$ direction at the scattering angle Θ, Φ. The velocities $v_1 = p_1/E_1, v_2 = p_2/E_2$, of the two mesons (or two leptons) are shown in the γN CM system, where they are in the θ_1, ϕ_1 and $\theta_2, \phi_2 = \pi + \phi_1$ directions. Note $\bar{\theta}, \bar{\phi}$ are the angles between the velocity difference vector $v_1 - v_2$ and the direction of the vector meson $q = q\hat{z}'$, all in the γN CM system. The decay plane is out of the scattering plane by an azimuthal angle $\bar{\phi}$, which equals ϕ_1.

frame, one can show that the angle $\bar{\theta}$ is related to the ρ rest frame angle θ_1 of the decay pion π_1 by $\tan \bar{\theta} = (E_\rho/m_\rho) \tan \theta_1$. The decay distribution is

$$W^V(\bar{\theta}, \bar{\phi}) = \mathrm{Tr}[M_\lambda \rho_f M_\lambda^\dagger], \qquad (12)$$

where M_λ describes the $\rho \to \pi_1 \pi_2$ decay amplitude which has the relativistic invariant form

$$M_\lambda = -ig_{\rho\pi\pi} \varepsilon^\mu(q, \lambda) (p_1 - p_2)_\mu , \qquad (13)$$

where $\varepsilon^\mu(q, \lambda)$ is the polarization vector of the ρ meson. Introducing the relative velocity of the decay products $v = v_1 - v_2$, M_λ equals a Wigner $\mathcal{D}$–function times a helicity independent kinematic factor

$$M_\lambda = ig_{\rho\pi\pi} \frac{2E_1 E_2}{E_1 + E_2} \, \varepsilon(q, \lambda) \cdot (v_1 - v_2) = ig_{\rho\pi\pi} \frac{2E_1 E_2}{E_1 + E_2} |v_1 - v_2| \mathcal{D}^1_{\lambda 0}(\bar{\theta}, \bar{\phi})^*,$$
$$(14)$$

which holds in the ρ rest frame and in the γN CM frame. For the $\bar{\theta}, \bar{\phi}$ angles see Fig. 1 In the ρ rest frame Eq. 14 reduces to

$$M_\lambda = ig_{\rho\pi\pi} \, 2p_1 \, \mathcal{D}^1_{\lambda 0}(\theta_1, \phi_1)^*.$$

Using the above relations and parity conservation, the distribution of decay pions for the case of an unpolarized beam and unpolarized target, with no measurement made of the recoil nucleon is: [2]

$$W^V(\bar{\theta}, \bar{\phi}) = \frac{1}{4\pi} \xi_V(\bar{\theta}) \left[1 - \sqrt{2} \sum_\mu T^V_{2\mu}(\Theta, \Phi) \, C^*_{2\mu}(\bar{\theta}\bar{\phi}) \right]$$

---

[2] Distributions for the case of polarized targets and beams and for lepton decay are presented in ref. [1].

$$= \frac{1}{4\pi}\xi_V(\bar{\theta})\left[1 - \sqrt{\frac{1}{2}}T_{20}^V(3\cos^2\bar{\theta} - 1) + \sqrt{3}\,T_{21}^V\sin 2\bar{\theta}\cos\bar{\phi}\right.$$
$$\left. -\sqrt{3}\,T_{22}^V\sin^2\bar{\theta}\cos 2\bar{\phi}\right]. \tag{15}$$

Here $C_{2\mu}^* \equiv \sqrt{\frac{4\pi}{5}}\,Y_{2\mu}^*$ is a spherical harmonic function. Note the above angular distribution is independent of the vector polarization of the vector meson, but depends on its tensor polarization. To understand how this property arises from parity and rotational symmetry, consider the density matrix folded in with the decay amplitude:

$$W = \sum_{\lambda,\lambda'} M_\lambda \rho_{\lambda,\lambda'} M_{\lambda'}^\dagger \propto \sum_{\lambda,\lambda'} \mathcal{D}_{\lambda 0}^{1*}\mathcal{D}_{\lambda' 0}^{1*}(-1)^{-\lambda'}\,\rho_{\lambda,-\lambda'}\,. \tag{16}$$

The Wigner function product is a symmetric function of the vector meson helicity labels λ, λ' whereas $(-1)^{-\lambda'}\,\rho_{\lambda,-\lambda'}$ is symmetric for the tensor τ and antisymmetric for vector part S; thus only the tensor polarization contributes to W.

The function

$$\xi_V(\bar{\theta}) = \frac{1}{(\sin^2\bar{\theta} + (\frac{E_\rho}{m_\rho})^2\cos^2\bar{\theta})^{5/2}}\,, \tag{17}$$

arises from the density of states and from the kinematical factor in M_λ for the γN CM system. Here E_ρ is the vector meson's energy; for $E_\rho \approx m_\rho$ clearly $\xi_V(\bar{\theta}) \to 1$, and the above expressions reduce to those obtained when the rest frame of the vector meson is used along with the amplitude in the γN CM frame. Therefore, $\xi_V(\bar{\theta})$ is the major effect of a consistent treatment of the decay and production, along with the use of the decay angle $\bar{\theta}$. From the expression in Eq. 15 one can extract the tensor spin observables.

For ρ decay into leptons, it is shown in ref. [1] that the vector polarization of the vector meson remains inaccessible when the lepton spins are not measured. For weak decays, vector polarization and tensor polarization (W and Z vector bosons decays) can be extracted from leptonic decay angular distributions because the parity violation operator $1 \pm \gamma_5$ yields S, P and D-wave interference terms in the angular distribution, which depend on the vector meson's vector polarization. The operator $\frac{1}{2}(1 \pm \gamma_5)$ also serves as a spin projection operator; hence, for strong interactions measuring a decay lepton's spin in the parity conserving ρ or ϕ leptonic decay creates a vector polarization contribution to the angular distribution.

5   Conclusion

In pionic and leptonic parity-conserving two-body decays of vector mesons, the final decay angular distribution yields only the tensor polarization of the vector meson. Access to the vector meson's vector polarization is possible if one measures the final state leptonic spin (or if one has access to a weak decay).

Hence, one of the simpler single spin observables, P_y^V is difficult to measure. One way to find P_y^V is to ascertain an interference mechanism with the background; another possibility, albeit remote, is to measure the final leptonic spin. Double, triple, and quadruple spin observables involving the vector polarization of the vector meson can also not be accessed from processes where the final vector meson decays into two mesons. The 16 single + double spin observables that are inaccessible, unless one measures the lepton spin in a leptonic decay mode, include one single spin observable: P_y^V, plus the 15 observables in Table 5.1.

Table 5.1. Inaccessible double spin observables

	$C_{xy}^{\gamma V}$		$C_{xx'}^{NV}$		$C_{xz'}^{NV}$	$C_{xx'}^{N'V}$	$C_{xz'}^{N'V}$
$C_{yx'}^{\gamma V}$		$C_{yz'}^{\gamma V}$		C_{yy}^{NV}		$C_{yy}^{N'V}$	
$C_{zx'}^{\gamma V}$		$C_{zz'}^{\gamma V}$	$C_{zx'}^{NV}$		$C_{zz'}^{NV}$	$C_{zx'}^{N'V}$	$C_{zz'}^{N'V}$

For an unpolarized photon beam, just by measuring the angular distribution of the decay mesons or leptons, 3 single tensor polarization observables of the vector meson (T_{20}, T_{21}, T_{22}) are accessible. For a polarized photon beam (γV), or a polarized target (NV), or a polarized recoil baryon $(N'V)$, one can access the double spin observables given in Table 5.2.

Table 5.2. Accessible double spin observables

$C_{x20}^{\gamma V}$	$C_{x21}^{\gamma V}$	$C_{x22}^{\gamma V}$		C_{x21}^{NV}	C_{x22}^{NV}		$C_{x21}^{N'V}$	$C_{x22}^{N'V}$	
	$C_{y21}^{\gamma V}$	$C_{y22}^{\gamma V}$	C_{y20}^{NV}	C_{y21}^{NV}	C_{y22}^{NV}	$C_{y20}^{N'V}$	$C_{y21}^{N'V}$	$C_{y22}^{N'V}$	
	$C_{z21}^{\gamma V}$	$C_{z22}^{\gamma V}$		C_{z21}^{NV}	C_{z22}^{NV}		$C_{z21}^{N'V}$	$C_{z22}^{N'V}$	

Acknowledgement. One author (F.T.) thanks the European Centre for Theoretical Studies in Trento and the workshop organizers for their gracious hospitality.

References

1. W.M. Kloet, W.T. Chiang, F. Tabakin: Phys. Rev. **C58**, 1086 (1998)

2. K. Gottfried, J.D. Jackson: Nuovo Cim. **33**, 309 (1964)

3. C.G. Fasano, F. Tabakin, B. Saghai: Phys. Rev. **C46**, 2430 (1992)

4. M. Pichowsky, Ç. Şavkli, F. Tabakin: Phys. Rev. **C53**, 593 (1996)

5. W.T. Chiang, F. Tabakin: Phys. Rev. **C55**, 2054 (1997)

6. G. Keaton, R. Workman: Phys. Rev. **C53**, 1434 (1996); R. Workman: these Proceedings

7. K. Schilling, P. Seyboth, G. Wolf: Nucl. Rev. **B15**, 397 (1970)

Few-Body Systems Suppl. 11, 316–326 (1999)

Few-
Body
Systems
© by Springer-Verlag 1999

Real and Virtual Compton Scattering (experiments)

Nicole d'Hose[a] *

for the VCS collaboration at MAMI

P. Bartsch[b], D. Baumann[b], J. Berthot[d], P.Y. Bertin[d], V. Breton[d], W.U. Boeglin[b], R. Böhm[b], T. Caprano[b], S. Derber[b], N. Degrande[c], M. Ding[b], M.O. Distler[b], J.E. Ducret[a], R. Edelhoff[b], I. Ewald[b], H. Fonvieille[d], J. Friedrich[b], J.M. Friedrich[b] †, R. Geiges[b], Th. Gousset[a], P.A.M. Guichon[a], H. Holvoet[c], Ch. Hyde-Wright[e], P. Jennewein[b], M. Kahrau[b], S. Kerhoas[a], M. Korn[b], H. Kramer[b], K.W. Krygier[b], V. Kunde[b], B. Lannoy[c], D. Lhuillier[a], A. Liesenfeld[b], C. Marchand[a], D. Marchand[a], J. Martino[a], H. Merkel[b], K. Merle[b], P. Merle[b], G. De Meyer[c], J. Mougey[a], R. Neuhausen[b], E. Offermann[b], Th. Pospischil[b], G. Quemener[d], O. Ravel[d], Y. Roblin[d], J. Roche[a] **, D. Rohe[b], G. Rosner[b], D. Ryckbosch[c], P. Sauer[b], H. Schmieden[b], S. Schardt[b], G. Tamas[b], M. Tytgat[c], M. Vanderhaeghen[a], L. Van Hoorebeke[c], R. Van de Vyver[c], J. Van de Wiele[f], P. Vernin[a], A. Wagner[b], Th. Walcher[b], S. Wolf[b]

a *CEA Saclay, DSM/DAPNIA/SPhN, F91191 Gif-sur-Yvette Cedex, France*
b *Institut für Kernphysik, Universität Mainz, Germany*
c *University of Gent, Belgium*
d *LPC, Univ. Blaise Pascal, IN2P3 Aubiere, France*
e *Old Dominium University, Virginia, U.S.A*
f *IPN, IN2P3 Orsay, France*

Abstract. This paper deals with Real and Virtual Compton Scattering off the proton at threshold and the way to deduce information about the nucleon polarizabilities.

1 Threshold regime of Compton Scattering and polarizabilities

Real Compton Scattering (RCS) off the proton refers to the reaction $\gamma p \to \gamma' p'$ with two real photons, while Virtual Compton scattering (VCS) is the generalization to the reaction $\gamma^* p \to \gamma' p'$ with γ^* denoting a virtual photon. This

---

* *E-mail address:* ndhose@cea.fr
† PhD student involved in the VCS experiment at the Mainz University
** PhD student involved in the VCS experiment at the CEA/Saclay

last reaction is performed by inelastic electron scattering on the proton. The best way to understand VCS at threshold [1] is to imagine that the final photon plays the role of a quasi-static external electromagnetic field applied to the proton. So the electron scattering measures the consequent deformation of the distributions of charge and current inside the nucleon, which is parameterized by the polarizabilities. They are of particular interest since they contain information about the complete excitation spectrum of the nucleon, in contrast to ground state properties like formfactors. The study of these observables as a function of Q^2 will be very sensitive to the evolution of both quark and pion degrees of freedom. Such information will be derived from a combined experimental study of RCS [2] and of VCS at MAMI (Q^2=0.33GeV^2) [3], Jefferson Lab. (Q^2=1 and 2 GeV^2) [4], and MIT-Bates (Q^2=0.05GeV^2) [5].

2 Real Compton Scattering at threshold

The amplitude for Real Compton Scattering off a proton of charge e and mass m can be written in terms of six structure dependent functions $A_i(q, \theta)$, i=1..6, with q = q$'$ denoting the photon energy in the c.m. frame and θ being the c.m. scattering angle

$$T = \sum_{i=1,6} A_i(q, \theta) \mathcal{F}_i(\vec{\epsilon}, \vec{\epsilon'}, \hat{q}, \hat{q'}, \vec{\sigma}). \tag{1}$$

Here $\vec{\epsilon}, \hat{q}$ ($\vec{\epsilon'}, \hat{q'}$) are the polarization vector and the direction of the incident (final) photon, while $\vec{\sigma}$ denotes the spin vector of the nucleon. The low energy expansion of the amplitude in powers of the photon energy q gives [6] [1]

$$\begin{aligned} T = & -\frac{e^2}{4\pi m}\vec{\epsilon}^{\,*} \cdot \vec{\epsilon'} + a_1(\theta)q\mathcal{F}'(\vec{\epsilon}, \vec{\epsilon'}, \hat{q}, \hat{q'}, \vec{\sigma}) \\ & +a_2(\theta)q^2 \mathcal{F}''(\vec{\epsilon}, \vec{\epsilon'}, \hat{q}, \hat{q'}) \\ & +a_3(\theta)q^3 \mathcal{F}'''(\vec{\epsilon}, \vec{\epsilon'}, \hat{q}, \hat{q'}, \vec{\sigma}) + 0(q^4). \end{aligned} \tag{2}$$

The first term is the Thomson amplitude, the second one is only dependent on the global properties of the proton (e, m, κ anomalous magnetic moment), while the next terms are related to the unknown internal structure of the nucleon. The scalar electric ($\bar{\alpha}$) and magnetic ($\bar{\beta}$) polarizabilities enter the amplitude at the order q^2 (with a function $\mathcal{F}''$ independent of the spin vector $\vec{\sigma}$) and the four spin polarizabilities ($\gamma_1, \gamma_2, \gamma_3, \gamma_4$) enter at the order q^3 (with a function $\mathcal{F}'''$ dependent of the spin vector $\vec{\sigma}$). The possibility of target spin gives an additional degree of freedom that responds to external electric and magnetic fields and gives rise to the spin polarizabilities.

The low energy expansion of the unpolarized cross section is then the sum of a model-independent Born contribution for scattering from a spin 1/2 point particle (with an anomalous magnetic moment), and of contributions of the order q^2, q^4, etc., containing the internal structure effects. The second term in the expansion, of quadratic order in the photon energy, containing the electric

---

[1] We use the Heaviside-Lorentz units with $\frac{e^2}{4\pi} = 1/137$.

($\bar{\alpha}$) and magnetic ($\bar{\beta}$) polarizabilities, corresponds to the interference between the Thomson term of order q^0 and the term of order q^2 in the amplitude. As far as the unpolarized cross section is concerned, the spin polarizabilities do not interfere with the leading Thomson term and they enter only in the third term of fourth order in the photon energy.

The dispersive relations and the optical theorem applied to the forward Compton scattering amplitude give fundamental sum rules [8],[9]

$$-\frac{e^2\kappa^2}{2m^2} = \frac{1}{4\pi^2}\int_{thr}^{\infty}\frac{\sigma_{1/2}-\sigma_{3/2}}{q}dq \qquad \text{GDH} \qquad (3)$$

$$\bar{\alpha}+\bar{\beta} = \frac{1}{2\pi^2}\int_{thr}^{\infty}\frac{\sigma_{tot}}{q^2}dq \qquad \text{Baldin} \qquad (4)$$

$$\gamma_1-\gamma_2-2\gamma_4 = \gamma_{\theta=0} = \frac{1}{4\pi^2}\int_{thr}^{\infty}\frac{\sigma_{1/2}-\sigma_{3/2}}{q^3}dq. \qquad (5)$$

The unpolarized differential cross section in the lab. can be written in the form

$$\frac{d\sigma}{d\Omega} = \frac{d\sigma^{Born}}{d\Omega} - \frac{e^2}{4\pi m}\left[\frac{q'_{lab}}{q_{lab}}\right]^2 q_{lab}q'_{lab}\left[\frac{\bar{\alpha}+\bar{\beta}}{2}(1+cos\theta)^2 + \frac{\bar{\alpha}-\bar{\beta}}{2}(1-cos\theta)^2\right] \qquad (6)$$

$$+0(q_{lab}^4).$$

where q_{lab} (q'_{lab}) is the initial (final) photon energy. Figure 1 shows the relative importance of the different contributions to the cross section. The relativistic "Klein-Nishina" cross section for the scattering on an ideal point particle of spin 1/2, differs significantly from the Born contribution, which takes into account the anomalous magnetic moment. However, this increase of the cross section is largely compensated by the effect of the $\bar{\alpha}$ and $\bar{\beta}$ polarizabilities (and the π pole term). Finally, all these effects due to the compositeness of the nucleon seem to yield only little information about the internal structure. The steep rise in the curve labeled "L'vov" is the complete cross section due to all higher order terms in q, calculated in a dispersive approach [7]. It is clear that just below the π^0 threshold, the deviation between the approximation to the quadratic order in $\bar{\alpha}$ and $\bar{\beta}$, and the complete estimation reaches 25%. This provides the main difficulty for the extraction of the scalar polarizabilities.

In spite of considerable experimental efforts, $\bar{\alpha}$ and $\bar{\beta}$ have not yet been independently determined to a good accuracy by Compton scattering. A "global average" on all the experiments is presented in the reference [2] and gives the results $\bar{\alpha}_p = (12.1\pm0.8\pm0.5)\ 10^{-4}\ fm^3$, $\bar{\beta}_p = (2.1\mp0.8\mp0.5)\ 10^{-4}\ fm^3$, the change of signs indicating that Baldin's sum rule has been used as a constraint.

A large effort by the MAMI, LEGS, SAL, Jefferson laboratories will be made in the near future to improve the determination of $\bar{\alpha}$ and $\bar{\beta}$ on the proton and to perform measurements of these quantities on the neutron (one possibility is photon scattering off the deuterium). Measurements of the proton

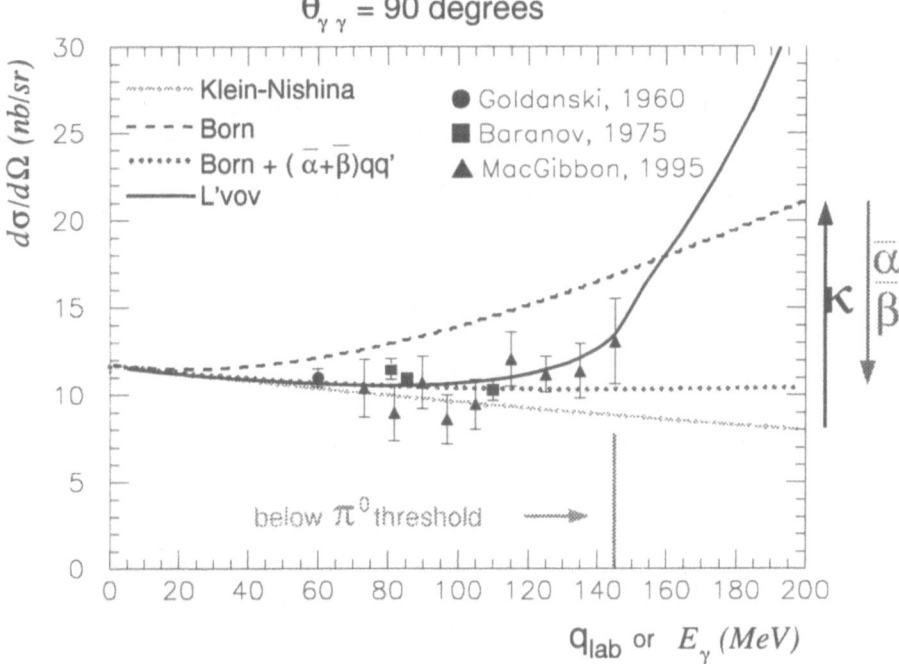

Figure 1. Differential cross section for a Compton scattering off the proton. See the comments in the text.

spin polarizabities will be investigated in double polarization experiments. The forward spin polarizability $\gamma_{\theta=0} = \gamma_1 - \gamma_2 - 2\gamma_4$ is the result of the study of $\sigma_{1/2} - \sigma_{3/2}$ evolution as a function of the photon energy (via the sum rule (5)). The backward spin polarizability $\gamma_{\theta=\pi} = \gamma_1 + \gamma_2 + 2\gamma_4$, which is estimated to be quite large could be directly accessed in the $\vec{\gamma}\vec{p} \to \gamma p$ reaction by measuring differential cross section angular distributions.

3   Virtual Compton Scattering at threshold

VCS ($\gamma^* p \to \gamma' p'$) can be realized experimentally by inelastic electron scattering ($ep \to e'p'\gamma'$), where the former process interferes with the Bethe-Heitler (BH) process, in which the real photon is emitted by the incoming or outgoing electron.

In addition to the two kinematical variables of RCS, e.g., the scattering angle θ and the energy q' of the outgoing real photon, one has to consider the modulus of the momentum of the virtual photon q in the photon-proton c.m. frame. Two other kinematical variables are also necessary: the linear polarization rate of the virtual photon ϵ and the angle between the leptonic and hadronic planes φ. The invariant quantities Q^2 and s are related to q and q' by the formula

$$q' = \frac{(s - m^2)}{2\sqrt{s}} \quad , \quad q = (Q^2 + \frac{(s - Q^2 - m^2)^2}{4s})^{\frac{1}{2}}. \tag{7}$$

The BH amplitude is exactly calculable from Quantum Electro-Dynamics (QED) if the proton form factors are well known, while the VCS amplitude can be split in two parts: the Born term containing only the nucleon and anti-nucleon contributions (exactly calculable) and the Non-Born term related to the excited states. The cross section has the form

$$\frac{d^5\sigma}{dk'_{lab}[d\Omega_e]_{lab}[d\Omega_p]_{CM}} = \frac{(2\pi)^{-5}}{32m} \left(\frac{k'_{lab}}{k_{lab}}\right) \frac{q'}{\sqrt{s}} \times \mathcal{M} \equiv \phi q'\mathcal{M}, \tag{8}$$

where $\mathcal{M}$ is the coherent sum of the different amplitudes

$$\mathcal{M} = \frac{1}{4} \sum_{spin} |T^{BH} + T^{VCS}|^2 = \frac{1}{4} \sum_{spin} |T^{BH} + T^{Born} + T^{NonBorn}|^2. \tag{9}$$

The low energy theorem from Low [10] states that in an expansion in powers of the real photon energy (q') (but fixed arbitrary q), the first term of the amplitudes T^{BH} and T^{Born} is of the order q'^{-1} (well-known infra-red divergence), while the first term of $T^{NonBorn}$ is of the order q'^1

$$T^{BH} + T^{Born} = \frac{b_{-1}(q,\theta)}{q'} + 0(q'^0), \tag{10}$$

$$T^{NonBorn} = b_1(q,\theta)q' + 0(q'^2). \tag{11}$$

The leading order term is parameterized by 6 generalized polarizabilities given for the electric and magnetic dipole radiation of the real photon, and constrained by selection rules for parity and angular momentum, plus crossing and charge-conjugation symmetries [11], [12]. The 6 polarizabilities are noted $P^{(\rho'L',\rho L)S}(q)$ where $L(L')$ denotes the initial (final) photon orbital angular momentum, $\rho(\rho')$ the type of multipole transition (0 for Electric, 1 for Magnetic) , and S distinguishes between non-spin-flip ($S = 0$) and spin-flip ($S = 1$) transitions at the nucleon side. One has 2 scalar ($S = 0$) and 4 spin or vector ($S = 1$) polarizabilities, functions of q or $Q^2|_{q'=0}$

$P^{(01,01)S=0,1}(q)$ corresponding to the E1/E1 transition

where $P^{(01,01)0}(0) = -\frac{4\pi}{e^2}\sqrt{\frac{2}{3}} \, \bar{\alpha}$,

and $P^{(01,01)1}(0) = 0$.

$P^{(11,11)S=0,1}(q)$ corresponding to the M1/M1 transition

where $P^{(11,11)0}(0) = -\frac{4\pi}{e^2}\sqrt{\frac{8}{3}} \, \bar{\beta}$,

and $P^{(11,11)1}(0) = 0$.

$P^{(01,12)1}(q)$ corresponding to the M2/E1 transition (N-D_{13} transition)

where $P^{(01,12)1}(0) = -\frac{4\pi}{e^2}\frac{\sqrt{2}}{3} \, \gamma_3$.

$P^{(11,02)1}(q)$ corresponding to the E2/M1 transition (N-Δ transition)
where $P^{(11,02)1}(0) - \frac{4\pi}{e^2}\sqrt{\frac{8}{27}} \ (\gamma_2 + \gamma_4)$.

The low energy expansion of $\mathcal{M}$ is the sum of the BH and Born contributions expressed in terms of the global properties of the nucleon (e, m, κ) and of contributions of the order q'^0, q'^1, etc containing the internal nucleon structure

$$\mathcal{M} = \mathcal{M}^{BH+Born} + (\mathcal{M}_0 - \mathcal{M}_0^{BH+Born}) + (\mathcal{M}_1 - \mathcal{M}_1^{BH+Born})q' + 0(q'^2). \quad (12)$$

The first term of order q'^0 corresponds to the interference between the term of order q'^{-1} in the BH+Born amplitude and the leading order term of order q'^1 in the Non-Born amplitude. According the equations (8) and (12), one has

$$d^5\sigma = d^5\sigma^{BH+Born} + \phi q'(\mathcal{M}_0 - \mathcal{M}_0^{BH+Born}) + 0(q'^2). \quad (13)$$

It is important to note all the differences between RCS and VCS, and to see the correspondence between the expansion terms of the two amplitudes. In RCS at threshold we consider the limit $(q = q') \to 0$ while in VCS, $q' \to 0$ at fixed arbitrary q.
1) The leading order terms are

$$T_{RCS}^{Born} \sim \vec{\epsilon}^* \cdot \vec{\epsilon'} \qquad , \qquad T_{VCS}^{Born} \sim \frac{\tilde{q}_0}{q'}\vec{\epsilon}^* \cdot \vec{\epsilon'}, \quad (14)$$

where $\tilde{q}_0 = m - \sqrt{m^2 + q^2}$ is the virtual photon energy. At the real Compton limit, these expressions are the same.
2) The term $b_1(q, \theta)q'$ in equation (11) which is parameterized by the generalized polarizabilities, corresponds to the term $a_2(\theta)q^2$ which generates the scalar polarizabilities (replace q^2 by qq') and also partially with the term $a_3(\theta)q^3$ which generates the spin polarizabilities (replace q^3 by q^2q').
3) The first order of the expansion of the Non Born cross section in RCS contains only the scalar polarizabilities while in VCS, the first order contains both scalar and spin polarizabilities.

4   The VCS experiment at MAMI

The first VCS experiment has been performed at MAMI [3] at $Q^2 = 0.33$ GeV^2 or q = 600 MeV/c. The scattered proton and electron are detected in coincidence in two magnetic spectrometers and the final photon is tagged by the reconstructed missing mass: $M_X^2 = 0$. Thanks to the excellent resolution of the facility, the separation between the two peaks "γ" and "π^0" in the missing mass is very pronounced as their FWHM's are about 17 times smaller than their absolute difference. This experiment benefits from the high duty cycle available at MAMI, and from a large luminosity of $\mathcal{L} = 4.10^{37}$ cm^{-2}s^{-1}.

Differential cross sections have been measured for 5 values of the real photon energy q' = 33.6, 45., 67.5, 90., 111.5 MeV in order to control the behavior of VCS at threshold. The polarization of the virtual photon ϵ is fixed at 0.62. As the two spectrometers are moving in the same plane, the angle between the leptonic and hadronic planes φ is close to 0 or 180°. Results are presented in Fig. 2, where the cross section is investigated in a large domain of the scattering angle θ far from the dominant Betle-Heitler peaks (note that one represents θ from $-\pi$ to π at $\varphi{=}0$.) The radiative corrections have been calculated for the first time by Vanderhaeghen *et al.* taking into account all the diagrams of order α_{em}^4 in the cross section [13].

$$d^5\sigma = d^5\sigma^{BH+Born} + \phi\, q'\, (M_0 - M_0^{BH+Born}) + \phi\, q'^2 \ldots$$

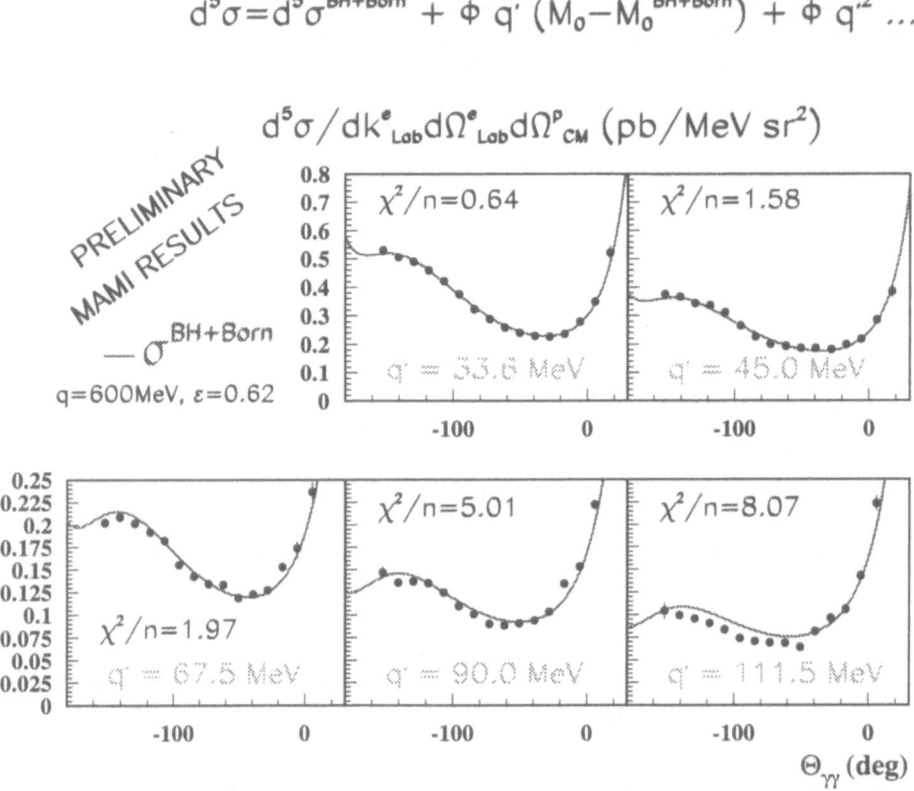

Figure 2. Differential cross sections as a function of θ for fixed q, ϵ, φ and for five different values of q'. The deviation of the calculated curve $d^5\sigma^{BH+Born}$ from the data is evaluated by a χ^2 calculation. When q' increases χ^2 increases also. This shows the effect of a structure effect beyond the "BH+Born" approximation: the polarizabilities

The experimental method to deduce information on the generalized polarizabilities is based on the expression (13). In Fig. 2, the deviation of the calculated curve $d^5\sigma^{BH+Born}$ from the data is evaluated by a χ^2 calculation (the statistical errors are close to 3% for all the data). The statement $d^5\sigma \rightarrow d^5\sigma^{BH+Born}$ when $q' \rightarrow 0$, is fully verified. It represents also a check of the experimental analysis method, the radiative corrections and the value of the proton form factors. The deviation at large real photon energies q′ indicates the effect of the generalized polarizabilities beyond the "BH+Born" approximation.

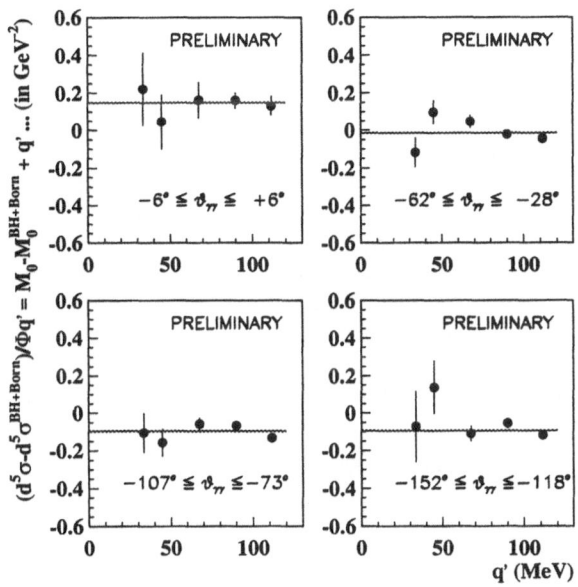

Figure 3. $(d^5\sigma - d^5\sigma^{BH+Born})/\phi q'$ studied as a function of the real photon energy q′ for ranges of scattering angles θ. The intercept at origin is $\mathcal{M}_0 - \mathcal{M}_0^{BH+Born}$. It is determined at each scattering angle θ by the mean value in the investigated real photon energy range (solid line) as no strong evolution with the real photon energy is revealed.

The quantity of interest $\mathcal{M}_0 - \mathcal{M}_0^{BH+Born}$ (for each scattering angle θ) is exactly the intercept at origin of $(d^5\sigma - d^5\sigma^{BH+Born})/\phi q'$ studied as a function of the real photon energy q′. Fig 3 shows the intrinsic difficulty of this experiment as the statistical errors increase when q′ decreases. Nevertheless, no strong evolution with the real photon energy is revealed, and the intercept at origin for all the scattering angles θ is preliminary determined by the mean value in the investigated real photon energy range.

In an unpolarized experiment, the difference $\mathcal{M}_0 - \mathcal{M}_0^{BH+Born}$ is

$$\mathcal{M}_0 - \mathcal{M}_0^{BH+Born} = v_{LL}(\theta, \varphi)\left(P_{LL}(q) - \frac{1}{\epsilon}P_{TT}(q)\right) + v_{LT}(\theta, \varphi)P_{LT}(q), \quad (15)$$

where $v_{LL}(\theta, \varphi)$ and $v_{LT}(\theta, \varphi)$ are known functions of θ and φ, $P_{LL}(q), P_{TT}(q), P_{LT}(q)$ are the following combinations of only 5 generalized polarizabilities (with G_E and G_M denoting the form factors applied at $Q^2|_{q'=0}$),

$$P_{LL}(q) = -2\sqrt{6}mG_E P^{(01,01)0}(q), \quad (16)$$

$$P_{TT}(\mathbf{q}) \;=\; 3G_M \mathbf{q}^2 \left(\sqrt{2}P^{(01,12)1}(\mathbf{q}) - \frac{1}{\tilde{q}_0}P^{(11,11)1}(\mathbf{q}) \right),$$

$$P_{LT}(\mathbf{q}) \;=\; \sqrt{\frac{3}{2}}\, m\frac{\mathbf{q}}{Q}G_E\, P^{(11,11)0}(\mathbf{q}) + \frac{3}{2}\frac{\mathbf{q}}{\tilde{q}_0}\tilde{Q}G_M\, P^{(01,01)1}(\mathbf{q}).$$

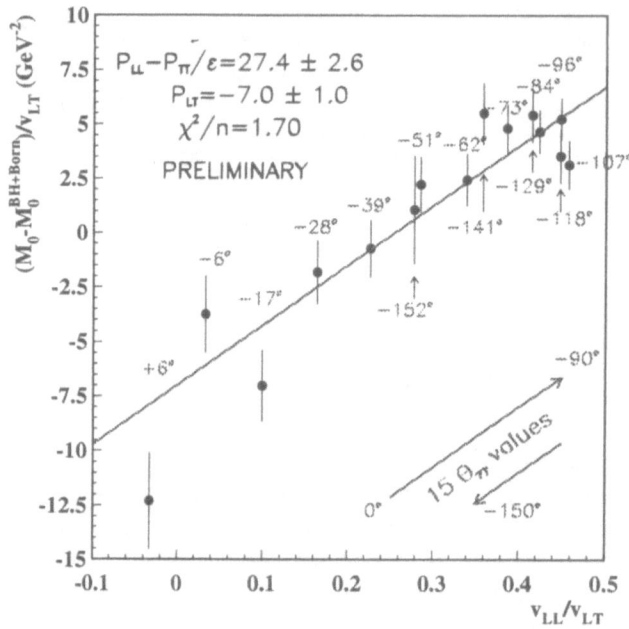

Figure 4. $\dfrac{\mathcal{M}_0 - \mathcal{M}_0^{BH+Born}}{v_{LT}(\theta,\varphi)} = \left(P_{LL}(\mathbf{q}) - \frac{1}{\epsilon}P_{TT}(\mathbf{q})\right)\dfrac{v_{LL}(\theta,\varphi)}{v_{LT}(\theta,\varphi)} + P_{LT}(\mathbf{q})$. All the data provided by the 15 scattering angles (from $+6°$ to $-152°$) are reasonably well aligned.

Now exploiting all the scattering angles θ, one should verify the expression

$$\frac{\mathcal{M}_0 - \mathcal{M}_0^{BH+Born}}{v_{LT}(\theta,\varphi)} = \left(P_{LL}(\mathbf{q}) - \frac{1}{\epsilon}P_{TT}(\mathbf{q})\right)\frac{v_{LL}(\theta,\varphi)}{v_{LT}(\theta,\varphi)} + P_{LT}(\mathbf{q}). \qquad (17)$$

in order to extract the two quantities $P_{LL}(\mathbf{q}) - \frac{1}{\epsilon}P_{TT}(\mathbf{q})$ and $P_{LT}(\mathbf{q})$. All the data provided by the 15 scattering angles (from $+6°$ to $-152°$) are reasonably well aligned (see Fig. 4): this surprising fact indicates that all the higher terms in the expansion of the differential cross section (formula 13) seem to be rather negligible.

The table 4.1 reports the preliminary results for the two quantities $P_{LL}(\mathbf{q}) - \frac{1}{\epsilon}P_{TT}(\mathbf{q})$ and $P_{LT}(\mathbf{q})$. The first error is the statistical one, while the second one

$Q^2 = 0.33$ (in GeV^2)	$P_{LL}(Q^2) - \frac{1}{\epsilon}P_{TT}(Q^2)$ (in GeV^2)	$P_{LT}(Q^2)$ (in GeV^2)
preliminary This experiment	27 ± 3 (± 12)	-7 ± 1 (± 4)
HBChPT	26.3	-5.7
LSM	10.9	0.
ELM	5.9	-1.9
NRQCM	17.0	-1.7

Table 4.1. The preliminary experimental results compared to theoretical predictions (HBChPT for Heavy Baryon Chiral Perturbation Theory, LSM for Linear Sigma Model, NRQCM for Non Relativistic Quark Constituent Model and ELM for Effective Lagrangian Model)

is due to our rough method of extraction and should be reduced in further studies. The two results are compared with different theoretical predictions:
- Guichon, Liu, and Thomas [14] made an estimation of the generalized polarizabilities in a non relativistic quark model (NRQCM).
- Vanderhaeghen [15] performed a calculation in an effective Lagrangian approach (ELM) including nucleon resonance effects.
- Metz and Drechsel [16] performed a one-loop calculation of the polarizabilities in the linear sigma model (LSM) in the limit of an infinite sigma mass. .
- A heavy-baryon chiral perturbation theory calculation (HBChPT) has also been performed to third order in the external momentum expansion by Hemmert, Holstein, Knöchlein and Scherer [17]. The results seem to notably favor this last model.

Although these results are still preliminary, VCS appears nevertheless to be a favorable case for the experimental study of polarizabilities as the first order of an expansion of the Non Born cross section, containing both scalar and spin polarizabilities is enough to reproduce the evolution with the real photon energy below pion threshold. This is in contrast to the RCS case, where the first term, containing only scalar polarizabilities, was not sufficient. The scheduled two experiments at MIT-Bates and Jefferson Lab. will also shed light on this problem by measuring the same combinations of generalized polarizabilities $P_{LL}(q) - \frac{1}{\epsilon}P_{TT}(q)$ and $P_{LT}(q)$ in different ranges of Q^2. To go further and to measure independently all the scalar and spin polarizabilities, it will be necessary to perform a double polarization experiment.

References

1. P.A.M. Guichon and M. Vanderhaeghen: Prog. Part. and Nucl. Phys. **41**, 121 (1998)

326

2. B.E. MacGibbon et al.: Phys. Rev. **C52**, 2097 (1995)

3. N. D'Hose and Th. Walcher: *Nucleon Structure Study by Virtual Compton Scattering*, MAMI proposal (1994)

4. P.Y. Bertin, P.A.M. Guichon and Ch. Hyde-Wright: *CEBAF proposal PR93-050* (1993)

5. J. Shaw and R. Miskimen: *Experimental Proposal at MIT-Bates* (1997)

6. Th. Hemmert et al.: Phys. Rev. **D57**, 5746 (1998)

7. A. L'vov : Int. J. Mod. Phys. **A8**, 5267 (1993)

8. S.B. Gerasimov: Yad. Fiz. **2**, 598 (1965); Sov. J. Nucl. Phys. **2**, 430 (1966); S.D. Drell and A.C. Hearn: Phys. Rev. Lett. **16**, 908 (1966)

9. A.M. Baldin: Nucl. Phys. **18**, 310 (1960)

10. F.E. Low, Phys. Rev. **96**, 1428 (1954); Phys. Rev. **110**, 974 (1958)

11. P.A.M. Guichon, G.Q. Liu, A.W. Thomas: Nucl Phys **A591**, 606 (1995)

12. D. Drechsel, G. Knöchlein, A. Metz, S. Scherer: Phys. Rev. **C55**, 424 (1997); D. Drechsel, G. Knöchlein, A. Yu Korchin, A. Metz, S. Scherer: Phys. Rev. **C57**, 941 (1998)

13. M. Vanderhaeghen, D. Lhuillier, D. Marchand, J. Van de Wiele: *in preparation*

14. G.Q.Liu, A.W.Thomas and P.Guichon: Aust. J. Phys. **49**, 905 (1996)

15. M. Vanderhaeghen: Phys. Lett. **B368**, 13 (1996)

16. A. Metz and D. Drechsel: Z. Phys. **A356**, 351 (1996); Z. Phys. **A359**, 165 (1997)

17. T.R. Hemmert, B.R. Holstein, G. Knöchlein, S. Scherer: Phys. Rev. **D55**, 2630 (1997); Phys. Rev. Lett. **79**, 22 (1997)

Few-Body Systems Suppl. 11, 327–334 (1999)

Few-
Body
Systems
© by Springer-Verlag 1999

Virtual Compton Scattering—Generalized Polarizabilities of Nucleons and Pions

S. Scherer*

Institut für Kernphysik, Johannes Gutenberg-Universität, J. J. Becher Weg 45, D-55099 Mainz, Germany

Abstract. Virtual Compton scattering off nucleons and pions at low energies is discussed. Predictions for the generalized polarizabilities of the nucleon are presented within the framework of heavy-baryon chiral perturbation theory and the linear sigma model. First results for the generalized polarizabilities of the charged pion in chiral perturbation theory at $\mathcal{O}(p^4)$ are shown.

1 Introduction

Real Compton scattering (RCS) off a stable particle has a long history as a means to study the dynamics responsible for the internal structure of a system. In this context, low-energy theorems (LET) play an important role in defining a reference point for the accuracy which has to be achieved in order to obtain information beyond global properties of the target and, thus, to potentially distinguish between different models. The famous low-energy theorem of Low [1] and Gell-Mann and Goldberger [2] for Compton scattering of real photons off a nucleon is based on the requirements of gauge invariance, Lorentz covariance, crossing symmetry, and the discrete symmetries. The low-energy amplitude is specified up to and including terms linear in the photon energy. The Taylor series coefficients are expressed in terms of the charge, mass, and magnetic moment of the target. Terms of second order in the frequency, which are not determined by this LET, can be parametrized in terms of two new structure constants, the electric and magnetic polarizabilities α and β (see, e.g., [3, 4]).

Recently, the investigation of low-energy virtual Compton scattering (VCS) as tested in, e.g., the reactions $e^-p \to e^-p\gamma$ [5, 6] and $\pi^-e^- \to \pi^-e^-\gamma$ [7], has attracted a lot of interest. The possibilities to investigate the structure of the target increase substantially, if virtual photons are used since (a) photon energy and momentum can be varied independently and (b) longitudinal

---

*This work was supported by the Deutsche Forschungsgemeinschaft (SFB 201) and a grant from NATO.

components of the transition current are accessible. For the nucleon, the model-independent properties of the low-energy VCS amplitude have been identified in [8, 9] whereas the spin-zero case has been discussed in Ref. [10]. In [8] the model-dependent part beyond the LET was analyzed in terms of a multipole expansion. Keeping only terms linear in the energy of the final photon, the corresponding amplitude was parametrized in terms of ten so-called generalized polarizabilities (GPs) which are functions of the three-momentum transfer of the virtual photon in the VCS process. The number of independent GPs reduces to six, if charge-conjugation invariance is imposed [11, 12]. Predictions for the generalized polarizabilities of the nucleon have been obtained in various frameworks [8, 13, 14, 15, 16, 17, 18] (for an overview, see Ref. [19]).

2   Kinematics and LET

The invariant amplitude consists of a Bethe-Heitler piece, where the real photon is emitted by the initial or final electrons, and the VCS contribution (see Fig. 1),

$$\mathcal{M} = \mathcal{M}_{\mathrm{BH}} + \mathcal{M}_{\mathrm{VCS}}. \tag{1}$$

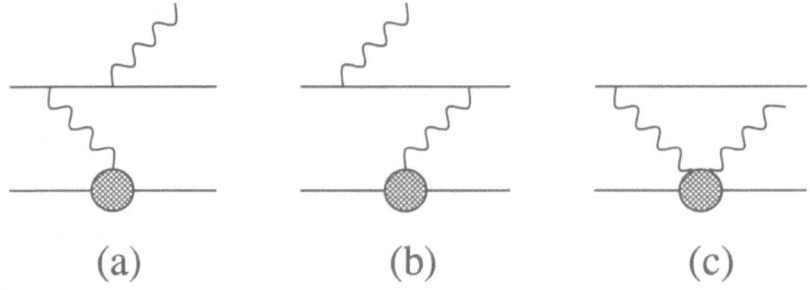

$$(a) \qquad\qquad (b) \qquad\qquad (c)$$

Figure 1. Bethe-Heitler diagrams (a) and (b). VCS diagram (c). The four-momenta of the virtual photons in the BH diagrams and the VCS diagram differ from each other.

In the following we will be concerned with the invariant amplitude for VCS,

$$\mathcal{M}_{\mathrm{VCS}} = -ie^2 \epsilon_\mu \epsilon_\nu'^* M^{\mu\nu} = -ie^2 \epsilon_\mu M^\mu = ie^2 \left(\epsilon_T \cdot M_T + \frac{q^2}{\omega^2} \epsilon_z M_z \right), \tag{2}$$

where $\epsilon_\mu = e\bar{u}\gamma_\mu u/q^2$ is the polarization vector of the virtual photon ($e > 0, e^2/4\pi \approx 1/137$), and where use of current conservation has been made. In the center-of-mass system, using the Coulomb gauge for the final real photon, the transverse (longitudinal) part of $\mathcal{M}_{VCS}$ for the nucleon can be expressed in terms of eight (four) independent structures [9, 16],

$$\epsilon_T \cdot M_T = \epsilon'^* \cdot \epsilon_T A_1 + \cdots, \qquad M_z = \epsilon'^* \cdot \hat{q} A_9 + \cdots, \tag{3}$$

where the functions A_i depend on three kinematical variables, $|q|$, $\omega' = |q'|$, $z = \hat{q} \cdot \hat{q}'$. For the spin zero case only one longitudinal and two transverse structures are required [10, 11].

Extending the method of Gell-Mann and Goldberger [2] to VCS, model-independent predictions for the functions A_i were obtained in [9]. For example, the result for A_1 up to second order in $|q|$ and ω' is

$$
\begin{aligned}
A_1 &= -\frac{1}{M} + \frac{z}{M^2}|q| - \left(\frac{1}{8M^3} + \frac{r_E^2}{6M} - \frac{\kappa}{4M^3} - \frac{4\pi\alpha}{e^2}\right)\omega'^2 \\
&\quad + \left(\frac{1}{8M^3} + \frac{r_E^2}{6M} - \frac{z^2}{M^3} + \frac{(1+\kappa)\kappa}{4M^3}\right)|q|^2.
\end{aligned}
\tag{4}
$$

To this order, all A_i can be expressed in terms of M, κ, G_E, G_M, r_E^2, α, and β. For $|q| = \omega'$, Eq. (4) reduces to the well-known RCS result.

In Ref. [10] the low-energy behavior of the VCS amplitude of $\pi^-(p_i) + \gamma^*(\epsilon, q) \to \pi^-(p_f) + \gamma(\epsilon', q')$ was found to be of the form

$$
\mathcal{M}_{\text{VCS}} = -2ie^2 F(q^2)\left[\frac{p_f \cdot \epsilon'^*(2p_i + q)\cdot\epsilon}{s - m_\pi^2} + \frac{p_i \cdot \epsilon'^*(2p_f - q)\cdot\epsilon}{u - m_\pi^2} - \epsilon\cdot\epsilon'^*\right] + \mathcal{M}_R,
\tag{5}
$$

where $F(q^2)$ is the electromagnetic form factor of the pion, $s = (p_i + q)^2$, and $u = (p_i - q')^2$. The residual term $\mathcal{M}_R$ is separately gauge invariant and at least of second order, i.e. $\mathcal{O}(qq, qq', q'q')$.

3   Generalized Polarizabilities

For the purpose of analyzing the model-dependent terms specific to VCS, the invariant amplitude $\mathcal{M}_{\text{VCS}}$ is split into a (generalized) pole piece $\mathcal{M}_A$ and a residual part $\mathcal{M}_B$. For the nucleon, the s- and u-channel pole diagrams are calculated using electromagnetic vertices of the form

$$
\Gamma^\mu(p', p) = \gamma^\mu F_1(q^2) + i\frac{\sigma^{\mu\nu}q_\nu}{2M}F_2(q^2), \quad q = p' - p,
\tag{6}
$$

where F_1 and F_2 are the Dirac and Pauli form factors, respectively. The corresponding amplitude $\mathcal{M}_A^{\gamma^*\gamma}$ contains all irregular terms as $q \to 0$ or $q' \to 0$ and is separately gauge invariant [8, 9]. For the pion, the situation is somewhat more complicated due to the fact that even for real photons the s- and u-channel pole diagrams are not separately gauge-invariant. A natural starting point is given by Eq. (5) with $\mathcal{M}_B \equiv \mathcal{M}_R$.

The generalized polarizabilities in VCS [8] result from an analysis of $\mathcal{M}_B^{\gamma^*\gamma}$ in terms of electromagnetic multipoles $H^{(\rho'L',\rho L)S}(\omega', |q|)$, where ρ (ρ') denotes the type of the initial (final) photon ($\rho = 0$: charge, C; $\rho = 1$: magnetic, M; $\rho = 2$: electric, E). The initial (final) orbital angular momentum is denoted by L (L'), and S distinguishes between non-spin-flip ($S = 0$) and spin-flip ($S = 1$) transitions. For the pion, only the case $S = 0$ applies. According to the LET for VCS, $\mathcal{M}_B^{\gamma^*\gamma}$ is at least linear in the energy of the real photon. A restriction to the lowest-order, i.e. linear terms in ω' leads to only electric and magnetic dipole radiation in the final state. Parity and angular-momentum selection rules (see Table 3.1) then allow for 3 scalar multipoles ($S = 0$) and 7 vector multipoles ($S = 1$). The corresponding ten GPs, $P^{(01,01)0}$, ..., $\hat{P}^{(11,2)1}$,

Table 3.1. Multipolarities of initial and final states

J^P	final real photon	initial virtual photon
$\frac{1}{2}^-$	E1	C1,E1
$\frac{3}{2}^-$	E1	C1,E1,M2
$\frac{1}{2}^+$	M1	C0,M1
$\frac{3}{2}^+$	M1	C2,E2,M1

are functions of $|\boldsymbol{q}|^2$, where mixed-type polarizabilities, $\hat{P}^{(\rho'L',L)S}(|\boldsymbol{q}|^2)$, have been introduced which are neither purely electric nor purely Coulomb type (see [8] for details). Only six of the above ten GPs are independent, if charge-conjugation symmetry is imposed [11, 12]. For example, for a charged pion, the constraint for the Compton tensor reads [10, 11]

$$\mathcal{M}_{\pi+}^{\mu\nu}(p_f, q'; p_i, q) \overset{C}{=} \mathcal{M}_{\pi-}^{\mu\nu}(p_f, q'; p_i, q) \overset{crossing}{=} \mathcal{M}_{\pi+}^{\mu\nu}(-p_i, q'; -p_f, q), \quad (7)$$

generating one relation between originally three GPs [11, 15].

Relations between the GPs at $|\boldsymbol{q}| = 0$ and the four spin-dependent RCS polarizabilities γ_i of Ref. [20] were discussed in [12]:

$$\gamma_3 = -\frac{e^2}{4\pi}\frac{3}{\sqrt{2}}P^{(01,12)1}(0), \quad \gamma_2 + \gamma_4 = -\frac{e^2}{4\pi}\frac{3\sqrt{3}}{2\sqrt{2}}P^{(11,02)1}(0), \quad (8)$$

i.e., only two of the four γ_i can be related to GPs at $|\boldsymbol{q}| = 0$.

4 Generalized Polarizabilities of the Nucleon

In the following we will discuss the predictions for the generalized polarizabilities of the nucleon obtained within the heavy-baryon formulation of chiral perturbation theory (HBChPT) and the linear sigma model.

In Ref. [16] the VCS amplitude was calculated using HBChPT to third order in the external momenta. At $\mathcal{O}(p^3)$, contributions to the GPs are generated by nine one-loop diagrams and the π^0-exchange t-channel pole graph (see [16]). For the loop diagrams only the leading-order Lagrangians are required,

$$\widehat{\mathcal{L}}_{\pi N}^{(1)} = \bar{N}_v(iv \cdot D + g_A S \cdot u)N_v, \quad \mathcal{L}_{\pi\pi}^{(2)} = \frac{F_\pi^2}{4}\text{Tr}\left[\nabla_\mu U(\nabla^\mu U)^\dagger\right], \quad (9)$$

where N_v represents a non-relativistic nucleon field, and $U = \exp(i\boldsymbol{\tau} \cdot \boldsymbol{\pi}/F_\pi)$ contains the pion field. The covariant derivatives $\nabla_\mu U$ and $D_\mu N_v$ include the coupling to the electromagnetic field, and u_μ contains in addition the derivative coupling of a pion. In the heavy-baryon formulation the spin matrix is given by $S^\mu = i\gamma_5 \sigma^{\mu\nu} v_\nu$, where v^μ is a four-vector satisfying $v^2 = 1, v_0 \geq 1$. Finally, for the π^0-exchange diagram one requires in addition to Eq. (9) the $\pi^0\gamma\gamma^*$ vertex provided by the Wess-Zumino-Witten Lagrangian,

$$\mathcal{L}_{\gamma\gamma\pi^0}^{(WZW)} = -\frac{e^2}{32\pi^2 F_\pi} \epsilon^{\mu\nu\alpha\beta} F_{\mu\nu} F_{\alpha\beta} \pi^0 , \tag{10}$$

where $\epsilon_{0123} = 1$ and $F_{\mu\nu}$ is the electromagnetic field strength tensor. At $\mathcal{O}(p^3)$, the LET of VCS is reproduced by the tree-level diagrams obtained from Eq. (9) and the relevant part of the second- and third-order Lagrangian,

$$\widehat{\mathcal{L}}_{\pi N}^{(2)} = -\frac{1}{2M} \bar{N}_v \left[D \cdot D + \frac{e}{2}(\mu_S + \tau_3 \mu_V) \epsilon_{\mu\nu\rho\sigma} F^{\mu\nu} v^\rho S^\sigma \right] N_v , \tag{11}$$

$$\widehat{\mathcal{L}}_{\pi N}^{(3)} = \frac{i e \epsilon_{\mu\nu\rho\sigma} F^{\mu\nu}}{8M^2} \bar{N}_v \left[\mu_S - \frac{1}{2} + \tau_3(\mu_V - \frac{1}{2}) \right] S^\rho D^\sigma N_v + h.c.. \tag{12}$$

The linear sigma model (LSM) represents a field-theoretical realization of chiral $SU(2)_L \times SU(2)_R$ symmetry. The dynamical degrees of freedom are given by a nucleon doublet Ψ, a pion triplet $\boldsymbol{\pi}$, and a singlet σ:

$$\mathcal{L}_S = i\bar{\Psi}\not{\partial}\Psi + \frac{1}{2}\partial_\mu\sigma\partial^\mu\sigma + \frac{1}{2}\partial_\mu\boldsymbol{\pi}\cdot\partial^\mu\boldsymbol{\pi}$$

$$-g_{\pi N}\bar{\Psi}(\sigma + i\gamma_5\boldsymbol{\tau}\cdot\boldsymbol{\pi})\Psi - \frac{\mu^2}{2}(\sigma^2 + \boldsymbol{\pi}^2) - \frac{\lambda}{4}(\sigma^2 + \boldsymbol{\pi}^2)^2, \tag{13}$$

$$\mathcal{L}_{s.b.} = -c\sigma, \tag{14}$$

where $\mathcal{L}_{s.b.}$ is a perturbation which explicitly breaks chiral symmetry. With an appropriate choice of parameters ($\mu^2 < 0, \lambda > 0$) the model reveals spontaneous symmetry breaking, $<0|\sigma|0> = F_\pi = 92.4$ MeV. The spectrum consists of massless pions, a massive sigma and nucleons with masses satisfying the Goldberger-Treiman relation $m_N = g_{\pi N} F_\pi$ with $g_A = 1$. The symmetry breaking of Eq. (14) generates the PCAC relation

$$\partial^\mu A_\mu^a = F_\pi m_\pi^2 \pi^a .$$

The interaction with the electromagnetic field is introduced via minimal substitution in Eq. (13). The generalized polarizabilities have been calculated in the framework of a one-loop calculation [15].

Numerical results for some of the generalized proton polarizabilities are shown in Fig. 2. In both ChPT and the LSM the electric polarizability α decreases considerably faster with $|q|^2$ than in the constituent quark model and the effective Lagrangian approach. Also, the chiral calculations show a distinct behavior for the slope of β near the origin. Note that at $\mathcal{O}(p^3)$, the results for the GPs are entirely given in terms of the pion mass m_π, the axial coupling constant g_A, and the pion decay constant F_π. Finally, it has been shown that ChPT and the LSM respect the relations between the GPs derived in Ref. [11, 12].

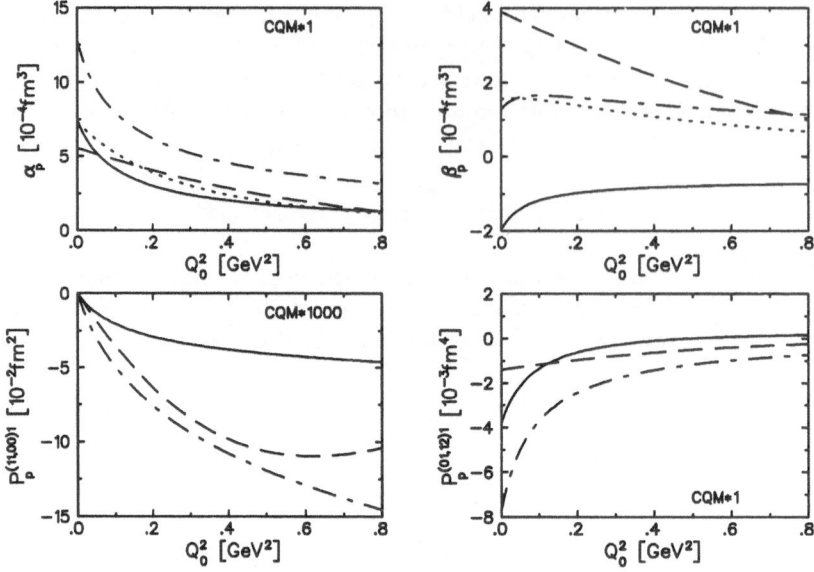

Figure 2. Some GPs of the proton. Solid curve: linear sigma model [15]. Dashed curve: Constituent quark model [8]. Dotted curve: Effective Lagrangian model [14]. Dashed-dotted curve: Chiral perturbation theory [16].

5   Generalized Polarizabilities of Pions

At the one-loop level, $\mathcal{O}(p^4)$, of chiral perturbation theory [21], the electromagnetic polarizabilities of the charged pion are entirely determined by an $\mathcal{O}(p^4)$ counter term [3],

$$\alpha = -\beta = \frac{e^2}{4\pi} \frac{2}{m_\pi F^2} (2l_5^r - l_6^r) = (2.68 \pm 0.42) \times 10^{-4}\,\mathrm{fm}^3, \qquad (15)$$

where the linear combination $2l_5^r - l_6^r$ is predicted through the decay $\pi^+ \to e^+ \nu_e \gamma$. Corrections to this result at $\mathcal{O}(p^6)$ were shown to be reasonably small, namely 12% and 24% of the $\mathcal{O}(p^4)$ values for α and β, respectively [22].

Presently, the pion VCS reaction is under investigation as part of the Fermilab E781 SELEX experiment, where a 600 GeV pion beam interacts with atomic electrons in nuclear targets [7]. In principle, the different behavior under the substitution $\pi^- \to \pi^+$ of $\mathcal{M}_{\mathrm{BH}}$ and $\mathcal{M}_{\mathrm{VCS}}$,

$$\mathcal{M}_{\mathrm{BH}}(\pi^-) = -\mathcal{M}_{\mathrm{BH}}(\pi^+), \quad \mathcal{M}_{\mathrm{VCS}}(\pi^-) = \mathcal{M}_{\mathrm{VCS}}(\pi^+), \qquad (16)$$

may be of use in identifying the contributions due to internal structure by comparing the reactions involving a π^- and a π^+ beam for the same kinematics:[1]

$$d\sigma(\pi^+) - d\sigma(\pi^-) \sim 4\mathrm{Re}\left(\mathcal{M}_{\mathrm{BH}}(\pi^+)\mathcal{M}^*_{\mathrm{VCS}}(\pi^+)\right). \qquad (17)$$

We have calculated the invariant amplitude for VCS at $\mathcal{O}(p^4)$ in the framework

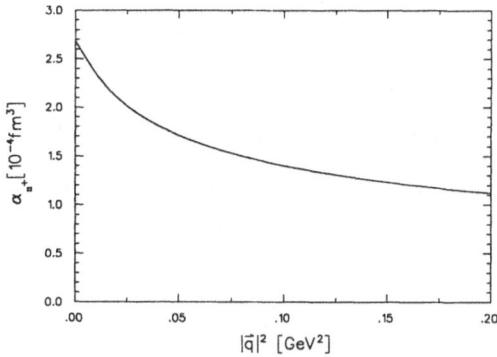

Figure 3. Generalized polarizability α of the charged pion as function of $|q|^2$ [23]. Note that $\alpha(|q|^2) = -\beta(|q|^2)$ at $\mathcal{O}(p^4)$.

of chiral perturbation theory. The result is in complete agreement with the LET of Ref. [10]. Using the procedure developed in Ref. [11] we find for the generalized polarizabilities of the charged pion (see Fig. 3)

$$
\begin{aligned}
\alpha(|\boldsymbol{q}|^2) &= -\beta(|\boldsymbol{q}|^2) \\
&= \frac{e^2}{8\pi m_\pi}\sqrt{\frac{m_\pi}{E_\pi}}\left[\frac{4(2l_5^r - l_6^r)}{F^2} - 2\frac{m_\pi - E_\pi}{m_\pi}\frac{1}{(4\pi F)^2}J^{(0)\prime}\left(2\frac{m_\pi - E_\pi}{m_\pi}\right)\right],
\end{aligned}
$$
$$(18)$$

where $E_\pi = \sqrt{m_\pi^2 + |\boldsymbol{q}|^2}$ and

$$J^{(0)\prime}(x) = \frac{1}{x}\left[1 - \frac{2}{x\sigma(x)}\ln\left(\frac{\sigma(x) - 1}{\sigma(x) + 1}\right)\right], \quad \sigma(x) = \sqrt{1 - \frac{4}{x}}, \quad x \leq 0.$$

The $|\boldsymbol{q}|^2$ dependence does not contain any $\mathcal{O}(p^4)$ parameter, i.e., it is entirely given in terms of the pion mass and the pion decay constant $F = 92.4$ MeV.

Acknowledgement. The author would like to thank D. Drechsel, H.W. Fearing, T.R. Hemmert, B.R. Holstein, G. Knöchlein, J.H. Koch, A.Yu. Korchin, A.I. L'vov, A. Metz, and C. Unkmeir for a pleasant and fruitful collaboration on various topics related to virtual Compton scattering. It is pleasure to thank J.M. Friedrich, N. d'Hose, M.A. Moinester, and A. Ocherashvili for useful discussions on experimental issues in VCS.

---

[1]This argument works for any particle which is not its own antiparticle such as the K^+ or K^0. Of course, one could also employ the substitution $e^- \to e^+$.

334

References

1. F.E. Low: Phys. Rev. **96**, 1428 (1954)

2. M. Gell-Mann, M.L. Goldberger: Phys. Rev. **96**, 1433 (1954)

3. B.R. Holstein: Comments Part. Nucl. Phys. **19**, 221 (1990); Comments Part. Nucl. Phys. **20**, 301 (1992)

4. A.I. L'vov: Int. J. Mod. Phys. **A8**, 5267 (1993)

5. G. Audit et al.: CEBAF Report No. PR 93-050, 1993; J.F.J. van den Brand et al., CEBAF Report No. PR 94-011, 1994; G. Audit et al.: MAMI proposal *"Nucleon structure study by Virtual Compton Scattering."* 1995; J. Shaw et al.: MIT–Bates proposal No. 97-03, 1997

6. N. d'Hose: *In these Proceedings*

7. M.A. Moinester, A. Ocherashvili: Private Communication

8. P.A.M. Guichon, G.Q. Liu, A.W. Thomas: Nucl. Phys. **A591**, 606 (1995)

9. S. Scherer, A.Yu. Korchin, J.H. Koch: Phys. Rev. **C54**, 904 (1996)

10. H.W. Fearing, S. Scherer: Few-Body Syst. **23**, 111 (1998)

11. D. Drechsel, G. Knöchlein, A. Metz, S. Scherer: Phys. Rev. **C55**, 424 (1997)

12. D. Drechsel, G. Knöchlein, A.Yu. Korchin, A. Metz, S. Scherer: Phys. Rev. **C57**, 941 (1998); Phys. Rev. **C58**, 1751 (1998)

13. G.Q. Liu, A.W. Thomas, P.A.M. Guichon: Austral. J. Phys. **49**, 905 (1996)

14. M. Vanderhaeghen: Phys. Lett. **B368**, 13 (1996)

15. A. Metz, D. Drechsel: Z. Phys. **A356**, 351 (1996); Z. Phys. **A359**, 165 (1997)

16. T.R. Hemmert, B.R. Holstein, G. Knöchlein, S. Scherer: Phys. Rev. **D55**, 2630 (1997); Phys. Rev. Lett. **79**, 22 (1997)

17. M. Kim, D.-P. Min: *hep-ph/9704381*

18. B. Pasquini, G. Salmè: Phys. Rev. **C57**, 2589 (1998)

19. P.A.M. Guichon, M. Vanderhaeghen: Prog. Part. Nucl. Phys. **41**, 125 (1998)

20. S. Ragusa: Phys. Rev. **D47**, 3757 (1993); Phys. Rev. **D49**, 3157 (1994)

21. J. Gasser, H. Leutwyler: Ann. Phys. (N.Y.) **158**, 142 (1984)

22. U. Bürgi: Phys. Lett. **B377**, 147 (1996); Nucl. Phys. **B479**, 392 (1996)

23. C. Unkmeir, S. Scherer, A.I. L'vov, D. Drechsel: *In Preparation*

Few-Body Systems Suppl. 11, 335–338 (1999)

Few-
Body
Systems
© by Springer-Verlag 1999

Dispersion Theoretical Analysis of the Nucleon Spin Polarizabilities

B. Pasquini [1] D. Drechsel[1], G. Krein [1], and A. Metz[2]

[1] Institut für Kernphysik, Universität Mainz, Germany
[2] Institut für Theoretische Physik, Universität Heidelberg, Germany

Abstract. The spin polarizabilities of the nucleon have been calculated from pion photoproduction data using forward dispersion relations. The feasibility of an experimental determination of these structure constants is discussed by focusing on polarization observables of the reaction $\vec{\gamma}\vec{p} \to \gamma p$.

1 Real Compton scattering below pion threshold

Compton scattering off the nucleon at low energy provides a very useful tool for investigating the structure of the nucleon. In general, the scattering amplitude can be divided into the Born terms and a residual contribution. Below pion threshold, both contributions are real and can be expanded in a power series of the incoming photon energy ω. According to low-energy theorems, the Born amplitude is entirely given by ground-state properties of the nucleon. Excitation effects enter only at $\mathcal{O}(\omega^2)$ in the residual amplitude through the well known electric (α) and magnetic (β) polarizabilities. At $\mathcal{O}(\omega^3)$ four new parameters, the socalled spin polarizabilities, show up in the residual amplitude [1]. The spin polarizabilities characterize the leading response of the nucleon due to its spin in the presence of an external electromagnetic field.

Up to now, experimental constraints exist only for those linear combinations of the γ_i which enter the forward and backward scattering amplitude, and only recently the first experiment for a direct measurement of the individual γ_i has been proposed [2]. An estimate for the forward spin polarizability, $\gamma = \gamma_1 - \gamma_2 - 2\gamma_4$, has been obtained by a phenomenological analysis using photoproduction multipoles [3] leading to the results $\gamma^{(p)} = -1.34 \times 10^{-4}\,\mathrm{fm}^4$ and $\gamma^{(n)} = -0.38 \times 10^{-4}\,\mathrm{fm}^4$ (the units $10^{-4}\,\mathrm{fm}^4$ are implicitly understood hereafter). Recently, also the backward spin polarizability, $\gamma_\pi = \gamma_1 + \gamma_2 + 2\gamma_4$, has been extracted from a fit to the unpolarized Compton cross section at $\theta = 180°$ with the result $\gamma_\pi^{(p)} = -27.1 \pm 2.2 \pm 2.8$ [4].

2 Dispersion calculation of the spin polarizabilities

To perform the dispersion calculation we follow the approach proposed by L'vov et al. [5]. The Compton scattering amplitude is parametrized by Lorentz-invariant functions $A_i(\nu, t)$, $i = 1, ...6$, which are free of both kinematic constraints and singularities. The dispersion relations for these functions read

$$\text{Re}\, A_i(\nu, t) = A_i^{\text{Born}}(\nu, t) + \frac{2}{\pi} \text{P} \int_{\nu_{\text{thr}}}^{\nu_{\text{max}}} \text{Im}\, A_i(\nu', t) \frac{\nu' d\nu'}{\nu'^2 - \nu^2} + A_i^{\text{as}}(t). \qquad (1)$$

According to eq. (1), the real part of the A_i is given by the sum of the Born terms, the contribution of the dispersion integral running from pion threshold ν_{thr} up to ν_{max}, and an asymptotic t-channel term. The asymptotic contribution is parametrized by the t-channel exchange of the π^0 in the case of A_2 and by the exchange of a "σ meson" in the case of A_1. For the remaining functions such an asymptotic contribution is quite negligible, at least for low energies. Using the relations between the γ_i and the functions $A_i(\nu, t)$ at $\nu = 0$ and $t = 0$, the desired dispersion results can be obtained [6].

In order to evaluate the dispersion integral in eq. (1) we saturate the imaginary part by one-pion photoproduction amplitudes using multipoles from the analysis of Hanstein, Drechsel and Tiator (HDT) [7]. For the upper limit of the integral we choose $\nu_{\text{max}} = 500$ MeV.

Table 2:1 Separate contributions to the spin polarizabilities of the proton: our dispersion theoretical predictions using the HDT multipoles [6] are shown in comparison with the results of HBChPT [8]. (See also text.)

	DR(HDT)			HBChPT		
$\gamma_i^{(p)}$	π^0-exchange	excitation	sum	WZW	loops	sum
$\gamma_1^{(p)}$	-22.5	$+5.1$	-17.4	-22.0	$+4.4$	-17.6
$\gamma_2^{(p)}$	0	-1.1	-1.1	0	-0.3	-0.3
$\gamma_3^{(p)}$	$+11.2$	-0.6	$+10.6$	$+11.0$	$+1.1$	$+12.1$
$\gamma_4^{(p)}$	-11.2	$+3.4$	-7.9	-11.0	$+1.3$	-9.7

Our results for the spin polarizabilities of the proton are listed in Table 2.1 and compared with a recent calculation in heavy baryon ChPT (HBChPT) [8]. For the t-channel contribution to $A_2^{\text{as}}(t)$ we use the parametrization of Ref. [5], which gives results very close to the Wess-Zumino-Witten term (WZW) in HBChPT. However, the contribution of the dispersion integral ("excitation") of our calculation and the pion-loop contribution of HBChPT show significant differences. The dependence of the dispersion calculation on the involved multipoles is extensively discussed in Ref. [6], where predictions from the HDT and the SAID-SP97K multipoles have been compared. These two multipole analyses mainly differ in the behaviour of the E_{0+} partial wave near threshold [6]. In the case of the forward spin polarizability, which is characterized by large cancellations between the dominant E_{0+} and M_{1+} multipoles, the HDT multipoles result in $\gamma^{(p)} = -0.6$, while the SAID multipoles lead to $\gamma^{(p)} = -1.2$. A further uncertainty of the dispersion approach is the t-channel term $A_2^{\text{as}}(t)$ which could also receive contribution from heavier meson. The most sensitive observable to clarify this uncertainty is γ_π. Our value of $\gamma_\pi^{(p)} = -34.3$ has to be compared with the above quoted experimental value of Ref. [4].

3 Polarization observables in $\vec{\gamma}\,\vec{p} \to \gamma\,p$ experiments

Experimentally the spin polarizabilities can be investigated by means of the $\vec{\gamma}\,\vec{p} \to \gamma\,p$ reaction. Below pion threshold, the cross section of this reaction can be expressed in terms of four response functions,

$$\frac{d\sigma}{d\Omega} = \Psi \left(W_{\text{T}} + \mathcal{P}\, P_z\, W_{\text{T}\text{T}'}^z + \mathcal{P}\, P_x\, W_{\text{T}\text{T}'}^x + \mathcal{L} \cos 2\Phi\, W_{\text{T}\text{T}}^0 \right). \tag{2}$$

In eq. (2) Ψ is the phase space factor. The degree of circular polarization of the photon is denoted by $\mathcal{P}$ while the degree of linear polarization is given by $\mathcal{L}$. The vector $\boldsymbol{P} = (P_x, P_y, P_z)$ specifies the polarization of the target nucleon. Our coordinate system is fixed by the photons in such a way that the incoming photon propagates along the z-axis and the momenta of both photons lie in the xz-plane. Finally, Φ is the angle between the linear polarization of the photon and the x-axis.

Most sensitive to the spin polarizabilities are the response functions with a circularly polarized photon, $W_{\text{T}\text{T}'}^z$ and $W_{\text{T}\text{T}'}^x$. The first one can be measured from the difference of the cross sections with parallel or antiparallel polarization of the nucleon with regard to the z-axis, $A_\parallel = 2\Psi W_{\text{T}\text{T}'}^z$. The latter one is obtained from the difference of the cross sections with the polarization of the proton along the positive or negative x-axis, $A_\perp = 2\Psi W_{\text{T}\text{T}'}^x$. From the low energy expansion of $A_\parallel$ and $A_\perp$ at different scattering angles, we obtain information about the individual spin polarizabilities: at $\theta = 90°$, $A_\perp$ and $A_\parallel$ receive, to leading order in ω, a contribution from γ_4 and $\gamma_1 + \gamma_3$, respectively, while γ and γ_π can be extracted from $A_\parallel$ at $\theta = 0°$ and $\theta = 180°$, respectively. In Fig. 4 we show $A_\parallel$ and $A_\perp$ calculated by expanding the scattering amplitude up to third order in ω. To study the influence of the γ_i we have plotted the results with and without the effects of the spin polarizabilities. Because of the large value of γ_π, the most significant effect shows up in $A_\parallel$ at $180°$, while in the forward direction the influence of the polarizabilities is small. At $\theta = 90°$, $A_\perp$ is most favourable, with effects up to 35%. The convergence of the low-energy expansion will be discussed in an upcoming publication [9]. A similar dispersion analysis with the SAID multipoles has been performed recently [10].

4 Summary

Predictions for the spin polarizabilities of the nucleon have been obtained from single-pion photoproduction multipoles using dispersion relations. The sensitivity of the results to different sets of multipoles has been discussed. Double

338

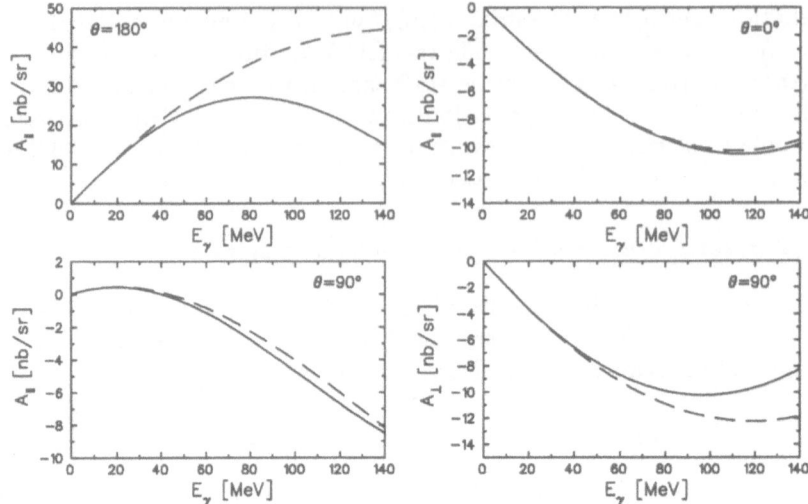

Figure 1. $A_{\parallel}$ and $A_{\perp}$ at different values of the scattering angle as function of the photon energy in the laboratory frame. The solid curves correspond to the low-energy expansion with the HDT dispersion results for the $\gamma_i^{(p)}$'s (Table 2.1). The dashed curves are the corresponding results with $\gamma_i^{(p)} = 0$, $i = 1, .., 4$.

polarization experiments at low energy give access to the spin polarizabilities. The calculated polarization observables indicate that high quality data are required to disentangle the individual polarizabilities.

Acknowledgement. The work of B.P. was supported by the Marie Curie-TMR Project.

References

1. S. Ragusa: Phys. Rev. **D47**, 3757 (1993); Phys. Rev. **D49**, 3157 (1994)

2. R. Miskimen: MIT-Bates design report (1996)

3. A.M. Sandorfi et al.: Phys. Rev. **D50**, R6681 (1994)

4. J. Tonnison et al.: Phys. Rev. Lett. **80**, 4382 (1998)

5. A.I. L'vov et al.: Phys. Rev. **C55**, 359 (1997)

6. D. Drechsel et al.: Phys. Lett. **B420**, 248 (1998)

7. O. Hanstein et al.: Nucl. Phys. **A632**, 561 (1998)

8. T.R. Hemmert et al.: Phys. Rev. **D57**, 5746 (1998)

9. D. Drechsel et al.: in preparation

10. D. Babusci et al.: Phys. Rev. **C58**, 1013 (1998)

Few-Body Systems Suppl. 11, 339–346 (1999)

The Spin of the Nucleon

F. Kunne

DAPNIA / SPhN, CEA Saclay, France

Abstract. We present here the status of the experimental studies on the spin structure of the nucleon. Recent results on spin structure functions measurements are shown. Global QCD analyses of all available data allow to determine the singlet and non singlet polarized parton distributions; they confirm the small contribution of quark spin to the nucleon spin. The results indicate a possible sizable contribution from the gluon (ΔG) to the spin of the nucleon. Direct measurements of ΔG foreseen at future experiments are described.

1 Spin structure function measurements

The total spin of the nucleon is usually written as $(1/2) = (1/2)\Delta\Sigma + \Delta G + L_z$, where $\Delta\Sigma$ represents the contribution from the spin of the quarks, ΔG from the spin of the gluons and L_z from the angular momentum. In order to study how the spin is distributed among the constituents of the nucleon, extensive measurements of spin structure functions have been performed at SLAC, CERN/EMC and SMC, and more recently at DESY/HERMES. Data were taken with proton, neutron and deuteron targets in a wide kinematic range. (Fig.1)

The spin structure functions are measured in inclusive deep inelastic scattering of polarized leptons off polarized nucleons. The two kinematic variables used to describe the process are Q^2, the four momentum of the virtual exchanged photon, and $x = Q^2/(2M_p.(E_\mu - E'_\mu))$; x represents the fraction of the proton momentum carried by the struck quark. The polarized cross section is written in terms of two spin structure functions g_1 (longitudinal) and g_2 (transverse) that can be extracted from the measurements of spin asymmetries. In the framework of the quark parton model, $g_1(x)$ has a simple interpretation : it can be written as $g_1(x) = (1/2)\sum_f e_f^2 \Delta q_f(x)$ where the index f runs over all quark flavors $u, \bar{u}, d, \bar{d}, s, \bar{s}$; $\Delta q_f(x)$ are the polarized quark distributions.

The results obtained for the longitudinal spin structure functions g_1 of the proton, the deuteron and the neutron [1, 2, 3] are shown in Fig.2. All the data, although measured at various Q^2, are evolved and shown at the common Q^2 of 5 GeV^2. There is a good agreement between all the data. They cover a

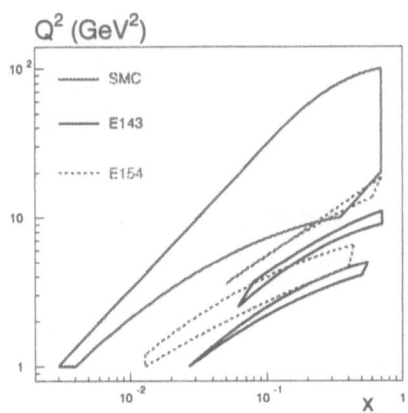

Figure 1. Kinematic x and Q^2 range covered by the CERN/SMC and SLAC/E143-E154 g_1 data.

Table 1. The experiments measuring g_1

	Beam GeV	Polarized target	Q^2 GeV^2	x_{min} $Q^2 > 1$	
SLAC	50	^3He(E142,E154) NH_3 ND_3(E143,E155) ^6LiD(E155)	3.6	0.014	high statistics
SMC	190	**p**- and **d**- butanol NH_3	10	0.003	low x, highQ^2
HERMES	28	^3He	2.5	0.023	pure target, particle id.

large kinematic domain in x, from the low value of $x = 10^{-3}$ for the SMC experiment, up to $x = 0.8$. The SLAC data (E143, E154) show high statistical precision. Additional data, still preliminary and not shown in the figure, were taken recently at SLAC (E155) on the proton with even more statistics. But even with this precision, the data never reach the precision obtained for the unpolarized structure function of the nucleon F_2. They require the measurement of spin asymmetries which are very small, due to the limited beam and target polarizations and, in the case of solid targets by the limited fraction of polarizable nucleons in the target material.

In order to compare the measurements with the Bjorken and Ellis-Jaffe sum rules predictions, one needs to calculate the first moment of g_1. For this, data are needed in the whole range of x and at a common Q_0^2, while they are measured in a limited region in x, and at Q^2 varying with x, even within a given experiment. So, two problems appear ; first the evolution of the data

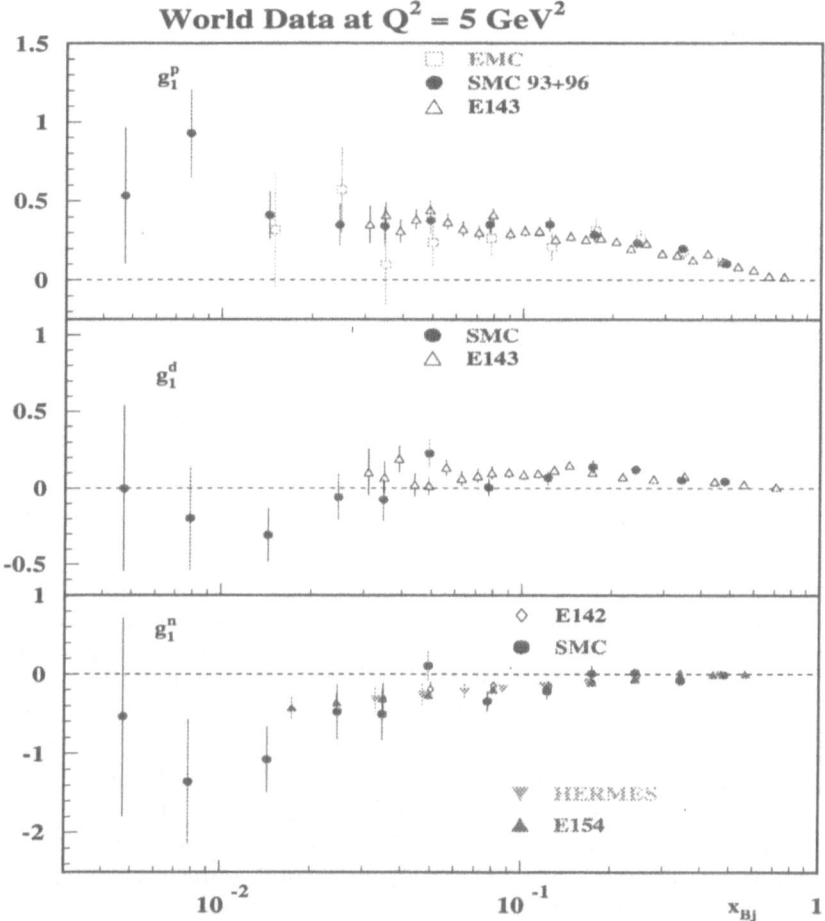

Figure 2. Data on the spin structure function g_1 of the proton, deuteron and neutron measured at CERN, SLAC and DESY. All data are evolved to $Q^2 = 5 GeV^2$

to a common Q_0^2 ; this will be discussed in the next section where the next to leading order QCD analysis for the Q^2 evolution of g_1 is explained. Then remains the problem of the extrapolation of g_1 to the unmeasured region of x. The high x region does not contribute much, while the contribution from the small x appears to be crucial. The behaviour at low x is not known, and when keeping only data with $Q^2 > 1 GeV^2$, the HERMES experiment can reach only the minimum value of $x = 0.023$, SLAC the value 0.014 and SMC 0.003. Up to 1997, a Regge behaviour of g_1 was assumed, with a constant value of g_1 down to $x = 0$. In general 100 % of error was assigned on this low x contribution to $\Gamma_1 = \int_0^1 g_1(x)dx$. This was not enough as can be seen from recent data on

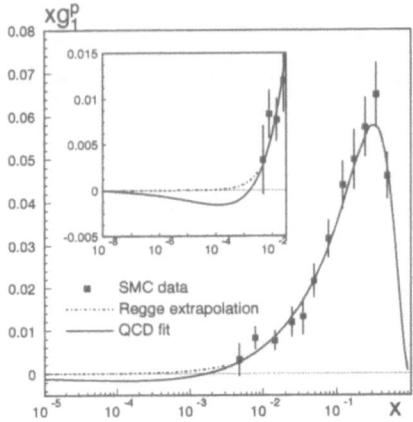

Figure 3. Low x behaviour of g_1^p : comparison of Regge type and QCD fits.

neutron from SLAC, or on proton from SMC. For instance for the neutron, SLAC[2] quotes for the first moment of g_1^n, a value of -0.041 when using a Regge extrapolation, and a value of -0.2 (5 times more) when using a power fit as $c.x^{-0.9}$. Looking now at the low x proton data (Fig.3), for the unmeasured part of the integral at low x, SMC finds a contribution of 0.002 ± 0.002 with a Regge type extrapolation, and a negative and much higher contribution of -0.011 ± 0.011 with a QCD fit [1]; here the error happens to be equal to the value only by chance. The latter contribution represents 10% of the total Γ_1^p value ! In conclusion the ignorance on the behaviour of g_1 at low x leads to uncertainties as high as 10 to 20 % of the value of the first moments; they are today the dominant uncertainties and are by far underestimated when taken equal to 100 % of the value calculated with a Regge fit.

2   Global QCD analysis of g_1 world data; determination of $\Delta\Sigma$

A major recent progress in the theory related to spin structure functions, is the calculation of polarized splitting functions, up to next-to-leading order (NLO) in α_s [4]. These functions are needed in the DGLAP equations which rule the Q^2 dependences of the polarized parton distributions $\Delta\Sigma$, ΔG and Δq_{NS} (quark non-singlet). Thus g_1, which is a combination of the above parton distributions can also evolve to different Q^2 values. Note that in the Q^2 evolution equations $\Delta\Sigma$ and ΔG are coupled, while Δq_{NS} evolves separately.

Global QCD analyses at NLO of all available g_1 data on proton, deuteron and neutron are performed by several authors. The aim is twofold. First, they provide the Q^2 evolution of g_1, needed to evolve all data to a common Q^2, in order to calculate the first moments and compare them with sum rules estimations. Second, they also give as an output a parameterization of the parton distributions. The analyses by different authors use similar methods and find similar results in general. A comparison of the different analyses and results is given in [5]. We give only here the results from one of the analyses (SMC [6]) which reflects the general trend.

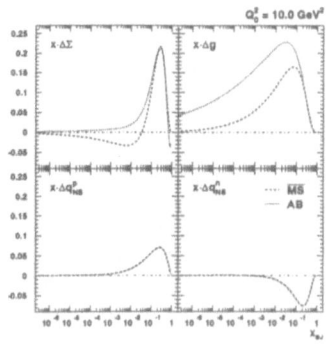

Figure 4. The polarized parton distributions $x\Delta\Sigma, x\Delta G$ and $x\Delta q_{Non-Singlet}$ from a global QCD fit of all g_1 data. The parameterizations for $\Delta\Sigma$ and ΔG depend on the chosen renormalization and factorization schemes

The problem which is encountered in the QCD analyses is their dependence on the renormalization and factorization schemes. The polarized splitting functions and the parton distributions depend on the choice of these schemes, while g_1 does not. We show here the results obtained when using two widely used schemes, the $\overline{MS}$ and the Adler-Bardeen (AB) schemes : ΔG contributes to the first moment of g_1 in the AB scheme, and not in the $\overline{MS}$ one. The consequence is that $\Delta\Sigma$ is not defined in the same way in both schemes. However, the relation between the two definitions is well established: $\Delta\Sigma_{\overline{MS}}(Q^2) = \Delta\Sigma_{AB} - (n_f\alpha_s/2\pi)\Delta G_{AB}(Q^2)$. The Fig.4 shows the polarized parton distributions obtained from the global QCD analysis of g_1 in both schemes. As expected the scheme dependence affects $\Delta\Sigma$. Both fits give a similar good description of g_1 data. This is illustrated in Fig.5 which shows the $\overline{MS}$ g_1 fit at the measured Q^2 of each data point. The values obtained from the global fit of world data performed by SMC, for the first moments of $\Delta\Sigma$ and ΔG at $Q^2 = 1GeV^2$ are given in Tab.2.

Table 2. First moments of $\Delta\Sigma$ and ΔG in the $\overline{MS}$ and AB schemes obtained from a global QCD NLO analysis of all g_1 data[6].

	$\overline{MS}$	AB
$\Delta\Sigma$	0.19 ± 0.05	0.38 ± 0.03
ΔG	$0.25^{+0.29}_{-0.22}$	$0.99^{+1.17}_{-0.31}$

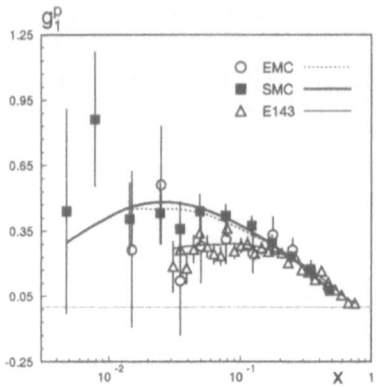

Figure 5. Global QCD fit of g_1 world data. Results of the fit are shown at the measured Q^2 of each g_1^p data point. The discrepancy between the SMC and E143 curves calculated at significantly different Q^2 values, reflects the scaling violation of spin structure functions.

From the above results, we conclude that $\Delta\Sigma_{\overline{MS}}$ and $\Delta\Sigma_{AB}$ are well determined. The data indicate also a possible large contribution of ΔG, which would increase with Q^2, from the value of 1 at $Q^2 = 1 GeV^2$ in the AB scheme to 2 at $Q^2 = 10$. However, the sensitivity to ΔG is very low. The x distribution of ΔG is not constrained by the g_1 data, and could be very different from the solutions shown in Fig.4.

3   Bjorken sum rule

The Bjorken sum rule is a fundamental result in pQCD that relates the first moments of g_1 for the proton and the neutron : $\Gamma_1^p - \Gamma_1^n = (1/6).(g_A/g_V).C_1^{NS}(Q^2)$. g_A/g_V is the axial vector coupling constant, and C_1^{NS} is a coefficient function known to fourth order in α_s [7]. To test the validity of the sum rule, the first moments of g_1^p and g_1^n are calculated from the QCD global analysis, but this time without any assumption on the value of g_A/g_V. The results from the global SMC analysis [6] is : $\Gamma_1^p - \Gamma_1^n = 0.174^{+0.024}_{-0.012}$ at $Q^2 = 5 GeV^2$ which is in excellent agreement with the theoretical value of 0.181 ± 0.003 [7].

4   Semi-inclusive measurements

In addition to *inclusive* polarized deep inelastic scattering measurements, *semi-inclusive* data $(\vec{\mu}\vec{p} \to \mu h^{\pm} X)$ where an outgoing charged hadron $h^{\pm}$ is detected, were also taken at SMC[1], and more recently at HERMES. In such a process, the outgoing leading hadron tags the flavor of the struck quark. Combining data from different polarized targets (p, d or n), one can extract separately the polarized valence and non-strange sea quarks distributions. Fig.6 shows the results for $x\Delta u_v(x), x\Delta d_v(x)$ and $x\Delta\bar{q}(x)$ obtained by SMC.

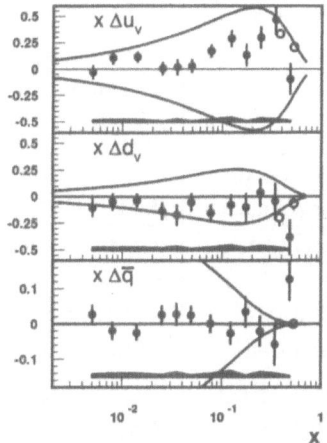

Figure 6. Polarized parton distributions $x\Delta u_v(x)$, $x\Delta d_v(x)$ and $x\Delta\bar{q}(x)$ measured by SMC. The full curves show the boundaries given by unpolarized distributions $\pm q(x)$.

5   Direct measurements of ΔG

We have seen that indirect determination of ΔG from QCD analyses of spin structure functions was not successful, given the present uncertainties on the g_1 data. The contribution of the gluon to the spin of the nucleon should then be accessed through direct measurements. This problem has been discussed and summarized in [8]. In the unpolarized case, the process of photon-gluon fusion ($\gamma G \to q\bar{q}$) has been used with success to probe the gluon distribution in the proton $G(x)$. If produced with initial polarized particles this process will be sensitive to ΔG. Several channels in the final state can be studied. For instance $c\bar{c}$ pairs will produce charmed mesons (D^0 or $\bar{D}^0$) identified by their decay into pions and kaons, or produce J/Ψ's decaying into muons or electrons pairs. $q\bar{q}$ pairs will hadronize in two jets. Calculations at leading order of the the polarized cross section of the elementary process $\gamma G \to q\bar{q}$, show that this reaction provides a significant analysing power for the measurement of $\Delta G/G$ with lepton beams of a few hundreds of GeV.

ΔG can also been accessed from other processes in $\bar{p}p$ collisions. As in the unpolarized case, several channels like the production of inclusive photons, of photons plus jet, or of two jets, will be used to probe the gluon contribution.

The Tab.3 gives a summary of all projects which aim at direct measurements of ΔG. Among them, only COMPASS [9] at CERN and RHIC at Brookhaven are approved. Both should start taking data around year 2000-2001.

Table 3. Projects for ΔG direct measurements

	process	x_g range	$\delta(\Delta G/G)$
RHIC	$\vec{p}\vec{p} \to \gamma$, jets	$0.01 < x < 0.4$	0.01 - 0.3
COMPASS	$\vec{\mu}\vec{p} \to$ charm	$< x > \simeq 0.1$	0.10
	high p_T hadrons	$0.04 < x < 0.3$	$\simeq 0.04$
SLAC(real γ)	$\vec{\gamma}\vec{p} \to$ charm	$0.1 < x < 0.4$	0.02
HERA	$\vec{e}\vec{p} \to 2$ jets	$0.0015 < x < 0.13$	0.007 - 0.09
(polarized p)	g_1 evolution	$0.0001 < x < 0.3$	$\delta(\Delta G) \simeq 0.15$
	$\vec{p}\vec{p}$ fixed target	$0.1 < x < 0.4$	0.03 - 0.4
HERA(real γ)	$\vec{\gamma}\vec{p} \to J/\psi$	$0.3 < x < 0.5$	0.15

6 Conclusions

The spin structure function g_1 has been measured extensively at CERN, SLAC and DESY, in a wide kinematic range. NLO QCD analyses of these data indicate a low value for the contribution of the quarks to the nucleon spin, and a possible large contribution of ΔG. RHIC and COMPAS experiments starting around year 2000, will provide direct measurements of $\Delta G/G$. Further projects are also proposed at HERA and SLAC.

References

1. D. Adams et al., (SMC Collaboration): Phys. Lett. **B396**, 338 (1997); Phys. Rev. **D56** 5330 (1997); B. Adeva et al.: Phys. Lett. **B412**, 414 (1997)

2. SLAC, K. Abe et al.: Phys. Lett. **B405**, 180 (1997)

3. HERMES, A. Airapetian et al.: hep-ex/9807015

4. R. Mertig and W.L. van Neerven: Z. Phys. **C170**, 637 (1996); W. Vogelsang: Nucl. Phys. **B475**, 47 (1996)

5. G.Altarelli, R.Ball, S.Forte and G.Ridolfi: hep-ph/9803237

6. SMC, B.Adeva et al.: Phys. Rev. **D58**, 112001 (1998); *ibid* **D58**, 112002 (1998)

7. S.Larin et al.: Phys. Rev. Lett. **66**, 862 (1991); Phys. Lett. **B259**, 345 (1991)

8. D.von Harrach, W.-D. Nowak and J.Soffer: hep-ph/9712207

9. COMPASS proposal, CERN; A. Bravar et al.:Phys. Lett. **B421**, 349 (1998)

Few-Body Systems Suppl. 11, 347–354 (1999)

Few-
Body
Systems
© by Springer-Verlag 1999

Nucleon Structure Functions and Light-Front Dynamics

M. Traini [a,*], Pietro Faccioli [a], Vicente Vento [b,†]

[a] Dipartimento di Fisica, Università degli Studi di Trento, and Istituto Nazionale di Fisica Nucleare, G.C. Trento, I-38050 POVO (Trento), Italy
[b] Departament de Física Teòrica, Universitat de València, and Institut de Física Corpuscular, Centre Mixt Universitat de València, Consejo Superior de Investigaciones Científicas, E-46100 Burjassot (València) Spain

Abstract. We present a quark-parton model to describe polarized and unpolarized nucleon structure functions. The twist-two matrix elements for the QCD evolution analysis of lepton-hadron scattering are calculated within a light-front covariant quark model. The relativistic effects in the three-body wave function are discussed for both the polarized and unpolarized cases. Predictions are given for the polarized gluon distributions as will be seen in future experiments.

1 Introduction

The description of deep inelastic lepton-hadron scattering data requires sophisticated and sometimes ad hoc parametrizations showing the complexity of the mechanisms involved in the description of the hadron structure. The research reported here analyzes how one can visualize these mechanisms in a scheme which unifies the description of low and high energy phenomena. We discuss a radiative approach which makes use of Quark Models (QM) to calculate the values of the twist two nucleon matrix elements occurring in the QCD analysis of lepton hadron scattering. QM require a reinterpretation in order to be used in conjunction with QCD perturbation theory. Jaffe and Ross [1] proposed that the quark model calculation of matrix elements give their values at a hadronic scale μ_0^2 and that for all larger Q^2 their coefficient functions evolve according to perturbative QCD.

Our formalism puts all these ingredients into a predictive scheme [2, 3]. We consider the nucleon to be consistent of valence quarks and gluons at the

---

*E-mail address: traini@science.unitn.it
†E-mail address: vicente.vento@uv.es

hadronic scale $Q^2 = \mu_0^2$ and generate the partonic content at $Q^2 \gg \mu_0^2$ dynamically via bremsstrahlung radiation of gluons and sea from the original system. The input distributions are explicitly related to the electromagnetic response of the constituent quark model which represents the non-perturbative part of the calculation. The investigations developed so far [2, 3, 4, 5, 6] have been based on non-relativistic quark model wave functions. Here we will demonstrate that the same approach can be used to incorporate realtivistic covariance in a rather transparent way and to this aim we develop a relativistic quark model making use of the light-front hamiltonian dynamics (for reviews cfr. refs. [7]).

2 Parton distributions at the hadronic scale μ_0^2

The parton distributions at the hadronic scale are assumed to be valence quarks and gluons, and their twist two component is determined by the quark momentum density (cfr. ref.[3])

In the light-front quark model the intrinsic momenta of the constituent quarks (k_i) can be obtained from the corresponding momenta (p_i) in a generic reference frame through a light-front boost $(k_i = \mathcal{L}_f^{-1}(P_{\text{tot}})\, p_i, \ P_{\text{tot}} \equiv \sum_{i=1}^{3} p_i)$ such that the Wigner rotations reduce to indentities. With the specific choice $\mathcal{L}_f^{-1}(P_{\text{tot}})\, P_{\text{tot}} = (M_0, 0, 0, 0)$, one has $\sum_{i=1}^{3} \mathbf{k}_i = 0$ and $M_0 = \sum_{i=1}^{3} \omega_i = \sum_{i=1}^{3} \sqrt{\mathbf{k}_i^2 + m_i^2}$. The nucleon state is characterized by isospin (and its third component), parity, light-front (non-interacting) angular momentum operators J and projection $J_{\hat{n}}$, where the unitary vector $\hat{n} = (0, 0, 1)$ defines the spin quantization axis. The nucleon state factorizes into $|N, J, J_n\rangle\, |\tilde{P}\rangle$ where $\tilde{P}$ is the total light-front nucleon momentum $\tilde{P} \equiv (P^+, \mathbf{P}_\perp) = \tilde{p}_1 + \tilde{p}_2 + \tilde{p}_3$. $P^+ = P^0 + \hat{n} \cdot \mathbf{P}$ and the subscript $\perp$ indicates the perpendicular projection with respect to the $\hat{n}$ axis. In order to achieve the ordinary composition rules, the intrinsic light-front angular momentum eigenstate $|N, J, J_n\rangle$ must be obatined from the *canonical* angular momentum eigenstate $|N, j, j_n\rangle$ by means of a unitary transformation which is a direct product of generalized Melosh rotations [8]. Finally the intrinsic part of the nucleon state, $|N, j, j_n\rangle$ is eigenstate of the mass operator $(M_0 + V)\, |N, j, j_n\rangle = M\, |N, j, j_n\rangle$, where the interaction term V must be independent on the total momentum P_{tot} and invariant under spatial rotations (cfr. refs.[7]).

In the present work we will discuss results of a confining mass equation of the following kind

$$(M_0 + V)\, \psi_{0,0}(\xi) \equiv \left(\sum_{i=1}^{3} \sqrt{\mathbf{k}_i^2 + m_i^2} - \frac{\tau}{\xi} + \kappa_l\, \xi \right) \psi_{0,0}(\xi) = M\, \psi_{0,0}(\xi)\,, \quad (1)$$

where $\xi = \sqrt{\rho^2 + \lambda^2}$ is the radius of the hypersphere in six dimension and ρ and λ are the intrinsic Jacobi coordinates $\rho = (\mathbf{r}_1 - \mathbf{r}_2)/\sqrt{2}$, $\lambda = (\mathbf{r}_1 + \mathbf{r}_2 - 2\,\mathbf{r}_3)/\sqrt{6}$ (solutions for non-relativistic reductions of Eq.(1) have been discussed by Ferraris *et al.* [9]).

The intrinsic nucleon state is antisymmetric in the color degree of freedom and symmetric with respect the orbital, spin and flavor coordinates. In particular, disregarding the color part, one can write $|N, J, J_n = +1/2\rangle = \psi_{0,0}(\xi)\, \mathcal{Y}^{(0,0)}_{[0,0,0]}(\Omega)\, [\chi_{MS}\phi_{MS} + \chi_{MA}\phi_{MA}]\,/\sqrt{2}$, where $\psi_{\gamma,\nu}(\xi)$ is the hyperradial wave function solution of Eq. (1), $\mathcal{Y}^{(L,M)}_{[\gamma,l_\rho,l_\lambda]}(\Omega)$ the hyperspherical harmonics defined in the hypersphere of unitary radius, and ϕ and χ the flavor and spin wave function of mixed $SU(2)$ symmetry. Let us note that, in order to preserve relativistic covariance, the spin wave functions have to be formulated by means of the appropriate Melosh transformation of the ith quark spin wave function:

We have solved the mass equation (1) numerically by expanding the hyperradial wave functions $\psi_{\gamma\nu}(\xi)$ on a truncated set of hyperharmonic oscillator basis states [10]. Making use of the Rayleigh-Ritz variational principle the HO constant has been determined and convergence has been reached considering a basis as large as 17 HO components. The parameters of the interaction, have been determined phenomenologically in order to reproduce the basic features of the (non strange) baryonic spectrum up to ≈ 1600 MeV, namely the position of the Roper resonace and the average value of the 1^- states[1]. We obtain: $\tau = 3.3$ and $\kappa_l = 1.8$ fm^{-2} [10] to be compared with the corresponding non-relativistic fit $\tau = 4.59$ and $\kappa_l = 1.61$ fm^{-2} [9]. The constituent quark masses have been chosen $m_u = m_d = m_q = M_N/3$.

As a result a huge amount of high momentum components is generated in solving the mass equation (cfr. Fig. 1.), and they play an important role in the evaluation of transitions and elastic form factors within light-front constituent quark models as discussed by Cardarelli et $al.$ [11] in connection with the solutions of the Isgur-Capstick model Hamiltonian.

The effects of the high momentum components on the unpolarized parton distributions at the hadronic scale are shown on the right panel of Fig. 1. Their important role to reproduce the behaviour of the structure functions for large value of the Bjorken variable x will be discussed in the next section. The relevant effects of relativistic covariance are even more evident looking at the polarized distributions [12]. In that channel the introduction of Melosh transformations results in a substantial suppression of the responses at large values of x and in an enhancement for $x \lesssim 0.15$ as it can be seen from Fig. 2.

3   Numerical results and comparison with the experimental data

The results we are going to comment are related to two scenarios according to the assumption on the gluon distribution at the hadronic scale $G(x, \mu_0^2)$:

i) scenario A: Quark model or extreme scenario, defined in such a way that only valence quarks exist at the hadronic scale (i.e. $G(x, \mu_0^2) = 0$). One has [4], $\mu_0^2 = 0.094$ GeV^2 at NLO ($[\alpha_s(\mu_0^2)/(4\pi)]_{\text{NLO}} = 0.142$).

ii) scenario B: Partonic scenario, characterized by the existence of valence quarks and gluons at the hadronic scale. A natural choice for the unpolarized gluon distribution within the present approach, has been discussed in refs.[3, 4]

---

[1]The well known problem of the energy location of the Roper resonance is solved, in the present case, by the use of $1/\xi$ potential, as discussed in the non-relativistic case by Ferraris et $al.$ [9].

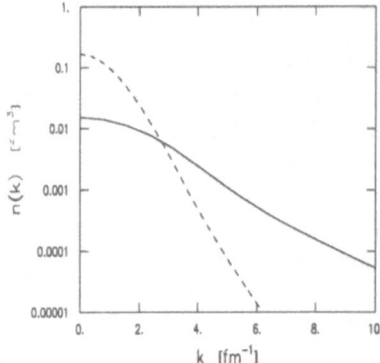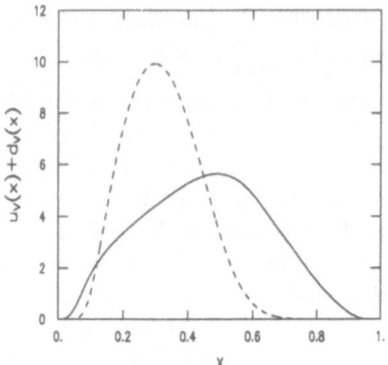

Figure 1. Left panel: $n(\mathbf{k}^2) = \sum_q [n_q^\uparrow(\mathbf{k}^2) + n_q^\downarrow(\mathbf{k}^2)]$ (the valence quark momentum distributions) as function of $|\mathbf{k}|$. Relativistic results: full curve, non-relativistic approximation: dashed curve. On the right panel the corresponding total valence distributions $u_V(x, \mu_0^2) + d_V(x, \mu_0^2)$, at the hadronic scale.

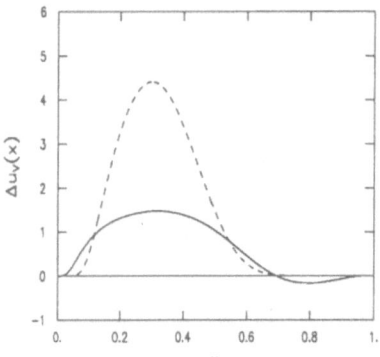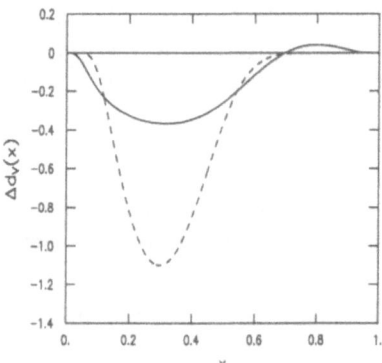

Figure 2. Left panel: the polarized distribution $\Delta u_V(x, \mu_0^2)$ as function of x within the relativistic (full curve) and non relativistic (dot-dashed curve) schemes. On the right panel the distribution $\Delta d_V(x, \mu_0^2)$ same notations as in Fig. 1.

and it assumes the *valence-like* form $G(x, \mu_0^2) = \mathcal{N}_g \left[u_V(x, \mu_0^2) + d_V(x, \mu_0^2) \right] / 3$. As a consequence $\int G(x, \mu_0^2) \, dx = 2$ and only 60% of the total momentum is carried by the valence quarks at the scale μ_0^2.

If the gluons were fully polarized one would have $|\Delta G(x, \mu_0^2)| = G(x, \mu_0^2)$, which reduces to $\Delta G(x, \mu_0^2) = f \, G(x, \mu_0^2)$ introducing the fraction f of polarized gluons, to be considered with the appropriate sign. As an example we discuss results followig a suggestion due to Jaffe [13]: $f \approx -0.35$ ($\int dx \, \Delta G(x, \mu_0^2) \approx -0.7$). In this case one obtains a consistent lower bound to the x-dependence of $\Delta G(x, \mu_0^2)$ [2]. In this case $\mu_0^2 = 0.220 \, \text{GeV}^2$ at NLO ($[\alpha_s(\mu_0^2)/(4\pi)]_{\text{NLO}} = 0.053$).

In Fig. 3 the results for the proton structure function $g_1^p(x, Q^2)$ are shown and compared with the experimental data within scenario A. The non relativistic approximation appears to reproduce rather poorly the experimental observations, a result already discussed in ref.[4] for other non-relativistic quark model wave functions. The introduction of relativistic covariance in the quark wave function, mainly due to spin dynamics induced by the Melosh rotations [12], leads to a suppression of the structure in the small-x region ($x \lesssim 0.5$), Such a large effect brings the theoretical predictions quite close to experimental data in the region $0.01 \leq x \lesssim 0.4$, even under the simple assumption of a pure valence component at the hadronic scale (scenario A). We stress that the calculation is parameter-free and the only adjustable parameters (τ and κ_l in Eq. (1)) have been fixed to reproduce the low-lying nucleon spectrum as already discussed.

Let us comment also on the comparison of LO versus NLO calculations. The differences shown in Fig. 3 indicate the relevance of higher order corrections in our parton model approach. The initial scale μ_0^2 is rather low, and NLO corrections have to be included.

In order to introduce gluons we evolve the unpolarized distributions predicted by the scenario A, up to the scale of scenario B where 60% only of the total momentum is carried by valence partons. At that scale the fraction of polarized gluons is chosen to be negative, according to the Jaffe result [13] (scenario B). Looking at the Figs. 3,4 one can conclude that the low-x data on g_1^p do not constrain the gluon strongly. If the fraction of polarized gluons varies from 35% to 100% the quality of the agreement is deteriorated in the region $0.01 \leq x \lesssim 0.4$ only slightly. For larger values of x the valence contribution plays a major role and the behaviour of the structure functions will depend largely on the potential model.

The comparison of the predicted neutron structure function with the data (Figs. 3,4) differs quite substantially according to the amount of polarized gluons at the hadronic scale.

Within scenario A the values of $x \, g_1^n(x, Q^2)$ remain quite small according to the fact that the mass operator (1) is $SU(6)$ symmetric, while the introduction of *negative* gluon polarization, as suggested by Jaffe, brings the predictions of

---

[2]In fact in ref.[13] it has been shown that $\int \Delta G(x, \mu_0^2) \, dx < 0$. Such inequality does not imply $\Delta G(x, \mu_0^2) < 0$ in the whole x-range. We are therefore investigating a lower bound to $\Delta G(x, \mu_0^2)$.

352

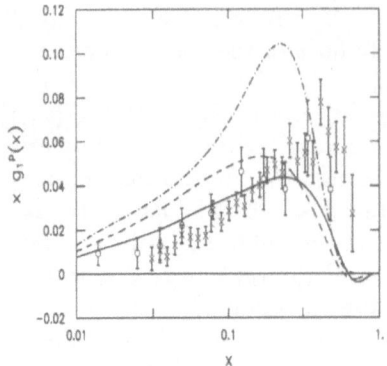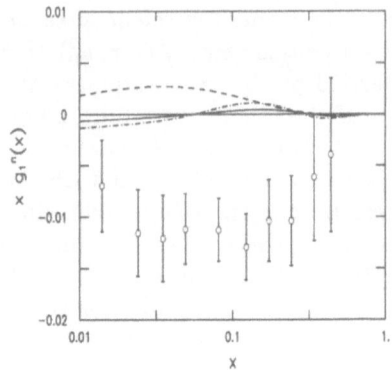

Figure 3. The proton (left) and neutron (right) polarized structure functions at $Q^2 = 3$ GeV^2, within scenario A. The full curves represent the relativistic results abtained by means of a complete NLO evolution; the dashed curves show the corresponding LO predictions. Dot - dashed curve: the (NLO) non relativitic calculation. Data are from the SMC and E143 experiments for the proton [14], and E154 for the neutron [15].

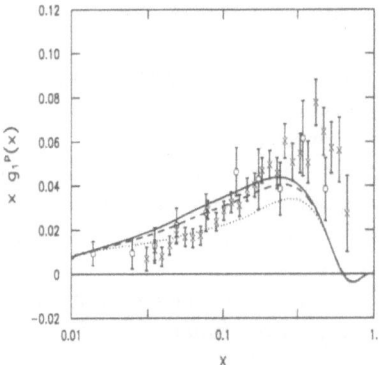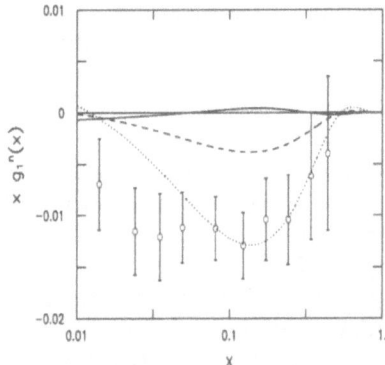

Figure 4. NLO $x\,g_1^{p,n}(x, Q^2 = 3$ GeV$^2)$ within different scenarios (see text). Scenario A $(f = 0)$: full curves Scenario B: for different values of the fraction of polarized gluons at the hadronic scale: $f = -0.35$ (dashed curves), $f = 1$ (dotted curves). Data as in Fig. 3.

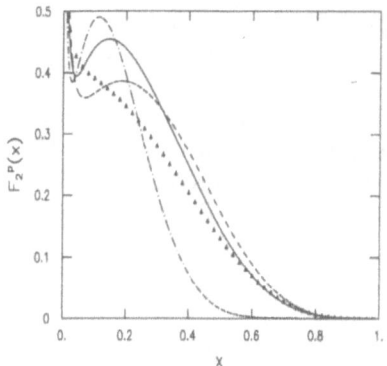
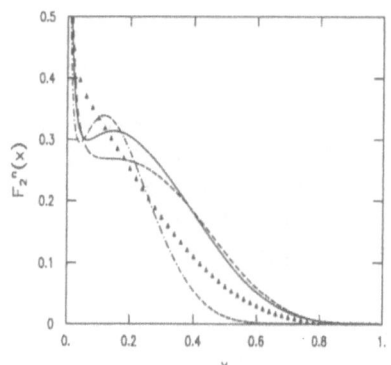

Figure 5. $F_2^{p,n}(x, Q^2 = 3 \text{ GeV}^2)$ within scenario A. Notations as in Fig. 3. Data fit (triangles) from ref.[16].

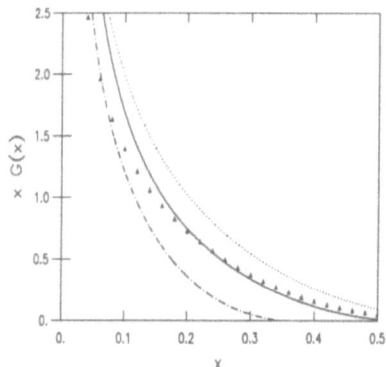
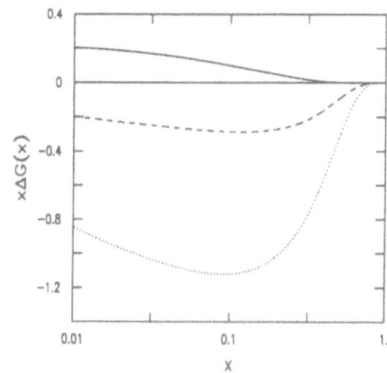

Figure 6. NLO unpolarized (left) and polarized (right) gluon distributions at $Q^2 = 3$ GeV^2. Left panel: scenario A: relativistic results (full curve), non relativistic approximation (dot-dashed curve). Scenario B: dotted curve (DIS factorization scheme). Right panel: scenario A: full curve; scenario B ($\overline{\text{MS}}$ factorization scheme): 35% polarization fraction (dashed curve), 100% polarization fraction (dotted curve).

the present relativistic quark model quite close to the experimental observations at least in the $x \gtrsim 0.1$ region. Larger *negative* fraction of gluon polarization is favored by the data in agreement with the large effect required to split the nucleon and Δ mass at the hadronic mass scales.

In Fig. 5 we show results for the unpolarized $F_2^{p,n}$ structure functions. The relativistic approach improves their description in the whole x-range, in particular in the region $x \gtrsim 0.4$ where high momentum components play a relevant role and the valence distribution dominate the response. In the complementary

354

range ($x \lesssim 0.4$) the distributions could be improved by considering the non-perturbative $q\bar{q}$ contributions [5] neglected at the present stage of development of our relativistic scheme.

Finally in Fig. 6 we show the gluon distributions both for polarized and unpolarized scattering. ΔG will be measured in dedicated experiments at CERN and BNL as discussed also during the workshop and our predictions show the sensitivity of that observable on the gluon polarization at the scale of the constituent quark model.

References

1. R.L. Jaffe and G.C. Ross: Phys. Lett. **B93**, 313 (1980)

2. M. Traini, L. Conci and U. Moschella: Nucl. Phys. **A544**, 731 (1992); M. Ropele, M. Traini and V. Vento: Nucl. Phys. **A584**, 634 (1995)

3. M. Traini, A. Zambarda and V. Vento: Mod. Phys. Lett. **10**, 1235 (1995); M. Traini, V. Vento, A. Mair and A. Zambarda: Nucl. Phys. **A614**, 472 (1997)

4. A. Mair and M. Traini: Nucl. Phys. **A624**, 564 (1997)

5. A. Mair and M. Traini: Nucl. Phys. **A628**, 296 (1998)

6. S. Scopetta, V. Vento and M. Traini: Phys. Lett. **B421**, 64 (1998); Phys. Lett. **B442**, 28 (1998)

7. B.D. Keister and W.N. Plolyzou: Adv. in Nucl. Phys. **20**, 225 (1991); F. Coester: Progr. Part. Nucl. Phys. **29**, 1 (1992)

8. H.J. Melosh: Phys. Rev. D **9**, 1095 (1974)

9. M. Ferraris, M.M. Giannini, M. Pizzo, E. Santopinto and L. Tiator: Phys. Lett. **B364**, 231 (1995)

10. P. Faccioli: thesis, University of Trento 1998, unpublished. M. Traini and P. Faccioli, in preparation.

11. F. Cardarelli, E. Pace, G. Salmè and S. Simula: Phys. Lett. **B357**, 267 (1995); Phys. Lett. **B371**, 7 (1996)

12. P. Faccioli, M. Traini and V. Vento: in preparation.

13. R.L. Jaffe: Phys. Lett. **B365**, 359 (1996)

14. D. Adams et al. (SMC): Phys. Lett. **B329**, 399 (1994); **B357**, 248 (1995); K. Abe et al. (E143): Phys. Rev. Lett. **74**, 346 (1995); **75**, 25 (1995)

15. K. Abe et al. (E154): Phys. Rev. Lett. **79**, 26 (1997)

16. H.L. Lai et al. (CTEQ collaboration): Phys. Rev. D **55**, 2862 (1997)

Few-Body Systems Suppl. 11, 355–358 (1999)

Few-
Body
Systems
© by Springer-Verlag 1999

Nucleon Structure Functions in a Constituent Quark Scenario *

Sergio Scopetta[a], Vicente Vento [a,b], and Marco Traini [c]

[a] Departament de Fisica Teòrica, Universitat de València
46100 Burjassot (València), Spain;
[b] Institut de Física Corpuscular, Consejo Superior de Investigaciones Científicas;
[c] Dipartimento di Fisica, Università di Trento, I-38050 Povo (Trento), Italy, and INFN, Gruppo Collegato di Trento.

Abstract. Using a simple picture of the constituent quark as a composite system of point-like partons, we construct the polarized parton distributions by a convolution between constituent quark momentum distributions and constituent quark structure functions. We achieve good agreement with experiments in the unpolarized as well as in the polarized case, though a good description of the recent polarized neutron data requires the introduction of one more paramater. When our results are compared with similar calculations using non-composite constituent quarks, the accord with the experiments of the present scheme is impressive. We conclude that DIS data are consistent with a low energy scenario dominated by composite constituents of the nucleon.

At low energies, the so called naive quark model accounts for a large number of experimental observations. At large energies, QCD sets the framework for an understanding of the Deep Inelastic Scattering (DIS) phenomena beyond the Parton Model. However, the perturbative approach to QCD does not provide absolute values for the observables. The description based on the Operator Product Expansion (OPE) and the QCD evolution requires the input of non-perturbative matrix elements. We have developed an approach which uses model calculations for the latter ingredients [1]. Moreover, in order to relate the constituent quark with the current partons of the theory, a procedure, hereafter called ACMP, has been applied [2, 3]. Within this approach, constituent quarks are effective particles made up of point-like partons (current quarks (antiquarks) and gluons), interacting by a residual interaction described as in

*Supported in part by DGICYT-PB94-0080 and TMR programme of the European Commission ERB FMRX-CT96-008

a quark model. The hadron structure functions are obtained by a convolution of the constituent quark model wave function with the constituent quark structure function. This idea has been recently used to estimate the pion structure function [4]. We summarize here our application to the unpolarized [3] and polarized [5] DIS off the nucleon. It will be found that DIS data are consistent with a low energy scenario dominated by composite constituents.

In our picture the constituent quarks are themselves complex objects whose structure functions are described by a set of functions Φ_{ab} that specify the number of point-like partons of type b, which are present in the constituents of type a with fraction x of its total momentum [2, 3]. In general a and b specify all the relevant quantum numbers of the partons, i.e., flavor and spin. Let us discuss first the unpolarized case for the proton [3].

The functions describing the nucleon parton distributions omitting spin degrees of freedom are expressed in terms of the independent $\Phi_{ab}(x)$ and of the constituent probability distributions u_0 and d_0, at the hadronic scale μ_0^2 [1], as

$$f(x, \mu_0^2) = \int_x^1 \frac{dz}{z} [u_0(z, \mu_0^2) \Phi_{uf}(\frac{x}{z}, \mu_0^2) + d_0(z, \mu_0^2) \Phi_{df}(\frac{x}{z}, \mu_0^2)] \tag{1}$$

where f labels the various partons, i.e., valence quarks (u_v, d_v), sea quarks (u_s, d_s, s), sea antiquarks ($\bar{u}, \bar{d}, \bar{s}$) and gluons g. The different types and functional forms of the structure functions for the constituent quarks are derived from three very natural assumptions [2]: *i)* The point-like partons are the quarks, antiquarks and gluons described by QCD; *ii)* Regge behavior for $x \to 0$ and duality ideas; *iii)* invariance under charge conjugation and isospin.

These considerations define the following structure functions [2]

$$\Phi_{qf}(x, \mu_0^2) = C_f x^{a_f} (1 - x)^{A-1} , \tag{2}$$

where $f = q_v, q_s, g$ for the valence quarks, the sea and the gluons, respectively. Regge phenomenology suggests: $a_{q_v} = -0.5$ (ρ meson exchange) and $a_{q_s} = a_g = -1$ (*pomeron* exchange). The other ingredients of the formalism, i.e., the probability distributions for each constituent quark, are defined according to the procedure of ref. [1] and shown in [3]. Our last assumption relates to the hadronic scale μ_0^2, i.e., that at which the constituent quark structure is defined. We choose $\mu_0^2 = 0.34$ GeV^2, as defined in Ref. [1], namely by fixing the momentum carried by the various partons. This choice of the hadronic scale determines all the parameters except one, which is fixed through the data [3]. To complete the process, the above input distributions are NLO-evolved in the DIS scheme to the experimental scale, where they are compared with the data.

We next generalize our previous discussion to the polarized parton distributions. As it is shown in ref. [5], using $SU(6)$ (spin-isospin) symmetry and other reasonable simplifying assumptions, it can be shown [5] that

$$\Delta f(x, \mu_0^2) = \int_x^1 \frac{dz}{z} [u_0(z, \mu_0^2) \Delta\Phi_{uf}(\frac{x}{z}, \mu_0^2) + d_0(z, \mu_0^2) \Delta\Phi_{df}(\frac{x}{z}, \mu_0^2)] , \tag{3}$$

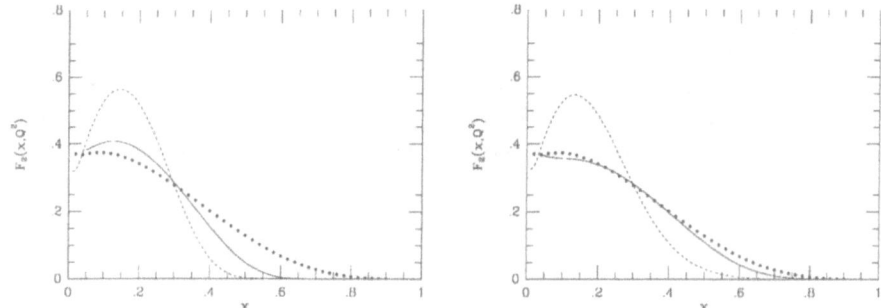

Figure 1. The proton $F_2(x, Q^2)$, obtained by NLO-evolution to $Q^2 = 10$ GeV^2 (full), compared to the data (dots) [10]. The result which would be obtained disregarding the constituent structure is also shown (dashed). Left (right) panel: constituent wave functions form ref. [8] (ref. [9]).

where f labels the various partons; it means that the $ACMP$ procedure can be extended to the polarized case just by introducing three additional structure functions for the constituent quarks: $\Delta\Phi_{qq_v}$, $\Delta\Phi_{qq_s}$ and $\Delta\Phi_{qg}$. In order to determine them we add two minimal assumptions: *iv)* factorization: $\Delta\Phi$ cannot depend upon the quark model used; *v)* positivity: the constraint $\Delta\Phi \leq \Phi$ is saturated for $x = 1$. In such a way we determine completely the $\Delta\Phi$'s. In fact, the QCD partonic picture, Regge behavior and duality imply that

$$\Delta\Phi_{qf} = \Delta C_f x^{-\Delta a_f}(1 - x)^{\Delta A_f - 1} \tag{4}$$

and $-\frac{1}{2} < \Delta a_f < 0$, for all $f = q_v, q_s, g$, as allowed by dominant exchange of the A_1 meson trajectory [7]. Moreover, the assumption that the positivity restriction is saturated for $x = 1$, in the spirit of ref. [6], implies that the $\Phi's$ and the $\Delta\Phi's$ have the same large x behavior, and that $\Delta C_f = C_f$, (the latter being introduced in (2)); it means that the partons which carry all of the momentum also carry all of the polarization. Let us stress that the change between the polarized functions and the unpolarized ones comes only from Regge behavior; as a matter of fact, it turns out that, *except for the exponent* Δa_f shown above, the $\Delta\Phi$'s, Eq. (4), are given by the unpolarized functions, Eq. (2). The other ingredients, i.e., the polarized distributions for each constituent quark, are defined according to the procedure of ref. [1] and they are shown in ref. [5]. Finally, the parton distributions at the hadronic scale are evolved to the experimental scale by performing a NLO evolution in the AB scheme [7]. Results are shown in Figs. 1 and 2. Fig. 1 refers to the unpolarized case. The structure function $F_2(x, Q^2)$, obtained evolving the parton distributions Eq. (1), calculated using Eq. (2) for the Φ_{qf}'s and two different models for u_o and d_o, describes successfully the data. The agreement becomes impressive if compared with a similar calculation with non-composite constituents. In the polarized case, it is found [5] that the constituent structure functions Eq. (4) give a good result for the proton, but they fail in reproducing the recent precise neutron data. This is to be ascribed to our naive input for the sea and to the symmetry for the u and d quarks [5]. In particular, it has been shown that, by

358

redefining the sea $\Delta\Phi$, changing *only one* parameter so that the experimental sea polarization is recovered, also the neutron is rather well described. Fig. 2 refers to this last scenario. The procedure is also able to predict successfully several observables, such as the nucleon axial charges [5]. It should be noticed that in this framework the *spin crisis*, as initially presented, does not arise.

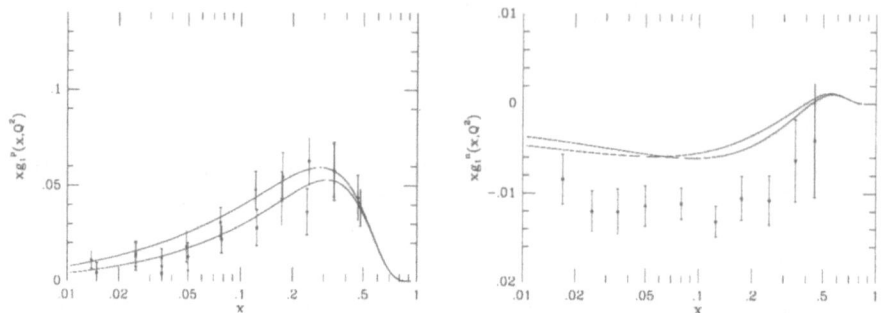

Figure 2. Left (Right): $xg_1(x, Q^2)$ for the proton (neutron) evolved at NLO to $Q^2 = 10\,(5)$ GeV^2, for the two extreme Regge behaviors mentioned in the text (full curves). The wave functions used are from ref. [9]. The data [10] are shown for comparison.

Summarizing, low energy models seem to be consistent with DIS data when a structure for the constituent is introduced. The crucial role played by the sea in the polarized case, as well as the implementation of Chiral Symmetry Breaking in our procedure, have to be more deeply investigated. It will be the subject of future work.

References

1. M. Traini, V. Vento, A. Mair and A. Zambarda: Nucl. Phys. **A614**, 472 (1997)

2. G. Altarelli , N. Cabibbo, L. Maiani and R. Petronzio: Nucl. Phys. **B69**, 531 (1974)

3. S. Scopetta, V. Vento and M. Traini: Phys. Lett. B 421, 64 (1998)

4. G. Altarelli, S. Petrarca and F. Rapuano: Phys. Lett. **B373**, 200 (1996)

5. S. Scopetta, V. Vento and M. Traini: Phys. Lett. **B442**, 28 (1998)

6. R. Carlitz and J. Kaur: Phys. Rev. Lett. **38**, 673 (1977)

7. G. Altarelli, R.D. Ball, S. Forte and G. Ridolfi, Nucl. Phys. **B496**, 337 (1997)

8. N. Isgur and G. Karl: Phys. Rev. **D18**, 4187 (1978)

9. R. Bijker, F. Iachello and A. Leviatan: Ann. Phys. **236**, 69 (1994)

10. EMC Collaboration: Nucl. Phys. **B328**, 1 (1989); SMC Collaboration, Phys. Rev. **D56** (1997) 5330; E154 Collaboration: Phys. Rev. Lett. **79**, 26 (1997)

Few-Body Systems Suppl. 11, 359–362 (1999)

Few-
Body
Systems
© by Springer-Verlag 1999

Strangeness and Charm Production with HERMES

M. J. Amarian*†

Istituto Nazionale di Fisica Nucleare, Sezione di Roma,
and
Gruppo Collegato Sanita', viale Regina Elena, I-00161 Roma, Italy

Abstract. Experimental results on strangeness and charm production with HERMES experiment at HERA are presented. Different scenarios to access polarization of strange quarks and gluons using polarized positron(electron) beam and polarized targets are discussed.

1 Introduction

The HERMES experiment is designed to study the spin structure of the nucleon using E=27.5GeV polarized positron(electron) beam of HERA incident on polarized H, D, ^{3}He and unpolarized gaseous targets. The target gas is fed into the center of 40-cm-long open-ended thin-walled storage cell located inside the storage ring. The magnetic spectrometer is fully described in [1]. The angular acceptance of the spectrometer extends over the range 40 mrad $< \theta <$ 220 mrad. Average angular resolution is better than 1 mrad and the average momentum resolution is better than 2%. Particle identification is accomplished with a lead glass calorimeter, a preshower counter, a transition radiation detector and a gas threshold Čerenkov counter.

According to the current understanding of the spin structure of the nucleon based on the results of polarized inclusive experiments only a small fraction of the spin is carried by quarks. In order to explain the rest it is necessary to measure contribution from strange quarks and gluons.

2 Strangeness Production

One possibility to access strange quark polarization in the nucleon is via beam-target double spin asymmetry measurements with leading kaons in final state. In order to have clean kaon identification new RICH detector has been installed

---

* *On leave from:* Yerevan Physics Institute, Alikhanian Br.2, 375036 Yerevan, Armenia
† *E-mail address:* amarian@axiss.iss.infn.it

360

at HERMES replacing gas threshold Čerenkov counters. Currently experiment is taking data in this mode.

Another unique possibilty to measure spin transfer from quarks to the final state hadron is measurement of the Λ hyperon polarization. In this case one can use polarized beam and unpolarized target with order of magnitude higher total luminosity than with polarized gaseous targets.

There are two different scenarios for lambda production in the current and target fragmentation region [2, 3].

In the current fragmentation region ($x_f > 0$, where x_f is the fraction of longitudinal momentum along the beam line transferred to the hadron in $\gamma^* N$ center of mass frame) polarized quark fragmentation results in polarization of Λ. Significant negative polarization of Λ's from Z^0 decay observed in e^+e^- collisions [4, 5] is due to the fragmentation of highly polarized s quarks. In deep inelastic scattering lambda's are produced predominantly from the scattering on u quarks, therefore lambda polarization measurement has an acess to the polarized fragmentation function of u quarks.

In the target fragmentation region $x_f < 0$ authors of Ref. [3] suggested that the remnant s quark polarization is negatively correlated with the non-zero polarization of struck quark, hence lambda polarization will give an insight into polarization of strange sea quarks.

In Fig. 1 Λ kinematical distributions from HERMES data are presented.

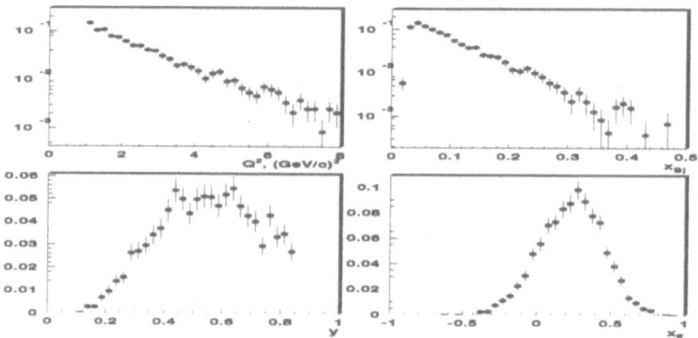

Figure 1. Λ kinematics: Q^2, x-Bjorken, y=ν/E_e-fraction of postron energy carried by virtual photon and x-Feynman distributions.

In 1999 an upgrade of the HERMES detector will greatly increase Λ detection acceptance [6]. Two round silicon detectors, called Lambda Wheels placed downstream of target area will detect pions which otherwise would have not been detected in standard acceptance. This will allow to access negative x_f with more than one order of magnitude higher statistics than at present setup.

In order to extract lambda polarization HERMES has been running with the positive and negative beam polarizations in 1996-97 running periods. According to the preliminary status of data analysis polarization transfer coefficient from the beam positron to Λ has been extracted

$$\sigma = D_y \tau_B = P_\Lambda/P_B = 0.1 \pm 0.17(stat.) \pm 0.08(syst.), \qquad (1)$$

where P_Λ, P_B are Λ and beam polarizations, D_y is depolarization factor, τ_B is polarization transfer. More detailed results will be published soon.

3   Charm Production

One of the most urgent tasks in high energy spin physics is a direct measurement of gluon polarization ΔG in the nucleon. Spin asymmetry measurement in polarized charm production via LO photon-gluon fussion is considered to be one of the cleanest ways to access polarized glue [7, 8]. In order to eval-

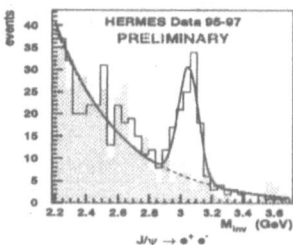 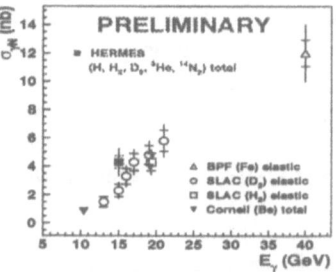

Figure 2. a)-J/ψ invariant mass distribution (left) and b)-production cross section compared to the world data (right).

uate Hermes capabilities to provide information on $\Delta G/G$ significant efforts have been devoted to extract charm signals from 1995-1997 data set. In Fig. 2a invariant mass spectrum of combined e^+e^- and $\mu^+\mu^-$ events is presented. About 140 events forming J/ψ over 40 of background is extracted. This allows to extract J/ψ production cross section near the threshold and compare with the world data mainly for elastic J/ψ production with real photon beams [9]. In Fig. 2b HERMES measurement along with existing data are presented ($\sigma_{\gamma p}[E_\gamma=15\text{GeV}]=(4.3\pm0.5)$ nb).

Extraction of gluon polarization via open charm production is preferrable due to the more definite theoretical description of fragmentation process. However, open charm particle reconstruction is a very difficult task due to the lack of particle identification and combinatorial background. In such conditions the best way to extract signal is with D*-D^0 tagging method. Due to the small binding energy of pion in D* the mass difference must peak around 146 MeV.

In Fig. 3 D*-D^0 missing mass from HERMES 1996-97 is presented. Shaded area was obtained with the wrong sign slow pion and D^0 combination.

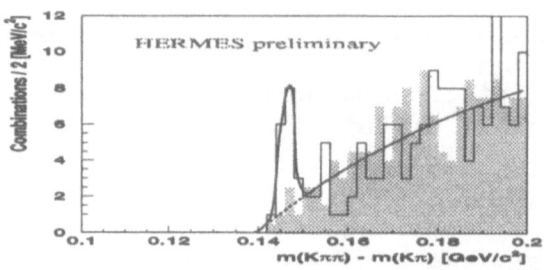

Figure 3. Missing mass spectrum M(K$\pi\pi$)-M(Kπ) for the process $D^* \to D^0(\to K\pi)\pi$

As one can see data sample is small, however with all upgrades of experimental setup and improved particle identification with partial reconstruction of final states like in semileptonic decay HERMES will collect enough statistics to put constrains on gluon polarization.

References

1. HERMES Collab., K. Ackerstaff et al.: Nucl. Instrum. Meth. **A417**, 230 (1998)

2. R. Jaffe: Phys. Rev. **D54**, 6581 (1996)

3. J. Ellis, D.Kharzeev and A.Kotzinian: Z. Phys.,**C69**, 467 (1996)

4. D. Buskulic et al.: Phys. Lett. **B374**, 319 (1996)

5. K. Ackerstaff et al.: CERN-PPE/97-104, 1997

6. M. Amaraian et al.: HERMES Internal Report 97-032, 1997

7. COMPASS Proposal: CERN/SPSLC 96-14, March 1996

8. M. Amarian et al.: HERMES Internal Report 97-001, 1997

9. U. Camerini et al.: Phys. Rev. Lett **35**, 483 (1975); A.R. Clark et al.: Phys. Rev. Lett. **43**, 187 (1979); B. Gittelman et al.: Phys. Rev. Lett. **35**, 1617 (1975); S.D. Holmes et al.: Ann. Rev. Nucl. Part. Sci. **35** 397 (1985)

Few-Body Systems Suppl. 11, 363–364 (1999)

Few-
Body
Systems
© by Springer-Verlag 1999

Questions on the Quark Model: Panel Discussion *

Nimai C. Mukhopadhyay[†] and R.M. Davidson **

Department of Physics, Applied Physics and Astronomy, Rensselaer
Polytechnic Institute, Troy, New York 12180-3590

In atomic, nuclear, and particle physics, one often deals with complicated potentials where the analytic solution to the bound state wave functions is not known. One general approach to this problem is to diagonalize the Hamiltonian in a finite, i.e., truncated, model space. While this method is generally very good in determining the low-lying energy levels, it is not without problems. For example, an operator that is known to commute with the Hamiltonian may not, in the model space, actually commute with the Hamiltonian. To deal with this problem, one tries to find effective operators in the model space that do satisfy the properties of the operators in the untruncated space.

A specific example of this problem is current conservation in the quark model, which was discussed at length at this workshop [1, 2]. One problem is that due to truncation of the space, the current is not conserved, i.e., $\partial_\mu j^\mu \neq 0$. Of course, one can always define a new current that is conserved, e.g., in momentum space,

$$\tilde{j}^\mu = j^\mu - \frac{K \cdot j}{K^2} K^\mu \, , \tag{1}$$

where j is the current and K is the photon four-momentum. However, this is not unique, which is not satisfactory since we want a microscopic model to give us an unique answer.

The degree of violation of current conservation in the quark model has been investigated by Bourdeau and Mukhopadhyay [3] and Drechsel and Giannini [4]. The general conclusion from these works is that there is a severe violation of current conservation, making predictions of small quantities, such as the $N - \Delta(1232)$ $E2$ transition, unreliable. Indeed, Buchmann [2] presented evidence at this workshop that two-body currents, neglected in most quark model calculations of $E2$, actually dominate this transition amplitude.

Presented by NCM at the Trento N workshop
[†] E-mail address : mukhon@rpi.edu
** E-mail address : davidr@rpi.edu

Close and Li [5] have proposed general effective operators to deal with this problem. Capstick and collaborators [6] have partially implemented these operators in their relativised quark model, but even Capstick admitted at this workshop that $E2$ calculations are still not very accuarte. The problem in these calculations is not just the truncation of the space. In fact, compared to the earliest quark model calculations, the model spaces now being used are huge. The additional problem one is now encountering is what to use for the electromagnetic transition operators. In other words, how do we *consistently* relativize these operators? In atomic physics, this may be done by means of a Foldy-Wouthuysen (FW) transformation. Unfortunately, in quark model calculations the FW series converges very slowly, if at all [7]. A specific example of this may be found in the classic text book by Close [8], where, in the FW scheme, the magnetic interaction energy turns out to be nearly as large as the Coulomb interaction energy.

In conclusion, this is a difficult problem that deserves serious attention. the good news is that it is being taken seriously as was evidenced by the presentations of Demetriou [1] and Buchmann [2] at this workshop.

We thank the participants of this panel for many stimulating discussions.

References

1. See P. Demetriou's contribution to this workshop.

2. A. J. Buchmann, E. Hernandez, and Amand Faessler: Phys. Rev. **C55**, 448 (1997); see also Buchmann's contribution to this workshop

3. M. Bourdeau and N. C. Mukhopadhyay: Phys. Rev. Lett. **58**, 976 (1987); *ibid.* **63**, 335 (1989)

4. D. Drechsel and M. M. Giannini: Phys. Lett. **143B**, 329 (1984)

5. F. E. Close and Z. P. Li: Phys. Rev. **D42**, 2194 (1990)

6. S. Capstick: Phys. Rev. **D46**, 2864 (1992)

7. L. Zhang and N. C. Mukhopadhyay: unpublished.

8. F. E. Close: *An Introduction to Quarks and Partons,* Academic Press, London (1979)

List of Participants

Shahin AGAEV
High Energy Physics Laboratory
Baku State University
Z. Khalilov st. 23, 370148 Baku
Azerbaijan
azhep@lan.ab.az

Moskov AMARIAN
Istituto Nazionale di Fisica Nucleare
Sezione di Roma
P.le Aldo Moro 2, I-00185 Roma
Italy
amarian@axiss.iss.infn.it

Gisela ANTON
Friedrich Alexander Universität
Physikalisches Institut
D-91058 Erlangen
Germany
anton@pib1.physik.uni-bonn.de

Nicola BIANCHI
Istituto Nazionale di Fisica Nucleare
Laboratori Nazionali di Frascati
I-00044 Frascati
Italy
nicola.bianchi@lnf.infn.it

Alfons BUCHMANN
Institute for Theoretical Physics
University of Tübingen
D-72076 Tübingen
Germany
alfons.buchmann@uni-tuebingen.de

Volker BURKERT
Jefferson Laboratory
Newport News
Virginia 23606
USA
burkert@jlab.org

Simon CAPSTICK
SCRI and Department of Physics
Florida State University
Tallahassee, Florida 32306
USA
capstick@scri.fsu.edu

Carl-Edwin CARLSON
Nuclear and Particle Theory Group
College of William and Mary
Williamsburg, VA 23187-8795
USA
carlson@physics.wm.edu

Richard DAVIDSON
Department of Physics
Rensselaer Polytechnic Institute
Troy, NY, 12180
USA
davidr@rpi.edu

Vivian P. DEMETRIOU
Dipartimento di Fisica Nucleare e Teorica
Università di Pavia and INFN Sezione di Pavia
Via Bassi 6, I-27100 Pavia
Italy
vivian@pv.infn.it

Jean-Pierre DIDELEZ
IN2P3
Institut de Physique Nucléaire
F-91406 Orsay
France
didelez@ipncls.in2p3.fr

Nicole D'HOSE
Service de Physique Nucléaire
Centre d'Etudes de Saclay
F-91191 Gif-sur-Yvette Cedex
France
dhose@hep.saclay.cea.fr

Alexander DONNACHIE
Department of Physics and Astronomy
University of Manchester
Manchester M13 9PL
United Kingdom
alexander.donnachie@cern.ch

Dieter DRECHSEL
Johannes Gutenberg-Universität
Institut für Kernphysik
J.J. Becherweg 45, D-55099 Mainz
Germany
klotter@kph.uni-mainz.de

Steven DYTMAN
Department of Physics and Astronomy
University of Pittsburgh
Pittsburgh, PA 15260
USA
dytman+@pitt.edu

Latifa ELOUADRHIRI
Christopher Newport University
1 University Place
VA 23606
USA
latifa@jlab.org

Mauro M. GIANNINI
Dipartimento di Fisica dell'Università
di Genova
INFN Sezione di Genova
via Dodecaneso 33, I-16164 Genova, Italy
giannini@genova.infn.it

Thomas HEMMERT
Institut für Kernphysik (Th)
FZ-Jülich
D-52425 Jülich
Germany
th.hemmert@fz-juelich.de

Barry HOLSTEIN
Department of Physics and Astronomy
University of Massachusetts
Amherst, MA 01002
USA
holstein@phast.umass.edu

Franco IACHELLO
Center for Theoretical Physics
Yale University
New Haven, CT 06520-8120
USA
peggy@nst.physics.yale.edu

Kyungseon JOO
Department of Physics
University of Virginia
Charlottesville, VA 22901
USA
kjoo@jlab.org

Norbert KAISER
Physik Department T39
Technische Universität München
James Franck Strasse, D-85747 Garching
Germany
nkaiser@physik.tu-muenchen.de

Sabit KAMALOV
Laboratory of Theoretical Physics
JINR Dubna
SU-101000 Moscow
Russia
kamalov@kph.uni-mainz.de

Vladimir A. KARMANOV
Lebedev Physical Institute
Leninsky Prospekt 53
117924 Moscow
Russia
karmanov@sci.lebedev.ru

Mariana KIRCHBACH
Institut für Kernphysik
Universität Mainz
J.J. Becherweg 45, D-55099 Mainz
Germany
mariana@kph.uni-mainz.de

Peter KROLL
Fachbereich Physik
Universität Wuppertal
D-42097 Wuppertal
Germany
kroll@theorie.physik.uni-wuppertal.de

Bernd KRUSCHE
II Physikalisches Institut
Universität of Giessen
Heinrich-Buff-Ring 16, D-35392 Giessen
Germany
bernd@piggy.physik.uni-giessen.de

Fabienne KUNNE
Service de Physique Nucléaire
Centre d'Etudes de Saclay
F-91191 Gif-sur-Yvette Cedex
France
fkunne@cea.fr

Harry T.-S. LEE
Physics Division
Argonne National Laboratory
Argonne, Illinois 60439
USA
lee@anlphy.phy.anl.gov

Kei-Feh LIU
Department of Physics and Astronomy
Kentucky University
Lexington, KY 40506-0055
USA
liu@ukcc.uky.edu

Mark MANLEY
Department of Physics and Center
for Nuclear Research
Kent State University
Kent, OH 44242, USA
manley@ksuvxa.kent.edu

Tetsuro MIZUTANI
Department of Physics
Virginia Polytechnic Institute and
State University
Blacksburg, VA 24061, USA
mizutani@vpihe1.phys.vt.edu

Victor MOKEEV
Skobelt'sin Nuclear Physics Institute
Moscow State University
Moscow 119899
Russia
mokeev@ge.infn.it

Valeria MUCCIFORA
Istituto Nazionale di Fisica Nucleare
Laboratori Nazionali di Frascati
I-00044 Frascati
Italy
valeria@lnf.infn.it

Nimai C. MUKHOPADHYAY
Department of Physics
Rensselaer Polytechnic Institute
Troy, New York 12180-3590
USA
mukhon@rpi.edu

Eulogio OSET
Departamento de Física Teórica and IFIC
Centro Mixto Universidad de Valencia - CSIC
46100 Burjassot (Valencia)
Spain
oset@evalvx.ific.uv.es

Barbara PASQUINI
Institut für Kernphysik
Universität Mainz
J.J. Becherweg 45, D-55099 Mainz
Germany
pasquini@kph.uni-mainz.de

Willibald PLESSAS
Institute for Theoretical Physics
University of Graz
Universitätsplatz 5, A-8010 Graz
Austria
plessas@bkfug.kfunigraz.ac.at

Steve POLLOCK
Department of Physics
University of Colorado
Boulder CO 80303
USA
pollock@lucky.colorado.edu

Marco RADICI
Istituto Nazionale di Fisica Nucleare
Sezione di Pavia
Via Bassi 6, I-27100 Pavia
Italy
radici@pv.infn.it

Anatoly RADYUSHKIN
Physics Department
Old Dominion University
Norfolk, VA 23529
USA
radyush@cebaf.gov

Marco RIPANI
Istituto Nazionale di Fisica Nucleare
Sezione di Genova
Via Dodecaneso 33, I-16146 Genova
Italy
ripani@genova.infn.it

Dan Olof RISKA
Department of Physics
University of Helsinki
P.O. Box 9, 00014 Helsinki
Finland
riska@pcu.helsinki.fi

Bijan SAGHAI
Service de Physique Nucléaire
Centre d'Etudes de Saclay
F-91191 Gif-sur-Yvette Cedex
France
bsaghai@cea.fr

Andy M. SANDORFI
Physics Deptartment
Brookhaven National Laboratory
Upton, NY 11973
USA
sandorfi@bnl.gov

Elena SANTOPINTO
Istituto Nazionale di Fisica Nucleare
Sezione di Genova
Via Dodecaneso 33, I-16146 Genova
Italy
santopinto@ge.infn.it

Vladimir SAULI
Nuclear Physics Institute
Academy of Sciences
Rez near Prague
Czech Republic
gemma@ujf.cas.cz

Carlo SCHAERF
Universitá di Roma "Tor Vergata" and
INFN Sezione di Roma II
Via della Ricerca Scientifica, I-00133 Roma
Italy
schaerf@roma2.infn.it

Stefan SCHERER
Institut für Kernphysik
Johannes Gutenberg-Universität
J.J. Becherweg 45, D-55099 Mainz
Germany
scherer@kph.uni-mainz.de

Reinhard SCHUMACHER
Department of Physics
Carnegie Mellon University
Pittsburgh, PA 15213
U.S.A.
schumacher@cmu.edu

Norberto N. SCOCCOLA
Physics Department
CNEA, Av. Libertador 8250
(1429) Buenos Aires
Argentina
scoccola@tandar.cnea.edu.ar

Sergio SCOPETTA
Departament de Fisica Teòrica
Universitat de València
46100 Burjassot (València)
Spain
scopetta@titan.ific.uv.es

Silvano SIMULA
Istituto Nazionale di Fisica Nucleare
Sezione di Roma III
Via della Vasca Navale 84, I-00146 Roma
Italy
simula@hpteo1.roma3.infn.it

Fl. STANCU
Institute of Physics
University of Liege
J Sart Tilman, B-4000 Liege 1
Belgium
fstancu@ulg.ac.be

Paul STOLER
Department of Physics
Rensselaer Polytechnic Institute
Troy, NY, 12180
USA
stolep@rpi.edu

Albert SVARC
Rudjer Bošković Institute
Bijenička c.54
10000 Zagreb
Croatia
svarc@rudjer.irb.hr

Frank TABAKIN
Department of Physics and Astronomy
University of Pittsburgh
Pittsburgh, Pennsylvania 15260
USA
tabakin@vms.cis.pitt.edu

Timo THONHAUSER
Institute for Theoretical Physics
University of Graz
Universitätsplatz 5, A-8010 Graz
Austria
timo.thonhauser@kfunigraz.ac.at

Marco TRAINI
Dipartimento di Fisica
Università degli Studi di Trento
I-38050 Povo (Trento)
Italy
traini@science.unitn.it

Robert F. WAGENBRUNN
Institute for Theoretical Physics
University of Graz
Universitätsplatz 5, A-8010 Graz
Austria
rfw@physik.kfunigraz.ac.at

Ron WORKMAN
Department of Physics
Virginia Tech
Blacksburg, VA 24061
USA
workman@clsaid.phys.vt.edu

Author Index

SpringerPhysics

B. Desplanques, K. Protasov,
B. Silvestre-Brac, J. Carbonell (eds.)

Few-Body Problems in Physics '98

Proceedings of the 16th European Conference on Few-Body
Problems in Physics, Autrans, France, June 1–6, 1998

1999. XIX, 548 pages. 217 figures and 1 Frontispiece.
Hardcover DM 220,–, öS 1540,–, sFr 198,–
Reduced price for subscribers to "Few-Body Systems":
Hardcover DM 198,–, öS 1386,–, sFr 179,–
(all prices are recommended retail prices)
ISBN 3-211-83280-7
Few-Body Systems, Supplement 10

The book contains invited and contributed talks presented
at the 16th European Conference on Few-Body Problems
in Physics, held in Autrans (France), June 1–6, 1998.
The conference was devoted to the description and the
properties of few-body systems in various fields of physics.
Contributions essentially concern the following topics: resolution methods for few-body problems, mathematical aspects,
relativity, few-body dynamics: atomic and mesoscopic
systems, threshold effects and stability limit, few-body dynamics: nuclear and particle systems.

SpringerWienNewYork

Sachsenplatz 4–6, P.O.Box 89, A-1201 Wien, Fax +43-1-330 24 26
e-mail: books@springer.at, Internet: http://www.springer.at
New York, NY 10010. 175 Fifth Avenue • D-14197 Berlin, Heidelberger Platz 3
Tokyo 113, 3–13, Hongo 3-chome, Bunkyo-ku

Springer-Verlag
and the Environment